Seedling Production and Field Performance of Seedlings

Seedling Production and Field Performance of Seedlings

Special Issue Editors

Johanna Riikonen
Jaana Luoranen

MDPI • Basel • Beijing • Wuhan • Barcelona • Belgrade

MDPI

Special Issue Editors
Johanna Riikonen
Natural Resources Institute Finland,
Finland

Jaana Luoranen
Natural Resources Institute Finland,
Finland

Editorial Office
MDPI
St. Alban-Anlage 66
4052 Basel, Switzerland

This is a reprint of articles from the Special Issue published online in the open access journal *Forests* (ISSN 1999-4907) from 2017 to 2018 (available at: https://www.mdpi.com/journal/forests/special_issues/seedling)

For citation purposes, cite each article independently as indicated on the article page online and as indicated below:

LastName, A.A.; LastName, B.B.; LastName, C.C. Article Title. *Journal Name* **Year**, *Article Number*, Page Range.

ISBN 978-3-03921-255-2 (Pbk)
ISBN 978-3-03921-256-9 (PDF)

Cover image courtesy of Jaana Luoranen.

Contents

About the Special Issue Editors

Johanna Riikonen (Dr) is a senior scientist at the Natural Resources Institute Finland, and is specialized in plant physiology and seedling production of forest tree species.

Jaana Luoranen (Dr) Senior scientist in the field of forest regeneration and production of forest tree seedling since 1995.

forests

MDPI

Editorial

Seedling Production and the Field Performance of Seedlings

Johanna Riikonen [1,*] and Jaana Luoranen [2]

[1] Natural Resources Institute Finland (Luke), 70200 Kuopio, Finland
[2] Natural Resources Institute Finland (Luke), 77600 Suonenjoki, Finland; jaana.luoranen@luke.fi
* Correspondence: Johanna.riikonen@luke.fi

Received: 19 November 2018; Accepted: 20 November 2018; Published: 27 November 2018

Abstract: The rapid establishment of seedlings in forest regeneration or afforestation sites after planting is a prerequisite for successful reforestation. The relationship between the quality of the seedling material and their growth and survival after outplanting has been recognized for decades. Despite the existence of a substantial amount of information on how to produce high-quality seedlings, there is still a need to develop practices that can be used in nurseries and at planting sites to be able to produce well-growing forest stands in ever-changing environments. This Special Issue of Forests is focused on seedling quality and how it can be manipulated in a nursery as well as how the quality of the seedlings affects their field performance after planting.

Keywords: cultural practice; field performance; nursery production; seedling quality; tree seedling

1. Use of High Quality Seedlings Is the Basis for Tree Planting Success

Seedling survival after outplanting is a complex process which can be affected by many nursery and silvicultural practices. The factors contributing to seedling quality have been comprehensively reviewed by Landis et al. [1] and Grossnickle and MacDonald [2]. Seedling quality can be assessed by measuring several morphological, physiological and performance attributes, the latter integrating the morphological and physiological attributes. However, in the end, the limiting factors on the outplanting site determine the most desirable morphological and physiological seedling attributes for improving the chances for increased growth and survival after the outplanting [3]. In this Special Issue, Grossnickle and MacDonald [4] review the historical development of the discipline of seedling quality, as well as where it is today. Because seedling quality consists of several features, such as the genetic source, morphological properties, nutritional status, stress resistance and the vitality of the seedlings, the seedling responses to different nursery practices may be variable in different tree species and under variable growth conditions [1,5]. In this Special Issue, Pinchot et al. [6] and Pinto et al. [7] consider the relationship between the initial size of the seedlings and their growth after outplanting. These studies highlight once more how the responses of the seedlings to different nursery practices are dependent on plant species and stock type.

The quality and germinability of seeds greatly influence the success of producing healthy and well-growing seedlings. Germinability and seedling health can be enhanced through different production methods [8]. In this issue, Kaliniewicz and Tylek [9] found that the quality of pedunculate oak acorns can be improved by different seed treatments prior to germination. They concluded that scarification and the elimination of infected acorns significantly increased the germination capacity of the acorns.

2. New and Existing Challenges along the Seedling Production Chain

Global change and development of technology provide new challenges and opportunities for influencing processes along the seedling production chain. According to the projections made by

Intergovernmental Panel on Climate Change [10], the global temperature will increase throughout the century. The world's forests play a key role as a carbon sink [11], and therefore, their responses to climate change may amplify or dampen atmospheric change at a regional and continental scale. During the last few years, the increased severity and frequency of summer heat waves and associated droughts have raised concerns about how climate change will interfere with forest regeneration processes. These climate extremes are projected to increase in the 21st century in many land areas [10] and they may eventually alter species compositions (as found by Vander Mijnsbrugge et al. [12] in this Special Issue), and even predispose some vulnerable species to disappearance from certain growth habitats (as found by Santos et al. [13] in this Special Issue).

Mining activity has a large impact on the surrounding landscape. It has caused significant forest losses and severe soil degradation worldwide. The post-mine areas are often reclaimed to non-forest land which results in a loss of biodiversity [14]. The reforestation of mined land would help mitigate the increase in atmospheric CO_2 concentrations and restore the potential for the land to provide forest ecosystem services and goods [15]. The restoration of forest on reclaimed post-mine land is often dependent on artificial regeneration [16]. Planted seedlings, however, are threatened by a variety of stresses, including low quality of rooting media, pre-existing competing vegetation and herbivory. In this issue, the first-year results from two experiments conducted in the reclaimed Appalachian surface mines are presented. Bell et al. [17] compared the survival and growth of native shortleaf pine to those of non-native loblolly pine (*Pinus taeda*). Hackworth et al. [18] studied herbivore damage in different tree species and how it could be reduced.

A current question in forest regeneration is how to transfer the gains from tree breeding programmes to forestry. One way to do this is to use vegetative propagation for producing somatic embryo plants. Somatic embryogenesis has been widely developed to mitigate shortages of regeneration material of a high breeding value in different conifer species ([19], and references therein). Fluctuation in the availability of genetically improved seed material of the Norway spruce has increased interest in developing the technology for the production of somatic embryos in Finland also. In this special issue, Tikkinen et al. [20] report that when state-of-the-art embryo storage and in vitro germination protocols were combined, somatic embryo plants can be grown and large-scale field testing can be initiated, although further development is still required to increase the cost-efficiency of the method.

Nursery production has traditionally focused on producing seedlings efficiently and economically. Nowadays, there is a growing interest in reducing the environmental impacts of seedling production. Sphagnum peat moss is widely used as a growth media in forest tree nurseries. However, due to its very long regeneration time, peat is no longer considered to be a renewable resource. Furthermore, peat extraction damages peatland ecosystems and reduces its capacity to act as a carbon sink ([21], and references therein). One way to reduce the C footprint of peat extraction is to develop an alternative growth media for Sphagnum peat moss. In this Special Issue, Dumroese et al. [22] evaluated different modes of biochar delivery to amend and replace Sphagnum peat moss in the production of nursery plants in containers.

In Fennoscandia, tree planting is the preferred method of stand regeneration. Most seedlings are planted manually in the regeneration sites. Economic pressure and labour shortages are pushing forest owners to manage their forests more intensively to increase wood production and profitability. Mechanized tree planting has been developed in Fennoscandia as an alternative to manual planting. It has been shown to be time efficient and to lead to high-quality regeneration when compared to manual planting [23]. However, due to its low cost-efficiency, the proportion of mechanically planted seedlings in Finland and Sweden has been only a few percentages of the total amount of plantings over the last few years [24,25]. In this issue, Ersson et al. [26] discuss the key factors that may affect the future growth of mechanized planting. They conclude that the cooperation between Sweden and Finland's forest industries and research institutes is an efficient way to enhance the mechanization level of Fennoscandian tree planting.

3. Conclusions

The papers included in this Special Issue cover a broad range of aspects, ranging from cultural practices in nurseries to the field performance of seedlings under challenging environmental conditions. Broader insights into how the existing and new information could be applied to the forest regeneration chain in the future were provided. We hope that the information in this Special Issue will be useful for the progress of science in the field of silviculture.

Conflicts of Interest: The authors declare no conflict of interest.

References

1. Landis, T.D.; Dumroese, R.K.; Haase, D.L. *The Container Tree Nursery Manual. Seedling Processing, Storage and Outplanting*; Agricultural Handbook 674; U.S. Department of Agriculture, Forest Service: Washington, DC, USA, 2010; Volume 7, 199p.
2. Grossnickle, S.C.; MacDonald, J.E. Why seedlings grow: Influence of plant attributes. *New For.* **2017**, *49*, 1–34. [CrossRef]
3. Ritchie, G.A.; Landis, T.D.; Dumroese, R.K. *Assessing Plant Quality. The Container Tree Nursery Manual. Volume 7: Seedling Processing, Storage, and Outplanting*; Agriculture Handbook 674, Chapter 2: Assessing Plant Quality; Landis, T.D., Dumroese, R.K., Haase, D.L., Eds.; USDA Forest Service: Washington, DC, USA, 2010; pp. 17–81.
4. Grossnickle, S.C.; MacDonald, J.E. Seedling Quality: History, Application, and Plant Attributes. *Forests* **2018**, *9*, 283. [CrossRef]
5. Simpson, D.G.; Ritchie, G.A. Does RGP predict field performance? A debate. *New For.* **1996**, *13*, 249–273. [CrossRef]
6. Pinchot, C.C.; Hall, T.J.; Saxton, A.M.; Schlarbaum, S.E.; Bailey, J.K. Effects of Seedling Quality and Family on Performance of Northern Red Oak Seedlings on a Xeric Upland Site. *Forests* **2018**, *9*, 351. [CrossRef]
7. Pinto, J.R.; McNassar, B.A.; Kildisheva, O.A.; Davis, A.S. Stocktype and Vegetative Competition Influences on *Pseudotsuga menziesii* and *Larix occidentalis* Seedling Establishment. *Forests* **2018**, *9*, 228. [CrossRef]
8. Himanen, K.; Nygren, M. Seed soak-sorting prior to sowing affects the size and quality of 1.5-year-old containerized *Picea abies* seedlings. *Silva Fenn.* **2015**, *49*, 1056. [CrossRef]
9. Kaliniewicz, Z.; Tylek, P. Influence of Scarification on the Germination Capacity of Acorns Harvested from Uneven-Aged Stands of Pedunculate Oak (*Quercus robur* L.). *Forests* **2018**, *9*, 100. [CrossRef]
10. IPCC. *Global Warming of 1.5 °C. An IPCC Special Report on the Impacts of Global Warming of 1.5 °C above Pre-Industrial Levels and Related Global Greenhouse Gas Emission Pathways, in the Context of Strengthening the Global Response to the Threat of Climate Change, Sustainable Development, and Efforts to Eradicate Poverty*; Masson-Delmotte, V., Zhai, P., Pörtner, H.O., Roberts, D., Skea, J., Shukla, P.R., Pirani, A., Moufouma-Okia, W., Péan, C., Pidcock, R., et al., Eds.; IPCC: Geneva, Switzerland, 2018; in press.
11. Settele, J.; Scholes, R.; Betts, R.A.; Bunn, S.; Leadley, P.; Nepstad, D.; Overpeck, J.; Taboada, M.A.; Fischlin, A.; Moreno, J.M.; et al. Terrestrial and Inland water systems. In *Climate Change 2014 Impacts, Adaptation and Vulnerability: Part A: Global and Sectoral Aspects*; Cambridge University Press: Cambridge, UK, 2015; pp. 271–360. [CrossRef]
12. Vander Mijnsbrugge, K.; Turcsán, A.; Maes, J.; Duchêne, N.; Meeus, S.; Van der Aa, B.; Steppe, K.; Steenackers, M. Taxon-Independent and Taxon-Dependent Responses to Drought in Seedlings from *Quercus robur* L., *Q. petraea* (Matt.) Liebl. and Their Morphological Intermediates. *Forests* **2017**, *8*, 407. [CrossRef]
13. Santos, M.M.; Borges, E.E.L.; Ataíde, G.M.; Souza, G.A. Germination of Seeds of *Melanoxylon brauna* Schott. under Heat Stress: Production of Reactive Oxygen Species and Antioxidant Activity. *Forests* **2017**, *8*, 405. [CrossRef]
14. Wickham, J.; Wood, P.B.; Nicholson, M.C.; Jenkins, W.; Druckenbrod, D.; Suter, G.W.; Strager, M.P.; Mazzarella, C.; Galloway, W.; Amos, J. The overlooked terrestrial impacts of mountaintop mining. *Bioscience* **2013**, *63*, 335–348. [CrossRef]
15. US Environmental Protection Agency (USEPA). *Mountaintop Mining/Valley Fills in Appalachia: Final Programmatic Environmental Impact Statement*; USEPA. Report No. EPA 9-03-R-05002; USEPA: Washington, DC, USA, 2005.

16. Zipper, C.E.; Burger, J.A.; Skousen, J.G.; Angel, P.N.; Barton, C.D.; Davis, V.; Franklin, J.A. Restoring Forests and Associated Ecosystem Services on Appalachian Coal Surface Mines. *Environ. Manag.* **2011**, *47*, 751–765. [CrossRef] [PubMed]

17. Bell, G.; Sena, K.L.; Barton, C.D.; French, M. Establishing Pine Monocultures and Mixed Pine-Hardwood Stands on Reclaimed Surface Mined Land in Eastern Kentucky: Implications for Forest Resilience in a Changing Climate. *Forests* **2017**, *8*, 375. [CrossRef]

18. Hackworth, Z.J.; Lhotka, J.M.; Cox, J.J.; Barton, C.D.; Springer, M.T. First-Year Vitality of Reforestation Plantings in Response to Herbivore Exclusion on Reclaimed Appalachian Surface-Mined Land. *Forests* **2018**, *9*, 222. [CrossRef]

19. Egertsdotter, U. Plant physiological and genetical aspects of the somatic embryogenesis process in conifers. *Scand. J. For. Res.* **2018**. [CrossRef]

20. Tikkinen, M.; Varis, S.; Aronen, T. Development of Somatic Embryo Maturation and Growing Techniques of Norway Spruce Emblings towards Large-Scale Field Testing. *Forests* **2018**, *9*, 325. [CrossRef]

21. Kern, J.; Tammeorg, P.; Shanskiy, M.; Sakrabani, R.; Knicker, H.; Kammann, C.; Tuhkanen, E.-M.; Smidt, G.; Prasad, M.; Tiilikkala, K.; et al. Synergistic use of peat and charred material in growing media—An option to reduce the pressure on peatlands? *J. Environ. Eng. Landsc. Manag.* **2017**, *25*, 160–174. [CrossRef]

22. Dumroese, R.K.; Pinto, J.R.; Heiskanen, J.; Tervahauta, A.; McBurney, K.G.; Page-Dumroese, D.S.; Englund, K. Biochar Can Be a Suitable Replacement for Sphagnum Peat in Nursery Production of *Pinus ponderosa* Seedlings. *Forests* **2018**, *9*, 232. [CrossRef]

23. Hallongren, H.; Laine, T.; Rantala, J.; Saarinen, V.-M.; Strandström, M.; Hämäläinen, J.; Poikel, A. Competitiveness of mechanized tree planting in Finland. *Scand. J. For. Res.* **2014**, *29*, 144–151. [CrossRef]

24. Ersson, B.T. Concepts for Mechanized Tree Planting in Southern Sweden. Ph.D. Thesis, SLU, Umeå, Sweden, 2014.

25. Laine, T.; Kärhä, K.; Hynönen, A. A survey of the Finnish mechanized tree-planting industry in 2013 and its success factors. *Silva Fenn.* **2016**, *50*, 1323. [CrossRef]

26. Ersson, B.T.; Laine, T.; Saksa, T. Mechanized Tree Planting in Sweden and Finland: Current State and Key Factors for Future Growth. *Forests* **2018**, *9*, 370. [CrossRef]

forests

MDPI

Article

Establishing Pine Monocultures and Mixed Pine-Hardwood Stands on Reclaimed Surface Mined Land in Eastern Kentucky: Implications for Forest Resilience in a Changing Climate

Geoffrey Bell [1], Kenton L. Sena [2,*], Christopher D. Barton [2] and Michael French [3]

[1] Department of Environment and Ecology, University of North Carolina, 3305 Venable Hall
    Campus Box 3275, Chapel Hill, NC 27599, USA; gwbell@email.unc.edu
[2] Department of Forestry, University of Kentucky, 218 T. P. Cooper Bldg, Lexington, KY 40546, USA;
    barton@uky.edu
[3] Green Forests Work, 6071 N. SR 9, Hope, IN 47246, USA; michael.french@greenforestswork.org
*   Correspondence: kenton.sena@uky.edu

Received: 13 September 2017; Accepted: 29 September 2017; Published: 3 October 2017

Abstract: Surface mining and mine reclamation practices have caused significant forest loss and forest fragmentation in Appalachia. Shortleaf pine (*Pinus echinata*) is threatened by a variety of stresses, including diseases, pests, poor management, altered fire regimes, and climate change, and the species is the subject of a widescale restoration effort. Surface mines may present opportunity for shortleaf pine restoration; however, the survival and growth of shortleaf pine on these harsh sites has not been critically evaluated. This paper presents first-year survival and growth of native shortleaf pine planted on a reclaimed surface mine, compared to non-native loblolly pine (*Pinus taeda*), which has been highly successful in previous mined land reclamation plantings. Pine monoculture plots are also compared to pine-hardwood polyculture plots to evaluate effects of planting mix on tree growth and survival, as well as soil health. Initial survival of shortleaf pine is low (42%), but height growth is similar to that of loblolly pine. No differences in survival or growth were observed between monoculture and polyculture treatments. Additional surveys in coming years will address longer-term growth and survival patterns of these species, as well as changes to relevant soil health endpoints, such as soil carbon.

Keywords: reforestation; shortleaf pine; restoration ecology; mine reclamation; Appalachia; loblolly pine

---

1. Introduction:

1.1. Surface Mine Reclamation and Reforestation

Surface mining is a major driver of land use change throughout Appalachia, including eastern Kentucky. While early surface mining reclamation practices often resulted in successful post-mining forest restoration, surface mines reclaimed prior to 1978 were often characterized by haphazardly-placed mine spoils that were prone to landslides and erosion, and significantly impaired water quality. Public Law 95-87, The Surface Mining Control and Reclamation Act of 1977 (SMCRA), ushered in a new era of surface mine reclamation, requiring a return of landforms to the approximate original contour, stabilized spoil placement to eliminate landslides, and establishment of herbaceous vegetation to control erosion. Revegetation was commonly performed by hydro-seeding competitive, fast-growing nonnative species such as tall fescue (*Schedonorus arundinaceus*) and lespedeza (*Lespedeza cuneata*). Surface mines reclaimed after SMCRA were often characterized by heavily compacted spoils with poor infiltration and aeration [1]. Aggressive groundcovers competed with planted and volunteer tree seedlings for nutrients, water, and light, and the compacted soils were often not conducive to vigorous tree growth. As a result, many mining companies began implementing hay/pastureland or wildlife habitat as post-mining land uses. These reclamation practices present challenges to subsequent reforestation of reclaimed mine sites.

An estimated 600,000 ha of previously forested Appalachian land was surface mined and reclaimed to non-forest land (termed "legacy mined land") [2], perpetuating negative landscape effects of surface mining, including forest fragmentation and spread of invasive species, as well as habitat and biodiversity loss [3]. In addition to these ecological challenges, this extensive land area is mostly unmanaged and economically unproductive. Thus, this broad area of unforested land presents opportunities for ecological improvement, including restoration of threatened and endangered forest species, habitat restoration, and carbon storage, as well as short-term (e.g., restoration industry jobs) and long-term economic opportunities (e.g., timber and non-timber forest products) [4].

A team of researchers, regulators, and industry practitioners have addressed the reforestation challenges on reclaimed mine sites by developing a set of recommendations known as the Forestry Reclamation Approach (FRA) [4,5]. When these guidelines are followed during initial mine reclamation, forest establishment can be successful, with high survival and hardwood growth rates similar to regenerating stands of high-quality forests [6–8]. Additionally, reclaimed surface mined lands that currently exist as grasslands or shrublands (legacy mines) can be rehabilitated using the FRA by controlling competing vegetation, mitigating soil compaction, and planting a diverse mix of native tree and shrub seedlings [9–11].

The FRA recommends planting both early- and late-successional species [5], however, the survival and growth of planted hardwoods on legacy mined land can be restricted by severe competition from grasses and shrubs, especially tall fescue, lespedeza, and autumn olive (*Elaeagnus umbellata*) [12]. In contrast, pines typically demonstrate high survival and growth rates on legacy sites [13], rapidly achieving canopy closure and shading out competitive invasive species in the understory. The potential for pines to act as a "nurse" crop on harsh legacy sites should also be evaluated. For example, pines could be planted in monoculture stands to improve soil quality through organic matter contribution and to eliminate invasive species from the understory, and subsequently underplanted with hardwoods, which could be released in stages. Alternatively, pines and hardwoods can be planted together initially, and pines can be selectively thinned as needed.

1.2. Shortleaf Pine Restoration

Shortleaf pine (*Pinus echinata*), an economically and ecologically valuable species native throughout the southeast US, is a potential candidate for mine reforestation. Shortleaf pine forest types have experienced significant declines throughout the southeast US due in part to insect and disease pressure, extensive timber harvesting, fire suppression and poor management [14–19]. Shortleaf pine is currently the focus of a major restoration effort (Shortleaf Pine Initiative: http://shortleafpine.net/) throughout its native range [20,21] because of the suite of ecosystem services they provide. Shortleaf pine restoration leads to increased levels of plant available nutrients over time [22], in spite of initial loss of nitrogen [23]. Shortleaf pine restoration also provides important habitat for the federally endangered red-cockaded woodpecker (*Picoides borealis*), and also positively impacts diversity and/or abundance of populations of taxa including butterflies, reptiles, amphibians [24], other birds [25,26] and small mammals [27]. Shortleaf pine stands, characterized by relatively frequent fire maintaining low basal area, also provide important habitat for endangered Indiana bats (*Myotis sodalis*) [28], as well as a number of other bat species [29].

Loblolly pine (*Pinus taeda*) is another economically valuable tree species that is distributed across the southeast US, although not native to Kentucky, generally preferring poorly drained, fine-textured soils. In mixed stands, loblolly pine is commonly associated with hardwoods (including white oak) and other pines (including shortleaf pine). Loblolly pine is shallow-rooted; the majority of lateral roots are found in the top 15–46 cm of soil, especially in shallow soils with a hardpan or high water table [30].

Shortleaf pine has a broader distribution throughout the southeast US, ranging much farther north than loblolly pine, and it tolerates a broader range of climate conditions. While shortleaf pine grows best on deep, well-drained floodplain soils, it is also competitive on dry, shallow ridgetop soils, and is commonly associated with a number of hardwood and other pine species. When found in mixed stands with loblolly pine, shortleaf pine tends to be dominant in drier ridgetop sites; this is commonly attributed to shortleaf pine preferring better soil aeration and being more tolerant of poor soil fertility than loblolly pine [30].

While techniques for establishing shortleaf pine in relatively high-quality sites, such as existing hardwood forests or agricultural fields [31–34], are relatively well-understood, establishment of shortleaf pine on compaction-mitigated legacy surface mines has not yet been rigorously evaluated [35,36]. Shortleaf pine is competitive on drier ridgetop sites with frequent fire [37], but legacy mine sites can be characterized by poor infiltration resulting in ponding, which may limit site suitability for shortleaf pine. In contrast, loblolly pine prefers poorly drained soils and is more tolerant of higher moisture conditions [37], and has demonstrated good growth and survival on legacy sites in Kentucky [13].

Over even larger spatial scales and longer temporal scales, climate change represents a major threat to forest tree species, especially for species already stressed by insects, disease, and management issues [38,39]. Because trees are sessile and have long generation times, they may be particularly susceptible to the effects of rapid climate change, less resilient to changing temperatures and moisture than animals or plants with shorter generation times [40]. An option for conservation and management of forest trees with respect to climate change may be assisted migration, intentionally planting species of interest in their projected future range under climate change. Shortleaf pine is an example of a species already under significant pressure, which may be particularly threatened by climate change. With climate change projections indicating that the distribution of loblolly pine will shift north over time into Kentucky [14], the species is likely to move into these sites whether planted or not, and may potentially outcompete native species such as shortleaf pine. Focusing shortleaf pine reforestation efforts in the northern part of its range, such as eastern Kentucky, may improve its resilience to climate change.

This project was initiated to evaluate growth and survival of shortleaf pine and loblolly pine on surface mined land in eastern Kentucky grown in monoculture and in polyculture with white oak (*Quercus alba*), northern red oak (*Quercus rubra*), and chestnut oak (*Quercus montana*). This paper presents first-year growth and survival data. Long-term project goals will be assessed by follow-up surveys 5–7 years after establishment, including species effects (i.e., shortleaf pine vs. loblolly pine) and planting effects (i.e., polyculture vs. monoculture) on reforestation success, including tree (e.g., growth and survival) and soil (e.g., carbon, pH, etc.) outcomes.

2. Methods and Materials

2.1. Plot Establishment and Data Collection

A 1.3 ha plot of legacy mined land in a portion of the University of Kentucky Robinson Forest (Breathitt County, KY) was selected for this experiment (Figure 1). Exotic shrubs, primarily autumn olive (*Elaeagnus umbellata*), were removed prior to ripping using a small bulldozer (John Deere 550G). Soil compaction was mitigated by cross-ripping (plowing) the ground with a Caterpillar D-9 bulldozer equipped with two, rear-mounted ripping shanks. The two shanks were spaced approximately 2.4 m apart on the tool bar so that the two shanks were located directly behind the bulldozer's tracks. Ripping shanks were immersed approximately 1 m deep into the soil and pulled through the ground, creating parallel rips across the entire site. The bulldozer operator then turned perpendicular to the first set of parallel rips and ripped the site a second time. The experiment was set up as a split-plot design with six whole plots, each measuring 39 m × 31.7 m. Three of the plots were randomly assigned to a shortleaf pine treatment and the other three to a loblolly pine treatment. Each whole plot was divided into two

22 m × 12.2 m subplots that were randomly assigned either the pine monoculture or pine-hardwood polyculture treatment (i.e., split plot factor) (Figure 2). One-year-old bare root seedlings sourced from the Kentucky Division of Forestry were planted in March of 2016. Seedlings were planted in rows on a 2 m spacing, with 45 pines per monoculture subplot, and 22 hardwoods (red oak, white oak, and chestnut oak) and 23 pines per polyculture subplot. The buffer space outside the border of the split plots but within the whole plots was planted with seedlings for the pine species assigned to the whole plot.

Height and ground-line diameter were recorded for each individual at time of planting (spring 2016), and measurements were repeated after one year (spring 2017). In addition, soil samples (composited from six subsamples) were collected in duplicate at random in each subplot both at planting and after one year, and samples were analyzed for the following parameters: soil pH, P, K, Ca, Mg, and Zn. Additional soil analyses conducted only in 2017 included the following: total N, sand, silt, clay, CEC, total C, and exchangeable K, Ca, Mg, and Na. Sand, silt, and clay were determined by the micropipette method [41]; pH was determined in a 1:1 soil:water solution [42]. P, K, Ca, and Mg, were analyzed by Mehlich-III extraction [43]. Cation exchange capacity was determined by the ammonium acetate method at pH 3 [44]. Exchangeable base concentration was evaluated after ammonium acetate extraction using ICP [43]. Total N (%) and total C (%) were determined on a LECO CHN-2000 Analyzer (LECO Corporation, St. Joseph, MI, USA).

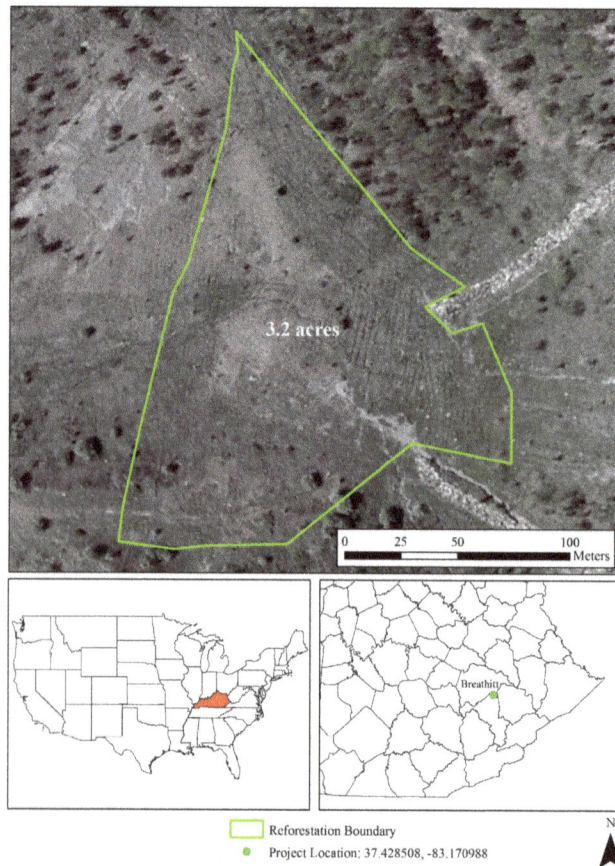

Figure 1. Plot location, Breathitt County, KY. (Figure credit: Kylie Schmidt).

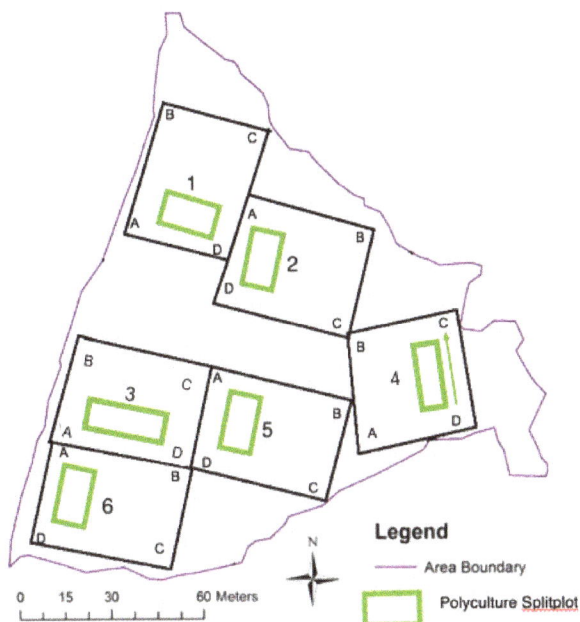

Figure 2. Whole plot (1–6) and subplot configuration of shortleaf pine and loblolly pine monoculture and pine/hardwood polyculture plantings in rehabilitated legacy mined land in eastern Kentucky.

2.2. Statistical Methods

Statistical analyses were conducted in SAS 9.3. Soils data collected in both 2016 and 2017 were analysed by repeated measures ANOVA using PROC MIXED, with subplot as the experimental unit. Planting mix (polyculture vs. monoculture) and species (loblolly pine vs. shortleaf pine), and their interaction, were modelled as fixed effects, replicate (each treatment replicated 3 times) modelled as a random effect, and year modelled in the repeated statement. Soils data collected in 2017 only were analysed by ANOVA using PROC GLM, with planting mix, species, and their interaction modelled as effects, with three replicates.

Tree height change was averaged by species for each subplot, and subplot means were treated as the experimental unit. Differences in change in tree height were detected by split-plot ANOVA using PROC GLM, with species, planting mix, and their interaction, modelled as effects. Tree survival was analysed using PROC GLIMMIX, with survival proportions calculated for each subplot as the experimental unit, and species, planting mix, and their interaction modelled as effects. Significant differences detected by all ANOVAs were followed up by a student's *t*-test to detect pairwise differences.

3. Results

Soil chemical and physical data are reported in Table 1. Of the soil chemical data assessed in both 2016 and 2017, only pH was significantly different, increasing slightly from 5.74 to 6.18 ($p < 0.05$). K, Mg, and Zn were significantly higher in loblolly pine than in shortleaf pine, and Zn was significantly higher in monoculture than polyculture ($p < 0.05$). Total N and exchangeable Mg were higher in loblolly pine than shortleaf pine plots ($p < 0.05$).

Table 1. Soil data (means ± SE) for soil samples collected from reforestation plots (three plots planted in loblolly pine and three plots planted in shortleaf pine) in Eastern Kentucky. Each plot was subdivided into pine-hardwood polyculture and pine-only monoculture subplots. Means with differing letters are significantly different, as detected by ANOVA and followed up by a student's *t*-test, at $p < 0.05$. "Exch" = "Exchangeable".

	Year		Pine		Planting Mix	
	2016	2017	Shortleaf Pine	Loblolly Pine	Monoculture	Polyculture
Soil pH	5.74b ± 0.31	6.18a ± 0.31	6.20 ± 0.42	5.72 ± 0.42	6.06 ± 0.42	5.86 ± 0.42
P (mg/kg)	6.92 ± 1.27	7.67 ± 1.27	6.79 ± 1.56	7.79 ± 1.56	7.83 ± 1.56	6.75 ± 1.56
K (mg/kg)	91.2 ± 6.24	78.6 ± 6.24	67.9b ± 6.58	101.9a ± 6.58	91.0 ± 6.58	78.8 ± 6.58
Ca (mg/kg)	996 ± 408	1409 ± 408	773 ± 529	1633 ± 529	1178 ± 529	1227 ± 529
Mg (mg/kg)	216.7 ± 16.5	206.1 ± 16.5	159.9b ± 22.4	262.9a ± 22.4	213.7 ± 22.4	209.1 ± 22.4
Zn (mg/kg)	3.09 ± 0.08	3.06 ± 0.08	2.28b ± 0.08	3.87a ± 0.08	3.39a ± 0.08	2.76b ± 0.08
Total N (%)	-	-	0.104b ± 0.014	0.196a ± 0.016	0.162 ± 0.023	0.138 ± 0.018
Sand (%)	-	-	62.7 ± 3	53.7 ± 4	58.0 ± 4	58.4 ± 4
Silt (%)	-	-	25.4 ± 2	32.7 ± 3	29.0 ± 3	29.0 ± 3
Clay (%)	-	-	12 ± 9	13.6 ± 1.2	12.9 ± 1.2	12.6 ± 0.9
CEC (meq/100 g)	-	-	7.46 ± 1.13	12.94 ± 1.20	10.84 ± 1.65	9.56 ± 1.13
Exch K (meq/100 g)	-	-	0.158 ± 0.02	0.308 ± 0.04	0.247 ± 0.04	0.219 ± 0.03
Exch Ca (meq/100 g)	-	-	3.58 ± 1.68	7.95 ± 2.09	6.63 ± 2.13	4.90 ± 1.85
Exch Mg (meq/100 g)	-	-	1.13b ± 0.16	2.26a ± 0.21	1.67 ± 0.23	1.72 ± 0.27
Exch Na (meq/100 g)	-	-	0.023 ± 0.004	0.026 ± 0.004	0.026 ± 0.005	0.023 ± 0.004
Total C (%)	-	-	0.022 ± 0.004	0.034 ± 0.002	0.029 ± 0.004	0.027 ± 0.003

After one growing season, most seedlings experienced positive growth in their height (77%) and diameter (72%). Negative height growth was related to deer and elk browse that sheared the tops off of the seedlings. Diameter growth did not differ between the two pine species, averaging 0.22 cm and ranging between −0.6 cm and 1.79 cm (Figure 3). Hardwood diameter growth was about half that of the pines with highest growth in white oaks (mean = 0.1 cm; range = −0.6 cm–1.1 cm) followed by chestnut oak (mean = 0.08 cm; range = −0.25 cm–0.5 cm), and red oak (mean = 0.06 cm; range = −0.8 cm–0.65 cm) (Figure 4). A similar species-specific pattern was observed in height growth. Individual loblolly pine seedling growth ranged from −11 cm to 69.3 cm and loblollies had the largest average height increase (16.02 cm), which was significantly greater than all the hardwoods but not shortleaf pine. Shortleaf pine height growth ranged from −19.8 cm to 72.5 cm with an average (10.51 cm) that was approximately 5.5 cm less than loblolly pine. Shortleaf pine height growth was not significantly different from loblolly pine growth but was significantly larger than two of the three hardwoods. White oak seedling height growth ranged from −32.2 cm to 52.5 cm. White oaks had the largest height growth among the hardwoods and was the only hardwood species to achieve a positive average height growth (5.65 cm). Although many red oak seedlings experienced positive height growth, ranging from −27.9 cm to 19 cm, their average was negative (−0.71 cm). Similarly, chestnut oak height growth ranged from −15.5 cm to 7.5 cm and averaged −1.19 cm. Despite the range of height growths among the hardwoods, none were significantly different from each other. Collectively, these results suggest that diameter growth was similar among all five species but that the pines grew taller than hardwoods, with the exception of no significant difference between shortleaf pines and white oaks.

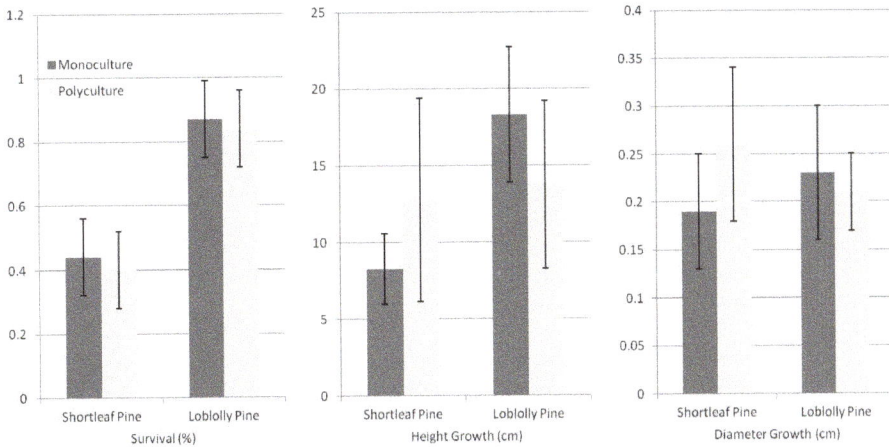

Figure 3. First-year growth and percent survival of shortleaf pine and loblolly pine planted in pine monoculture and pine-hardwood polyculture on rehabilitated legacy mine land in eastern Kentucky.

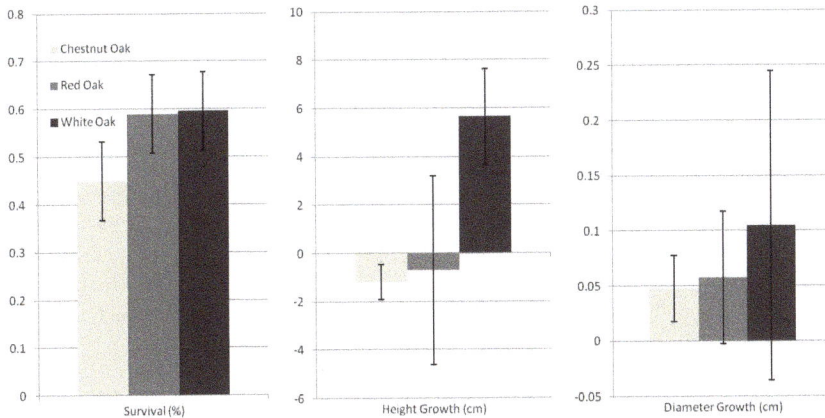

Figure 4. First-year growth and percent survival of white oak, chestnut oak, and northern red oak planted in pine-hardwood polyculture on rehabilitated mine land in eastern Kentucky.

4. Discussion

Planting mix (polyculture vs. monoculture) did not significantly influence tree growth or survival; however, growth and survival varied with species. Shortleaf pine survival (42%) was similar to that of planted hardwoods, but lower than that of loblolly pine (85%) ($p < 0.05$). Survival of shortleaf pine was lower than first-year survival reported by Angel (2008) of mixed hardwoods planted into mixed mine spoils with no vegetative competition in eastern Kentucky (69–98%) [6,45], lower than survival (65–75%) of seedlings planted into spoils seeded with groundcover species [46], and lower than first-year chestnut survival in legacy mined land in eastern Kentucky (72–97%) [9]. However, shortleaf pine survival was similar to survival of seedlings planted into mine spoil seeded with groundcover species (56%) [47], and greater than survival of shagbark hickory (*Carya ovata*) reported in the same study (24%) [47]. In contrast to relatively low shortleaf pine survival, loblolly pine survival (85%) was greater than first-year loblolly pine survival reported in Oklahoma by Dipesh et al. (2015) (76%) [48] and fourth-year survival of another loblolly pine planting near Robinson Forest

(77%) [13]. While first-year survival of shortleaf pine was relatively low, this planting could still be successful if ongoing mortality rates are low; subsequent surveys will be necessary to evaluate long-term suitability of shortleaf pine on these sites. Also, consistent with previous studies at Robinson Forest and elsewhere, first-year loblolly pine survival was high, supporting continued use of loblolly pine in mine reforestation efforts.

While first-year survival was lower, first-year growth of shortleaf pine (10 cm) was similar to that of loblolly pine (16 cm). Growth of loblolly pine dramatically outpaced growth of northern red oak in an adjacent site in legacy mined land in eastern Kentucky [49]. In that study, loblolly pine rapidly overgrew competing vegetation and shaded it out (in 4–8 yrs), leading to a bare understory characterized by a thick pine needle litter layer [13]. In contrast, the northern red oaks in Hansen et al. (2015) struggled against competing vegetation and were not successful in achieving canopy closure even after 10 years [13]. While loblolly pine has demonstrated its ability to rapidly outcompete nonnative vegetation in these conditions, this has yet to be seen with shortleaf pine. Further monitoring of our plots over the next several years should demonstrate whether shortleaf pine can compensate for its lower survival and be a reasonable candidate for reforestation on reclaimed surface mined land. Heavy competition from *Miscanthus* spp. and other herbaceous species (lespedeza and fescue) in these plots will likely be the most significant impediment to shortleaf pine survival and growth. Hardwood growth in this study was low, even negative in two species (chestnut oak and red oak), likely due to browse by deer and elk. Browse was observed on this site, as on many other similar plantings, and can significantly affect growth and survival [9,50,51]. Regardless of browse, hardwood growth tends to be low during the first 2–3 years after planting [45], with growth rates increasing after this 2–3 yr establishment period [6].

Higher survival of loblolly pine than shortleaf pine on our site is likely due to loblolly pine being favored by site soil moisture and chemistry conditions. Loblolly pine is more tolerant of poorly drained soils than shortleaf pine [37]. Large portions of the project site exhibited poor drainage and even standing water (which can frequently be the case on these sites [52,53]), suggesting that overall soil moisture conditions may be more favorable for loblolly pine than shortleaf pine. Chemically, soils were favorable across treatments, with pH, particle size distribution, nutrient levels, and CEC similar to those observed on soils favorable for tree growth and survival in another eastern Kentucky study [6]. However, the soils in loblolly pine plots in this study were chemically more favorable than the soils in shortleaf pine plots, with higher total N and exchangeable Mg.

The current study continues to provide support for the use of loblolly pine in surface mine restoration plantings; however, low first-year survival of shortleaf pine is concerning. Additional studies investigating survival and growth over time will provide additional valuable information about the potential for surface mines as shortleaf pine restoration sites. Also, the unique design of this project presents opportunity for investigation of more complex restoration ecology questions, specifically (1) whether pines planted in mixtures with hardwoods experience greater growth and survival than pines planted in monoculture and (2) whether rapid pine establishment can sufficiently reduce invasive species competition and improve soil health so as to act as a "nurse crop" for subsequent high-value hardwood species release. Finally, this experiment presents an opportunity for the long-term comparison of loblolly pine and shortleaf pine that will offer insights into restoration strategies involving these species, especially under climate change. Specifically, insights on whether restoration practitioners should consider species not historically native to a state or region (e.g., loblolly pine in Kentucky) suitable for restoration, given that climate change will shift their distributions.

Acknowledgments: This project was partially funded by a Sustainability Challenge Grant from the Tracey Farmer Institute for Sustainability and the Environment (TFISE); additional funds were provided by the National Fish and Wildlife Foundation (NFWF) through the Appalachian Forest Renewal Initiative. The authors gratefully acknowledge the assistance of the following University of North Carolina students, who assisted in developing the experimental design, conducting the planting, and collecting data: Joshua Dickens, Helen Drotor, Saideep Gona, Veronica Kapoor, Peter Oliveira, Levi Rolles, Alicia Wood, Jacob Baldwin, Emma Bogerd, Caroline Durham,

Megan Lott, Tyler Niles, and Victoria Triana. Participation by undergraduate students from University of North Carolina was generously funded by Brad Stanback and Shelli Lodge-Stanback.

Author Contributions: K.L.S., M.F., and C.D.B. obtained funding; C.D.B., M.F., and G.B. designed and implemented the planting; G.B., C.D.B., and K.L.S. collected data; K.L.S. analyzed data and prepared the manuscript.

Conflicts of Interest: The authors declare no conflict of interest.

References

1. Haering, K.C.; Daniels, W.L.; Galbraith, J.M. Appalachian mine soil morphology and properties: Effects of weathering and mining method. *Soil Sci. Soc. Am. J.* **2004**, *68*, 1315–1325. [CrossRef]
2. Zipper, C.E.; Burger, J.A.; McGrath, J.M.; Rodrigue, J.A.; Holtzman, G.I. Forest restoration potentials of coal-mined lands in the eastern United States. *J. Environ. Qual.* **2011**, *40*, 1567–1577. [CrossRef] [PubMed]
3. Wickham, J.; Wood, P.B.; Nicholson, M.C.; Jenkins, W.; Druckenbrod, D.; Suter, G.W.; Strager, M.P.; Mazzarella, C.; Galloway, W.; Amos, J. The overlooked terrestrial impacts of mountaintop mining. *Bioscience* **2013**, *63*, 335–348. [CrossRef]
4. Zipper, C.E.; Burger, J.A.; Skousen, J.G.; Angel, P.N.; Barton, C.D.; Davis, V.; Franklin, J.A. Restoring forests and associated ecosystem services on Appalachian coal surface mines. *Environ. Manag.* **2011**, *47*, 751–765. [CrossRef] [PubMed]
5. Davis, V.; Burger, J.A.; Rathfon, R.; Zipper, C.E.; Miller, C.R. Chapter 7: Selecting Tree Species for Reforestation of Appalachian Mined Lands. In *The Forestry Reclamation Approach: Guide to Successful Reforestation of Mined Lands*; Adams, M.B., Ed.; Gen. Tech. Rep. NRS-169; U.S. Department of Agriculture, Forest Service, Northern Research Station: Newtown Square, PA, USA, 2017; pp. 7-1–7-10.
6. Sena, K.; Barton, C.; Hall, S.; Angel, P.; Agouridis, C.; Warner, R. Influence of spoil type on afforestation success and natural vegetative recolonization on a surface coal mine in Appalachia, United States. *Restor. Ecol.* **2015**, *23*, 131–138. [CrossRef]
7. Wilson-Kokes, L.; Emerson, P.; DeLong, C.; Thomas, C.; Skousen, J. Hardwood tree growth after eight years on brown and gray mine soils in West Virginia. *J. Environ. Qual.* **2013**, *42*, 1353–1362. [CrossRef] [PubMed]
8. Cotton, C.; Barton, C.; Lhotka, J.; Angel, P.N.; Graves, D. Evaluating reforestation success on a surface mine in eastern Kentucky. In *Tech. Coords. National Proceedings: Forest and Conservation Nursery Associations-2011*; USDA Forest Service, Rocky Mountain Research Station: Fort Collins, CO, USA, 2012; pp. 16–23.
9. Sena, K.; Angel, H.; Barton, C. Influence of tree shelters and weed mats on growth and survival of backcrossed chestnut seedlings on legacy minelands in eastern Kentucky. *J. Am. Soc. Min. Reclam.* **2014**, *3*, 41–63. [CrossRef]
10. Fields-Johnson, C.W.; Burger, J.A.; Evans, D.M.; Zipper, C.E. Ripping improves tree survival and growth on unused reclaimed mined lands. *Environ. Manag.* **2014**, *53*, 1059–1065. [CrossRef] [PubMed]
11. Skousen, J.; Gorman, J.; Pena-Yewtukhiw, E.; King, J.; Stewart, J.; Emerson, P.; Delong, C. Hardwood tree survival in heavy ground cover on reclaimed land in West Virginia: Mowing and ripping effects. *J. Environ. Qual.* **2009**, *38*, 1400–1409. [CrossRef] [PubMed]
12. Barton, C.D.; Sweigard, R.J.; Marx, D.; Barton, W. Evaluating spoil amendment use and mycorrhizal inoculation on reforestation success in the eastern and western Kentucky coalfields. In Proceedings of the American Society of Mining and Reclamation, Richmond, VA, USA, 14–19 June 2008; Barnhisel, R.I., Ed.; pp. 98–111.
13. Hansen, E.; Barton, C.; Drayer, A. Challenges for native forest establishment on surface mines in a time of climate change. *Reclam. Matters* **2015**, *Spring*, 36–39.
14. Will, R.; Stewart, J.; Lynch, T.; Turton, D.; Maggard, A.; Lilly, C.; Atkinson, K. *Strategic Assessment for Shortleaf Pine*; Oklahomoa Forestry Services: Washington, OK, USA, 2013; p. 58.
15. Oswalt, C.M. Spatial and Temporal Trends of the Shortleaf Pine Resource in the Eastern United States. In Proceedings of the Shortleaf Pine Conference, East Meets West, Huntsville, AL, USA, 20–22 September 2011; Kush, J., Barlow, R.J., Gilbert, J.C., Eds.; 2012; p. 33.
16. Campbell, W.A.; Copeland, J.O.L. *Littleleaf Diseases of Shortleaf and Loblolly Pines*; USDA Circular No. 940; USDA: Washington, DC, USA, 1954; Volume 41.

17. Clarke, S.R.; Riggins, J.J.; Stephen, F.M. Forest management and southern pine beetle outbreaks: A historical perspective. *For. Sci.* **2016**, *62*, 166–180. [CrossRef]
18. Coyle, D.R.; Klepzig, K.D.; Koch, F.H.; Morris, L.A.; Nowak, J.T.; Oak, S.W.; Otrosina, W.J.; Smith, W.D.; Gandhi, K.J. A review of southern pine decline in North America. *For. Ecol. Manag.* **2015**, *349*, 134–148. [CrossRef]
19. Coyle, D.R.; Green, G.T.; Barnes, B.F.; Klepzig, K.D.; Nowak, J.T.; Gandhi, K.J. Landowner and manager awareness and perceptions of pine health issues and southern pine management activities in the southeastern United States. *J. For.* **2016**, *114*, 541–551. [CrossRef]
20. Liechty, H.O.; Luckow, K.R.; Daniel, J.S.; Marion, D.A.; Spetich, M.; Guldin, J.M. Shortleaf pine ecosystem restoration: Impacts on soils and woody debris in the Ouachita Mountains of the southern United States. In Proceedings of the 16th International Conference, Society for Ecological Restoration, Victoria, BC, Canada, 24–26 August 2004; p. 5.
21. Stewart, J.F.; Will, R.E.; Crane, B.S.; Nelson, C.D. The genetics of shortleaf pine (*Pinus echinata* mill.) with implications for restoration and management. *Tree Genet. Genom.* **2016**, *12*, 98. [CrossRef]
22. Liechty, H.O.; Luckow, K.R.; Guldin, J.M. Soil chemistry and nutrient regimes following 17–21 years of shortleaf pine-bluestem restoration in the Ouachita Mountains of Arkansas. *For. Ecol. Manag.* **2005**, *204*, 345–357. [CrossRef]
23. Hubbard, R.M.; Vose, J.M.; Clinton, B.D.; Elliott, K.J.; Knoepp, J.D. Stand restoration burning in oak–pine forests in the southern Appalachians: Effects on aboveground biomass and carbon and nitrogen cycling. *For. Ecol. Manag.* **2004**, *190*, 311–321. [CrossRef]
24. Perry, R.W.; Rudolph, D.C.; Thill, R.E. Reptile and amphibian responses to restoration of fire-maintained pine woodlands. *Restor. Ecol.* **2009**, *17*, 917–927. [CrossRef]
25. Thill, R.E.; Rudolph, D.C.; Koerth, N.E. Shortleaf pine-bluestem restoration for red-cockaded woodpeckers in the Ouachita Mountains: Implications for other taxa. In *Red-Cockaded Woodpecker: Road to Recovery*; Costa, R., Daniels, S.J., Eds.; Hancock House Publishers: Blaine, WA, USA, 2004; pp. 657–671.
26. Conner, R.N.; Shackelford, C.E.; Schaefer, R.R.; Saenz, D.; Rudolph, D.C. Avian community response to southern pine ecosystem restoration for red-cockaded woodpeckers. *Wilson Bull.* **2002**, *114*, 324–332. [CrossRef]
27. Masters, R.E.; Lochmiller, R.L.; McMurry, S.T.; Bukenhofer, G.A. Small mammal response to pine-grassland restoration for red-cockaded woodpeckers. *Wildl. Soc. Bull.* **1998**, 148–158.
28. O'Keefe, J.M.; Loeb, S.C. Indiana bats roost in ephemeral, fire-dependent pine snags in the southern Appalachian Mountains, USA. *For. Ecol. Manag.* **2017**, *391*, 264–274. [CrossRef]
29. Cox, M.R.; Willcox, E.V.; Keyser, P.D.; Vander Yacht, A.L. Bat response to prescribed fire and overstory thinning in hardwood forest on the Cumberland Plateau, Tennessee. *For. Ecol. Manag.* **2016**, *359*, 221–231. [CrossRef]
30. Burns, R.M.; Honkala, B.H. Tech. coords. 1990. Silvics of North America: 1. Conifers. In *Agriculture Handbook 654*; USDA Forest Service: Washington, DC, USA, 1990; Volume 1, p. 675.
31. Pile, L.S.; Waldrop, T. *Shortleaf Pine and Mixed Hardwood Stands: Thirty-Four Years after Regeneration with the Fell-and-Burn Technique in the Southern Appalachian Mountains*; US Department of Agriculture Forest Service, Southern Research Station: Asheville, NC, USA, 2016; Volume SRS-56, pp. 1–7.
32. Smalley, G.W.; Bower, D.R. *Site Index Curves for Loblolly and Shortleaf Pine Plantations on Abandoned Fields in Tennessee, Alabama, and Georgia Highlands*; USDA Forest Service, Southern Research Station: Asheville, NC, USA, 1971.
33. Stambaugh, M.C.; Guyette, R.P.; Dey, D.C. *What Fire Frequency Is Appropriate for Shortleaf Pine Regeneration and Survival?* USDA Forest Service Northern Research Station: Newtown Square, PA, USA, 2007; Volume NRS-P-15, pp. 121–128.
34. Stewart, J.F.; Will, R.E.; Robertson, K.M.; Nelson, C.D. Frequent fire protects shortleaf pine (*Pinus echinata*) from introgression by loblolly pine (*P. taeda*). *Conserv. Genet.* **2015**, *16*, 491–495. [CrossRef]
35. Ashby, W.C.; Baker, M.B. Soil nutrients and tree growth under black locust and shortleaf pine overstories in strip-mine plantings. *J. For.* **1968**, *66*, 67–71.
36. Walker, R.; West, D.; McLaughlin, S.; Amundsen, C. Performance of loblolly, virginia, and shortleaf pine on a reclaimed surface mine as affected by *Pisolithus tinctorius* ectomycorrhizae and fertilization. In Proceedings of the Biennial Southern Silvicultural Research Conference, Atlanta, GA, USA, 15–17 April 1985.

37. Mattoon, W.R. *Life History of Shortleaf Pine*; US Department of Agriculture Bulletin 244; US Department of Agriculture: Washington, DC, USA, 1915.

38. Butler, P.R.; Iverson, L.; Thompson, F.R.; Brandt, L.; Handler, S.; Janowiak, M.; Shannon, P.D.; Swanston, C.; Karriker, K.; Bartig, J. *Central Appalachians Forest Ecosystem Vulnerability Assessment and Synthesis: A Report from the Central Appalachians Climate Change Response Framework Project*; U.S. Department of Agriculture, Forest Service, Northern Research Station: Newtown Square, PA, USA, 2015.

39. McNab, W.H.; Spetich, M.A.; Perry, R.W.; Haywood, J.D.; Laird, S.G.; Clark, S.L.; Hart, J.L.; Torreano, S.J.; Buchanan, M.L. Climate-induced migration of native tree populations and consequences for forest composition. In *Climate Change Adaptation and Mitigation Management Options: A Guide for Natural Resource Managers in Southern Forest Ecosystems*; CRC Press, Taylor & Francis Group: Boca Raton, FL, USA, 2014; pp. 307–378.

40. Zolkos, S.G.; Jantz, P.; Cormier, T.; Iverson, L.R.; McKenney, D.W.; Goetz, S.J. Projected tree species redistribution under climate change: Implications for ecosystem vulnerability across protected areas in the eastern United States. *Ecosystems* **2015**, *18*, 202–220. [CrossRef]

41. Miller, W.; Miller, D. A micro-pipette method for soil mechanical analysis. *Commun. Soil Sci. Plant Anal.* **1987**, *18*, 1–15. [CrossRef]

42. Thomas, G. Soil ph and soil acidity. In *Methods of Soil Analysis Part 3—Chemical Methods*; Soil Science Society of America, American Society of Agronomy: Madison, WI, USA, 1996; pp. 475–490.

43. Soil and Plant Analysis Council. *Soil Analysis Handbook of Reference Methods*; CRC Press: Boca Raton, FL, USA, 2000.

44. Summer, M.E.; Miller, W.P. Cation exchange capacity and exchange coefficients. In *Methods of Soil Analysis. Part 3. Chemical Methods*; Sparks, D., Bartels, J.M., Eds.; Soil Science Society of America, American Society of Agronomy: Madison, WI, USA, 1996.

45. Angel, P.; Barton, C.; Warner, R.; Agouridis, C.; Taylor, T.; Hall, S. Forest establishment and water quality characteristics as influenced by spoil type on a loose-graded surface mine in eastern Kentucky. In Proceedings of the American Society of Mining and Reclamation, Richmond, VA, USA, 14–19 June 2008; pp. 28–65.

46. Fields-Johnson, C.W.; Zipper, C.E.; Burger, J.A.; Evans, D.M. First-year response of mixed hardwoods and improved american chestnuts to compaction and hydroseed treatments on reclaimed mine land. In Proceedings of the American Society of Mining and Reclamation, Billings, MT, USA, 30 May–5 June 2009; Barnhisel, R.I., Ed.; pp. 413–432.

47. Koropchak, S.; Zipper, C.; Burger, J.; Evans, D. Native tree survival and herbaceous establishment on an experimentally reclaimed Appalachian coal mine. *J. Am. Soc. Min. Reclam.* **2013**, *2*, 32–55. [CrossRef]

48. Dipesh, K.; Will, R.E.; Lynch, T.B.; Heinemann, R.; Holeman, R. Comparison of loblolly, shortleaf, and pitch x loblolly pine plantations growing in Oklahoma. *For. Sci.* **2015**, *61*, 540–547. [CrossRef]

49. Michels, A.; Barton, C.; Cushing, T.; Angel, P.; Sweigard, R.; Graves, D. Evaluation of low spoil compaction techniques for hardwood forest establishment on an eastern Kentucky surface mine. In Proceedings of the American Society of Mining and Reclamation, Gillette, WY, USA, 2–6 June 2007; Barnhisel, R.I., Ed.; pp. 2–6.

50. Stange, E.E.; Shea, K.L. Effects of deer browsing, fabric mats, and tree shelters on *Quercus rubra* seedlings. *Restor. Ecol.* **1998**, *6*, 29–34.

51. Robertson, D.J. Trees, deer, and non-native vines: Two decades of northern piedmont forest restoration. *Ecol. Restor.* **2012**, *30*, 59–70. [CrossRef]

52. Clark, E.V.; Zipper, C.E. Vegetation influences near-surface hydrological characteristics on a surface coal mine in eastern USA. *Catena* **2016**, *139*, 241–249. [CrossRef]

53. Evans, D.M.; Zipper, C.E.; Hester, E.T.; Schoenholtz, S.H. Hydrologic effects of surface coal mining in Appalachia (US). *J. Am. Water Resour. Assoc.* **2015**, *51*, 1436–1452. [CrossRef]

Article

Germination of Seeds of *Melanoxylon brauna* Schott. under Heat Stress: Production of Reactive Oxygen Species and Antioxidant Activity

Marcone Moreira Santos [1,*], Eduardo Euclydes de Lima e Borges [1], Glauciana da Mata Ataíde [2] and Genaina Aparecida de Souza [3]

[1]   Department of Forest Engineering, Federal University of Viçosa, 36570-000 Viçosa, Brazil; elborges@ufv.br
[2]   Department of Crop Sciences, Federal University of São João Del Rei, 35701-970 Sete Lagoas, Brazil; glaucianadamata@yahoo.com.br
[3]   Department of Plant Biology, Federal University of Viçosa, 36570-000 Viçosa, Brazil; genainasouza@yahoo.com.br
*   Correspondence: marconemoreirasantos@hotmail.com; Tel.: +55-31-997-505-923

Received: 9 August 2017; Accepted: 9 October 2017; Published: 25 October 2017

Abstract: In this article, the authors aimed to analyze the physiological and biochemical alterations in *Melanoxylon brauna* seeds subjected to heat stress. For this, seed germination, electric conductivity (EC), the production of reactive oxygen species (ROS), and the activity of antioxidant enzymes were assessed. Seeds were incubated at constant temperatures of 25, 35, and 45 °C. Independent samples were first incubated at 35 and 45 °C and then transferred to 25 °C after the intervals of 24, 48, 72, and 96 h. To evaluate EC, seeds were soaked for 0, 24, 48, and 72 h, at 25, 35, and 45 °C and then transferred to Erlenmeyer flasks containing 75 mL of deionized water at 25 °C, for 24 h. ROS production and enzyme activity were assessed every 24 h in seeds soaked at the aforementioned temperatures. Germination did not occur at 45 °C. Seeds soaked at 35 °C for 72 h and then transferred to 25 °C showed higher percentages of germination and a higher germination speed. Seed soaking at 45 °C increased peroxide production, which compromised the antioxidant enzyme system due to a reduction in the activity of enzymes APX, POX, and CAT, thus ultimately also compromising the cell membrane system.

Keywords: climate change; seeds; physiological quality; antioxidant enzymes

1. Introduction

The projections from the Brazilian Panel on Climate Change show that the global temperature will increase throughout the century. Such change might range from 1 to 5 °C until the end of this time period [1]. Considering the possibility of this temperature increase in the next years, the following questions remain: how will species adapt to such change and how can we interfere so that they do not disappear?

Melanoxylon brauna (Fabaceae-Caesalpinioideae), also known as brauna, is a native species to the Atlantic Forest, occurring in the Brazilian states of Bahia, São Paulo, Minas Gerais, Espírito Santo, Pará, and Rio de Janeiro [2]. The wood species is dense and highly used in the sailing industry, as well as in construction and the manufacture of light poles and furniture [3]. The species also has ornamental features, being used in afforestation and landscaping projects, as well as in folk medicine [2,4].

Brauna is currently included in the "Official List of Species from the Brazilian Flora Threatened with Extinction", under the 'vulnerable' category, according to the Brazilian Ministry of Environment [5]. In view of these factors, studies approaching seed physiology and germination represent starting points for the development of new strategies to preserve the brauna species [6].

Seed germination is influenced by environmental factors such as temperature, which can be manipulated to optimize the percentage, speed, and uniformity of germination, resulting in more vigorous seedlings and lower production costs [7,8]. Temperature affects water absorption by the seed and the biochemical reactions that regulate the entire seed metabolic process [9]. The temperature range in which germination occurs varies amongst species, and thus each species may have a base and an optimal germination temperature. Generally, the range of 20 to 30 °C is adequate for the germination of many subtropical and tropical species [10–12]. In brauna, for instance, the range between 25 and 30 °C is considered optimal for seed germination [13].

Heat stress increases the production and accumulation of reactive oxygen species (ROS) in seeds [14]. ROS include free radicals such as the superoxide anion ($O_2^{\bullet-}$), hydroxyl radical ($^{\bullet}OH$), and molecules that are not considered free radicals, like hydrogen peroxide (H_2O_2) and singlet oxygen ($^1O_2$) [15]. ROS are formed either due to excess energy in plants, specifically in chloroplasts, mitochondria, and plasma membranes; or as byproducts of metabolic pathways in different cell compartments [16]. Excess ROS are highly damaging, and when the levels of these molecules exceed the capacity of the defense mechanisms which scavenge them, cells undergo oxidative stress [17].

Plant cells have efficient enzymatic mechanisms for ROS removal, which enables them to remain undamaged by intoxication. Temperature affects the removal capacity of ROS, as it determines the activation and action of the enzymes superoxide dismutase (SOD), catalase (CAT), ascorbate peroxidase (APX), and peroxidase (POX), which are the main responsible agents for ROS scavenging [18].

Considering the ecological and economic importance of brauna and the influence of environmental conditions on seed germination, we aimed to evaluate the physiological and biochemical alterations that occur during the germination of *Melanoxylon brauna* seeds subjected to heat stress.

2. Materials and Methods

The experiments were performed between February and August 2016. *Melanoxylon brauna* fruits were collected in the Leopoldina municipality, in the State of Minas Gerais, southeastern Brazil (21°31′55″ S and 42°38′35″ W), in September 2015. Fruits were dried in the sun until opening and seeds were then extracted manually.

Seeds were incubated in water, in petri dishes, at 25, 35, and 45 °C under constant light. Another test was performed aiming to evaluate possible damage to the seeds after exposure to stressful temperatures. For that, independent samples were first incubated at 35 and 45 °C under the same previously described conditions, and then transferred to 25 °C after 24, 48, 72, and 96 h, after which they were evaluated for the germination percentage and germination speed index (GSI).

Seeds were considered germinated when the primary root emerged. GSI was calculated by Maguire's equation [19], with replicates of 20 seeds per treatment.

To evaluate electric conductivity (EC), seeds were soaked for 0, 24, 48, and 72 h at 25, 35, and 45 °C and then transferred to Erlenmeyer flasks containing 75 mL of deionized water at 25 °C for 24 h. EC of the solution was determined by a MICRONAL conductivity-meter, as described by Woodstock [20]. The variable was assessed in five replicates of 20 seeds and the results were expressed in $\mu S \, cm^{-1} \, seed^{-1}$.

The effect of temperature on ROS production, lipid peroxidation, and enzyme activity was evaluated throughout germination. The analyses were performed on the embryo axis of seeds soaked for 0, 24, 48, and 72 h at 25, 35, and 45 °C.

Superoxide was analyzed as described by Mohammadi and Kar [21]. Superoxide anion production was evaluated by determining the amount of accumulated adrenochrome [22], using a coefficient of molar absorptivity of $4.0 \times 10^3 \, M^{-1}$ [23].

Samples of 50 mg of embryonic axis and micropylar endosperm used to quantify hydrogen peroxide were crushed and homogenized in 2.0 mL of 50 mM potassium phosphate buffer, followed by centrifugation at $8400 \times g$ for 15 min at 4 °C, after which the supernatant was collected [24]. Aliquots of 100 µL of the supernatant were added to the reaction medium, which consisted of 250 µM ferrous ammonium sulfate, 25 mM sulfuric acid, 250 µM xylenol orange, and 100 mM sorbitol, in a final volume of 2 mL [25]. The mixture was then homogenized and kept in the dark for 30 min. Absorbance was determined by a spectrophotometer at 560 nm. Contents of H_2O_2 were quantified based on the calibration curve, using the peroxide concentration as a standard. Plant extracts were obtained from samples while analytical blanks were prepared in parallel.

Lipid peroxidation was evaluated by determining the TBA (thiobarbituric acid) concentration [26]. The results were expressed as mg MDA g^{-1} FW, after absorbance conversion [27]. Three replicates were used per treatment.

To evaluate enzyme activity, seeds were soaked at 25, 35, and 45 °C as previously described in the germination section, and samples were collected from seeds every 24 h. The embryonic axis was extracted, frozen in liquid nitrogen, and lyophilized. These samples were stored in a freezer (-20 °C) until analysis.

The enzyme extracts used to determine the activities of superoxide dismutase (SOD), ascorbate peroxidase (APX), and catalase (CAT) were obtained following the method described by Hodges [28], with adaptations. Samples of 50 mg were crushed and homogenized with 2.0 mL of a solution of 50 mM phosphate buffer pH 7.8 and 1% (*w/v*) polyvinylpolypyrrolidone (PVPP). Then, the extract was centrifuged at 19,000 g for 30 min at 4 °C and the supernatant was used as an enzyme extract. The entire procedure was conducted at 4 °C.

SOD activity: Superoxide dismutase activity was determined by an assay using 30 µL of extract and 2.97 mL of a reaction mixture comprised of 1500 µL of 100 mM phosphate buffer pH 7.5, 780 µL of 50 mM methionine, 225 µL of 1 mM p-nitro blue tetrazolium (NBT), 60 µL of 5 mM EDTA, 60 µL of 2 µM riboflavin, and 345 µL of distilled water [29]. The reaction was conducted at 25 °C in a reaction chamber under fluorescent light (15 W). After five min of light exposure, the blue formazan produced by NBT photoreduction was measured at 560 nm and the reading obtained at 560 nm was retrieved from the illuminated sample [30]. The absorbance at 560 nm of a reaction mixture equal to the other one, yet which was kept in the dark for an equal period, was used as the control. One SOD unit was defined as the amount of enzyme necessary to inhibit NBT photoreduction by 50% [31].

APX activity: Ascorbate peroxidase activity was determined by an assay adapted from Ramalheiro [32], using 100 µL of enzyme extract and 1400 µL of a reaction mixture comprised of 700 µL of 50 mM phosphate buffer pH 7.8, 400 µL of 0.25 mM ascorbic acid containing 0.1 mM EDTA, and 300 µL of 0.3 mM H_2O_2. Enzyme activity was calculated based on the molar extinction coefficient of 2.8 mM^{-1} cm^{-1} [33]. One activity unity (U) was defined as the amount of enzyme needed to convert 1 nmol of substrate into product per min, per mL, under the assay conditions.

CAT activity: Catalase activity was determined by an assay adapted from Hodges et al. [34], using 100 µL of enzyme extract and 1400 µL of a reaction mixture constituted by 900 µL of 50 mM phosphate buffer pH 7.8 and 500 µL of 0.97 M H_2O_2. Enzyme activity was calculated using the molar extinction coefficient of 36 M^{-1} cm^{-1} [35]. One activity unit was defined as the amount of enzyme needed to convert 1 µmol of substrate into product per min, per mL, under the assay conditions.

POX activity: Peroxidase activity was determined by adding 30 µL of crude enzyme extract to 2.97 mL of a reaction mixture constituted by 25 mM potassium phosphate buffer pH 6.8, 20 mM pyrogallol, and 20 mM H_2O_2 [36]. Purpurogallin production was determined in a spectrophotometer by the increase in absorbance at 420 nm, at 25 °C, until the second minute of the reaction. Enzyme activity was calculated using the molar extinction coefficient of 2.47 mM^{-1} cm^{-1} [37].

Enzyme activities were expressed as specific activity (SOD: U SOD min^{-1} mg $protein^{-1}$; APX: nmol Asc min^{-1} µg $protein^{-1}$; CAT: µmol H_2O_2 min^{-1} mg $protein^{-1}$; POX: µmol min^{-1} mg $protein^{-1}$).

The protein concentration in samples was determined by the Bradford method [38], with a standard curve constructed using bovine serum albumin (BSA) at 2.5 to 50 µg protein.

For all determinations, the statistical design was entirely randomized with five replicates. The data of germination was submitted to a variance analysis using the SAS statistical software (version 9.2; SAS Institute, Inc., Cary, NC, USA) [39] and the averages obtained for the treatments were compared by the Tukey test as a 5% significance. The data of EC, ROS, and enzyme activity were submitted to a regression analysis ($p < 0.05$).

Pearson correlation analysis was performed (SAS statistical software (version 9.2; SAS Institute, Inc., Cary, NC, USA)) [39] on the evaluated variables. The results were interpreted as suggested by Mukaka [40], under the following criteria: a correlation coefficient of 0.9 to 1.0 (positive or negative) indicates strong correlation (***), of 0.7 to 0.9 (positive or negative) indicates high correlation (**), of 0.5 to 0.7 (positive or negative) indicates moderate correlation (*), of 0.3 to 0.5 (positive or negative) indicates low correlation, and of 0 to 0.3 (positive or negative) indicates negligible correlation.

3. Results

3.1. Germination and GSI

A significant difference was detected between the mean values of germination as a function of temperature. In general, seeds incubated at 35 °C for 24 and 72 h and then transferred to 25 °C and seeds incubated constantly at 25 °C showed higher germination values: 88%, 84%, and 83%, respectively. However, constant incubation at 45 °C caused seed death. Moreover, temperature increase prior to the transfer of seeds to 25 °C caused a loss of vigor. A significant difference in GSI was also observed. The soaking of seeds at 35 °C for 24, 48, and 72 h followed by their ulterior transfer to 25 °C favored germination speed (Figure 1). After a 96 h soaking at 45 °C, all seeds died.

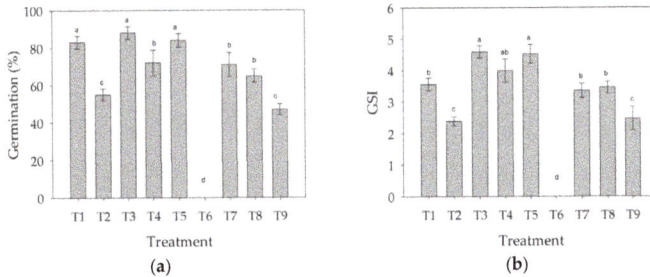

Figure 1. Germination percentage (a) and germination speed index (b) of *Melanoxylon brauna* seeds under different temperatures (T1: 25 °C; T2: 35 °C; T3: 35 °C/24 h; T4: 35 °C/48 h; T5: 35 °C/72 h; T6: 45 °C; T7: 45 °C/24 h; T8: 45 °C/48 h; T9: 45 °C/72 h). Vertical bars = ±SE, *n* = 5.

3.2. Eletric Condutivity

The interaction between temperature and soaking time was significant for EC, being highest at the highest temperature (Figure 2). EC at 25 °C only differed from that at 35 °C with a 72 h soaking period, with the former showing a clear decrease.

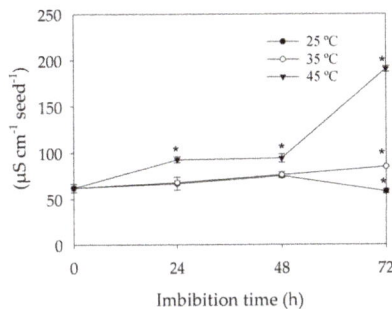

Figure 2. Electric conductivity in *Melanoxylon brauna* seeds under different temperatures and soaking periods. * Indicates statistical difference between means. Vertical bars = ±SE, *n* = 5.

3.3. Superoxide Anion and Hydrogen Peroxide

Superoxide anion was not detected by the adopted method under the tested conditions. The H_2O_2 concentration decreased during the first 24 h of soaking and increased from 48 h. At 45 °C, the embryonic axis and micropylar endosperm showed the highest H_2O_2 levels at all soaking times (Figure 3). At 25 and 35 °C, no difference in peroxide concentration in the embryonic axis was observed after any of the analyzed soaking times.

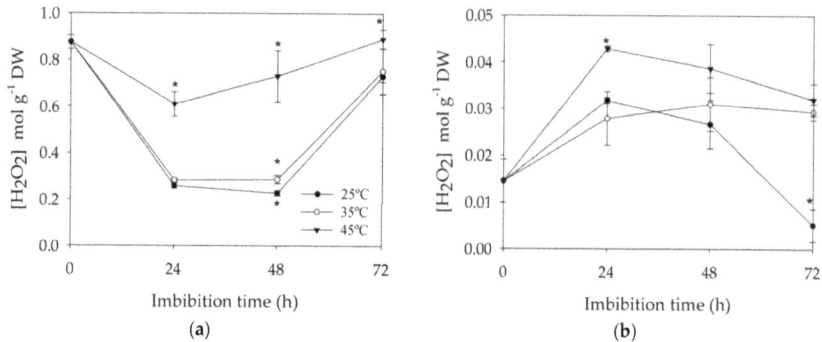

Figure 3. Hydrogen peroxide concentration in the embryonic axis (**a**) and micropylar endosperm (**b**) of *Melanoxylon brauna* seeds soaked at 25, 35, and 45 °C. * Indicates statistical difference between means. Vertical bars = ±SE, $n = 5$.

3.4. Lipid Peroxidation

Lipid peroxidation at 25 °C showed a decrease in the first 48 h of soaking, followed by an increase. At 35 and 45 °C, peroxidation increased during the first 24 h, with posterior reduction. The highest peroxidation levels were observed at 45 °C (Figure 4).

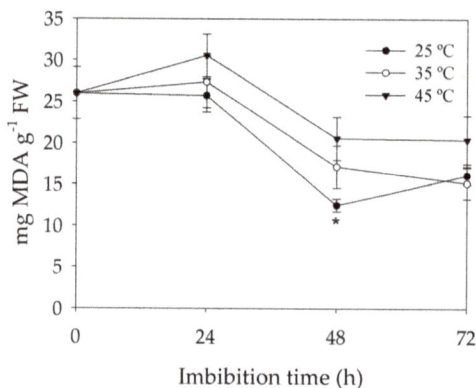

Figure 4. Malondialdehyde (MDA) concentration in *Melanoxylon brauna* seeds soaked at 25, 35, and 45 °C. * Indicates statistical difference between means. Vertical bars = ± SE, $n = 5$.

3.5. Specific Activity of Antioxidant Enzymes

A significant interaction was detected among the specific activity of enzymes APX, POX, SOD, and CAT at the different temperatures and soaking times, indicating that both of these factors influenced enzyme activity in the embryo axis during the germination of brauna seeds.

The highest values of SOD activity occurred in seeds subjected to 45 °C. At all temperatures, enzyme activity decreased after 48 h of soaking (Figure 5a). The differences between the two other temperatures were small, and such differences might have occurred due to a sampling effect. It is worth noting that the seeds showed a wide range of maturation levels during the harvest period.

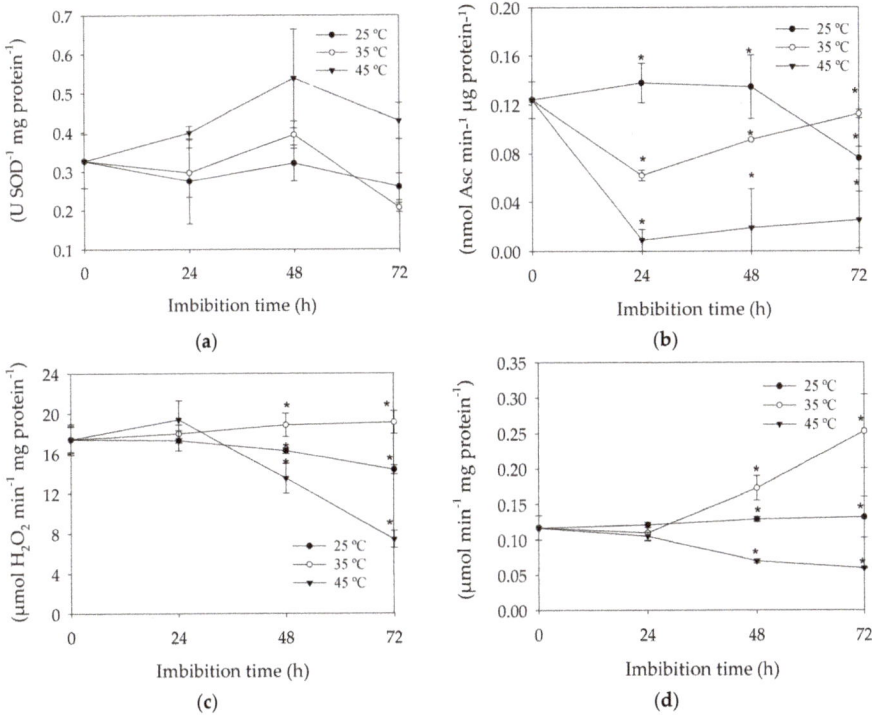

Figure 5. Specific activities of enzymes superoxide dismutase (SOD) (**a**), ascorbate peroxidase (APX) (**b**), catalase (CAT) (**c**), and peroxidase (POX) (**d**) in the embryonic axis of *Melanoxylon brauna* seeds during the germination period, after soaking at 25, 35, and 45 °C. * Mean statistical difference between means. Vertical bars = ±SE, *n* = 5.

Regarding APX specific activity at 25 °C, a slight increase was detected during the first 24 h, followed by a decrease. At 35 and 45 °C, a decrease in enzyme activity was observed during the first 24 h. After that period, an increase was observed at both temperatures, but more intensely in seeds incubated at 35 °C (Figure 5b).

CAT activity in the embryonic axis decreased after 24 h of soaking at 25 and 45 °C. At 35 °C, a small increase in enzyme activity was observed in the embryo during seed hydration. The differences in CAT activity among the three soaking temperatures were clearly highest after 72 h of hydration (Figure 5c).

POX activity was constant at 25 °C, but at 35 °C, it increased after 24 h of soaking. The opposite behavior was observed in seeds incubated at 45 °C, in which enzyme activity decreased during soaking (Figure 5d).

Pearson correlation was assessed for the following variables: EC, H_2O_2 concentration in the embryo and micropylar endosperm, and the activity of enzymes POX, SOD, APX, and CAT in *Melanoxylon brauna* seeds during soaking at 25, 35, and 45 °C. The obtained coefficients allowed for detecting significant correlations, both positive and negative, among the evaluated variables at all tested temperatures (Table 1).

Table 1. Pearson correlation coefficients for means of electric conductivity (EC), concentration of hydrogen peroxide in the embryo and micropylar endosperm, and activity of the enzymes peroxidase (POX), superoxide dismutase (SOD), ascorbate peroxidase (APX), and catalase (CAT) in *Melanoxylon brauna* seeds germinated after soaking at 25, 35, and 45 °C.

Temperature (°C)		Electric Conductivity	[H_2O_2] Embryo
25	[H_2O_2] embryo	−0.82 **	
	[H_2O_2] micropyle		−0.85 **
	POX	0.06	−0.22
	SOD	0.51	0.05
	APX	0.75 **	−0.59 *
	CAT	0.38	−0.19
35	[H_2O_2] embryo	−0.06	
	[H_2O_2] micropyle		−0.70 **
	POX	0.95 ***	0.24
	SOD	−0.47	−0.47
	APX	0.07	0.90 ***
	CAT	0.97 ***	−0.26
45	[H_2O_2] embryo	0.36	
	[H_2O_2] micropyle		−0.78 **
	POX	−0.80 **	−0.21
	SOD	0.25	−0.29
	APX	−0.48	0.60 *
	CAT	−0.88 **	−0.61 *

* Moderate correlation (correlation coefficient of 0.5 to 0.7 (positive or negative)); ** high correlation (correlation coefficient of 0.7 to 0.9 (positive or negative)); *** strong correlation (correlation coefficient of 0.9 to 1.0 (positive or negative)) (following Mukaka [37]).

4. Discussion

The observed variation in the germination percentages of brauna seeds with different temperatures is in accordance with what was described by Flores et al. [12]. These authors verified that germination in this species occurs between 12.3 and 42.5 °C and that 27 °C is the optimal temperature for germination. Similar to what we observed in the present study, they also verified that no germination occurs at 45 °C, and increasing the soaking time of seeds at this temperature and then transferring them to 25 °C reduces seed germination potential.

The soaking of seeds at 35 °C for 24, 48, and 72 h followed by their transfer to 25 °C yielded higher percentages of germination and GSI. Studies on the germination physiology of other species indicate that temperatures near 35 °C provide higher GSI values than those obtained at 25 °C, even when the latter temperature yields higher germination percentages. The appropriate temperature for germination is different from the appropriate temperature for germination speed [41]. The same pattern has been observed in *Torresia acreana* and *Cecropia glaziovii* seeds. This phenomenon occurs because water absorption and biochemical reactions occur more quickly at higher temperatures [42,43].

Besides being a determining factor of seed germination, temperature affects the EC of seeds during soaking. The EC value is associated with the number of leaked electrolytes in the solution, thus being directly related to the integrity of the cell membrane. Thus, high EC values indicate a high leakage of solutes due to alteration in the integrity of the cell membranes, and thereby represent reduced seed vigor. Consequently, EC has been proposed as a parameter to be used in the assessment of seed physiological quality [44].

The observed EC values in brauna seeds incubated at 45 °C were higher than those from seeds incubated at 25 and 35 °C. This indicates that the damage to cell membranes was higher in seeds soaked at 45 °C. At elevated temperatures, membrane selective permeability is lost due to the inability of the membrane to resume its functions because of the disorganization of the lipid bilayer [13].

During soaking, there is an increase in H_2O_2 concentration in the micropylar endosperm during the first 24 h (Figure 3). Such an increase is due to the resumption of respiration. Increased amounts of H_2O_2 leads to the weakening of the wall of seed coat cells, thus enhancing germination. The roles that ROS play in plants also include cell signaling, the promotion of programed cell death, and an increase in the expression of genes that encode oxidative stress enzymes. However, at elevated concentrations, these free radicals may attack the cell membrane system, causing its disruption [45,46].

Regarding ROS production, our results showed that under the stress conditions of elevated temperatures, there is an increased production of H_2O_2. The increased concentration of this molecule might lead to the occurrence of lipid peroxidation and to an ulterior disruption of the cell membrane, as indicated by the increased leakage at 45 °C, which caused a gradual decrease in seed viability. Similar results were found in *Dalbergia nigra* seeds, which showed a gradual loss in viability at 45 °C [18]. With increasing stress, the formation of ROS is intensified, and their elimination must be constant to avoid oxidative stress. Therefore, the synchronized action of enzymes responsible for ROS removal provides a higher stress tolerance to plants subjected to elevated temperatures.

The higher levels of H_2O_2 at 35 and 45 °C led to an increased production of malondialdehyde, which in turn is an indicator of high rates of lipid peroxidation. An increased peroxidation of lipids, mediated by free radicals and peroxides, is a possible reason for the loss of viability in seeds soaked at 45 °C.

Superoxide dismutase (SOD), the first enzyme of the antioxidant system to act, doing so by dismutating superoxide radical ($O_2{}^-$) into H_2O_2, showed low activity. However, such activity was higher in seeds soaked at 45 °C, indicating a possible detoxification. This is one of the possible explanations for the higher H_2O_2 concentration detected at that temperature. Our results are in accordance with what was reported by Flores et al. [13], who observed increased SOD activity in brauna seeds subjected to high temperatures. The behavior of SOD at 25 °C is similar to that found in *Picea omorika* seeds, in which enzyme activity remained constant during germination at 25 °C [47]. Similar results have also been reported for SOD activity in *Medicago sativa* seeds, which showed constant behavior at 22 °C [48]. During the soaking of *Dalbergia nigra* seeds, SOD activity is higher at 45 °C than at 25 °C [18].

Nevertheless, Kumar et al. [49] observed an increase in SOD activity until 40 °C in maize and rice genotypes, followed by a decrease after 45 °C. Although the production of SOD is one of the first responses to abiotic stress, the action of this enzyme must not be evaluated individually, since APX and/or CAT, for instance, eliminate H_2O_2, which permeates the membrane easily and is toxic to cells [50]. APX and CAT belong to different classes of antioxidant enzymes due to their different affinities for H_2O_2, in the orders of μM and mM, respectively. While APX is responsible for the refined modulation of ROS for cell signaling, CAT is responsible for removing the excess ROS generated during stress conditions [51,52].

The activity of SOD is stimulated at 45 °C in 48 h, when the highest H_2O_2 concentrations were detected. SOD activity is lower at 25 and 35 °C, indicating that peroxide levels at those temperatures are safe, being sufficiently low as to not be detected by the enzyme.

The activity of the enzymes APX, POX, and CAT decreased at 45 °C, which led to an increased concentration of H_2O_2 and consequent damage to the cell membranes (Figure 2), which ultimately affected seed germination (Figure 1).

There was no correlation between APX activity and H_2O_2 scavenging at 35 °C. The higher activity of this enzyme at 25 °C kept H_2O_2 concentrations at low levels. The decreased enzyme activity after 48 h and increased concentrations of H_2O_2 might be due to the affect of this compound on the weakening reactions of the wall of micropyle cells [45]. On the other hand, APX might also act on different organelles where H_2O_2 is produced at unsafe levels.

APX can scavenge H_2O_2 from cells using ascorbate as an electron donor for the reaction [53]. Sun et al. [54], after evaluating seeds and seedlings of wild plants and of two mutant linages of *Nicotiana tabacum* for the ATtAPX genes, observed that seeds from the wild genotype had lower percentages of germination at 42 °C, thus proving the importance of APX under stress conditions.

Hence, we suggest that APX is dependent on temperatures near the one that is ideal for germination of brauna seeds.

CAT activity at 25 °C was constant during the first 24 h of soaking, with a slight decrease during this period. At 35 °C, the activity of this enzyme increased as a function of soaking time, indicating that CAT acts under pre-stress conditions, avoiding ROS accumulation. However, at 45 °C, a slight decrease in CAT activity occurred after 24 h of soaking. Moreover, at 35 °C, both POX and CAT showed increased activities, especially the former. In the case of POX, the breakdown of storage lipids in peroxisomes results in the increase of H_2O_2 concentration. Additionally, even increased respiration in mitochondria, which results in an increased concentration of H_2O_2 [13], may determine an increase in the activity of both CAT and POX at 35 °C.

These results demonstrate that CAT has an important role in regulating ROS levels, acting in accordance with other metabolic cycles, such as that of ascorbate/glutathione. At 25 °C, APX is apparently responsible for H_2O_2 degradation, since at that temperature, neither CAT nor POX is important in such a process. This is justified by the fact that CAT acts at elevated H_2O_2 concentrations, which is not the case during the periods of 24 and 48 h. Despite acting on a micromolar scale, CAT seems to have not been essential in determining the existing H_2O_2 levels, which in turn did not cause major damage to seed cell metabolism.

Our results indicate that the low ROS production and the antioxidant enzyme activity at 25 °C maintained the physiological quality of seeds, thus favoring the occurrence of germination. Pre-soaking at 35 °C followed by posterior transfer to 25 °C increased seed metabolism and did not compromise seed viability, therefore enabling quicker and even more germination. Soaking at 45 °C, which stressed seeds, compromised antioxidant enzyme activity and membrane systems, resulting in increased H_2O_2 production and causing vigor loss and the absence of germination in *Melanoxylon brauna* seeds.

Table 1 shows that EC only had a significant inverse correlation with H_2O_2 concentration in the embryo at 25 °C. In that sense, it seems reasonable to presume that such a correlation is associated with the production of ROS and their impact on cell membrane integrity. An increase in ROS contents would lead to damage to the cell membranes, while a reduction in those contents would not affect the membrane structure. Hence, at 25 °C, the H_2O_2 levels and EC values were safe, thus causing neither oxidation damage nor damage to the membrane system (Figures 2 and 3).

The H_2O_2 concentrations in the micropyle and embryo showed a significant inverse correlation at all temperatures. Peroxide contents increased with leakage from the embryonic axis. Such leakage may possibly act on weakening the micropyle by means of the Fenton reaction [55]. This phenomenon occurs due to the high capacity of H_2O_2 to cross cell membranes, through protein channels that have important physiological roles in the capture, translocation, sequestration, and extrusion of this molecule [56]. Transport of H_2O_2 through protein channels might occur in response to the increased concentration of this ROS in the micropyle [56]. The increase in H_2O_2 concentration from the occurrence of respiratory activity in the embryo during germination is explained by the kinetic features of this ROS. Such features enable the binding of the substrate to aquaporins, due to the molecule dipole moment of 2.26×10^{-18} esu (H_2O_2) vs. 1.85×10^{-18} esu (H_2O), the dielectric constant of 73.1 (H_2O_2) vs. 80.4 (H_2O), a molecular diameter of 0.25–0.28 nm (H_2O_2) vs. 0.275 nm (H_2O), and the capacity to form hydrogen bonds [57]. The role played by H_2O_2 is not restricted to oxidative stress; during seed germination, this ROS is also responsible for softening micropyle tissues and signaling both apoptosis and cell proliferation [55].

Regarding the enzymes, no relevant correlation of SOD, CAT, or POX activity with EC and H_2O_2 levels was detected in the embryo or micropyle at 25 °C. APX showed a strong correlation with H_2O_2 concentration in the micropyle, as well as with EC. Although APX activity was constant during the first 48 h of soaking, such activity was enough to keep H_2O_2 concentrations at safe levels. These results reinforce the hypothesis that APX has its activity potentialized at temperatures near the optimal temperature for germination.

At 35 °C, POX and CAT activities showed a strong correlation with EC, while APX showed a higher correlation with H_2O_2 in the micropylar endosperm (Table 1). Our results indicate an initial oxidative stress and compromise of the membrane system. Thus, the increased enzyme activity at 35 °C enabled the elimination of excess H_2O_2 and the consequent reestablishment of metabolic routes related to seed germination. In general, heat stress becomes evident at 35 °C; however, at that temperature, the antioxidant system and the cell membrane system were not entirely compromised, which explains the occurrence of germination under those conditions (Figure 1).

A strong correlation of POX and CAT activities with EC was observed at 45 °C. APX showed a strong negative correlation with H_2O_2 levels in the embryo (Table 1). The decreased enzyme activity at 45 °C might have compromised the antioxidant system due to heat stress. Consequently, there was a higher accumulation of H_2O_2, the excess of which attacked the cell membrane system, causing its disruption due to increased EC and ultimately leading to seed death.

Unlike existing studies, we tried to analyze the seed tissues separately. By evaluating the behavior of the micropillary region and the embryo as such, we were able to interpret changes in each tissue separately.

5. Conclusions

Soaking at 35 °C for 72 h followed by the transfer to 25 °C favors germination and germination speed in brauna seeds.

At 45 °C, on the other hand, there is an increased production and accumulation of H_2O_2, and the antioxidant system at that temperature is then compromised. Catalase is the enzyme with the highest activity among peroxidases. The temperature of 45 °C had a deleterious effect on the peroxidases, while it was a stimulant for SOD. In general, enzyme activities vary between the temperatures and during the period of germination.

Consequently, EC is significantly increased, as the membrane system of seed cells is compromised with these conditions. As a result, no germination occurs at 45 °C. Additionally, soaking seeds at this temperature leads to seed death.

Acknowledgments: Authors are thankful to the Department of Forest Engineering at the Federal University of Viçosa, FAPEMIG, and CNPq for financing the project and productive scholarship.

Author Contributions: Marcone Santos, Eduardo Borges, Glauciana Ataíde, and Genaína Souza conceived and designed the experiments; Marcone Santos and Glauciana Ataíde performed the experiments; Marcone Santos, Genaína Souza, and Eduardo Borges analyzed the data; Marcone Santos, Glauciana Ataíde, and Genaína Souza wrote the paper.

Conflicts of Interest: The authors declare no conflict of interest.

References

1. Painel Brasileiro de Mudanças Climáticas. Available online: http://memoria.ebc.com.br/agenciabrasil/noticia/2013-12-24/painel-brasileiro-de-mudancas-climaticas-projeta-clima-mais-quente-para-este-seculo (accessed on 21 August 2017).
2. Lorenzi, H. *Árvores Brasileiras: Manual de Identificação e Cultivo de Plantas Arbóreas Nativas do Brasil*, 5th ed.; Instituto Platarum: Nova Odessa, Brazil, 2008; 368p.
3. Carvalho, F.A.; Nascimento, M.T.; Braga, J.M.A. Estrutura e composição florística do estrato arbóreo de um remanescente de Mata Atlântica submontana no município de Rio Bonito, RJ, Brasil (Mata Rio Vermelho). *Rev. Árvore* **2007**, *31*, 717–730. [CrossRef]
4. Silva, M.S.; Borges, E.E.L.; Leite, H.G.; Corte, V.B. Biometria de frutos e sementes de *Melanoxylon brauna* Schott. (Fabaceae-Caesalpinioideae). *Rev. Cerne* **2013**, *19*, 517–524. [CrossRef]
5. Ministério do Meio Ambiente (MMA). Lista Oficial das Espécies da Flora Brasileira Ameaçadas de Extinção. In *Instrução Normativa n° 6, de 23 de Dezembro de 2008*; MMA: Brazilia, Brazil, 2014. Available online: http://www.mma.gov.br/estruturas/ascom_boletins/arquivos/8319092008034949.pdf (accessed on 18 October 2016).

6. Ataíde, G.M.; Borges, E.E.L.; Gonçalves, J.F.C.; Guimarães, V.M.; Flores, A.V. Alterações fisiológicas durante a hidratação de sementes de *Dalbergia nigra* ((Vell.) Fr. All. ex Benth.). *Cienc. Florest.* **2016**, *26*, 615–625. [CrossRef]

7. Ferreira, A.G.; Borghetti, F. *Germinação: Do Básico ao Aplicado*, 1st ed.; Artmed: Porto Alegre, Brazil, 2004; 323p.

8. Pacheco, M.V.; Matos, V.P.; Ferreira, R.L.C.; Feliciano, A.L.P.; Pinto, K.M.S. Efeito de temperaturas e substratos na germinação de sementes de *Myracrodruon urundeuva* Fr. All. (Anacardiaceae). *Rev. Árvore* **2006**, *30*, 359–367. [CrossRef]

9. Bewley, J.D.; Bradford, K.; Hilhorst, H.; Nonogaki, H. *Seeds: Physiology of Development and Germination and Dormancy*, 3rd ed.; Springer: New York, NY, USA, 2013; 329p.

10. Lone, A.B.; Souza, G.R.B.; Oliveira, K.S.; Takahashi, L.A.S.; Faria, R.T. Temperatura e substrato para germinação de sementes de flor-de-maio (*Schlumbergera truncata* (Haw.) Moran). *Rev. Ceres* **2010**, *57*, 367–371. [CrossRef]

11. Oliveira, F.N.; França, F.D.; Torres, S.B.; Nogueira, N.W.; Freitas, R.M.O. Temperaturas e substratos na germinação de sementes de pereiro-vermelho (*Simira gardneriana* M.R. Barbosa & Peixoto). *Rev. Cienc. Agron.* **2016**, *47*, 658–666. [CrossRef]

12. Silva, M.L.M.; Alves, E.U.; Bruno, R.L.L.; Santos-Moura, S.S.; Neto, A.P.S. Germinação de sementes de *Chorisia glaziovii* O. Kuntze submetidas ao estresse hídrico sob diferentes temperaturas. *Cienc. Florest.* **2016**, *26*, 999–1007. [CrossRef]

13. Flores, A.V.; Borges, E.E.L.; Guimarães, V.M.; Ataíde, G.M.; Castro, R.V.O. Germinação de sementes de *Melanoxylon brauna* schott em diferentes temperaturas. *Rev. Árvore* **2014**, *38*, 1147–1154. [CrossRef]

14. Suzuki, N.; Rizhsky, L.; Liang, H.; Shuman, J.; Mittler, R. Enhanced tolerance to environmental stresses in transgenic plants expressing the transcriptional co-activator MBF1. *Plant Physiol.* **2005**, *139*, 1313–1322. [CrossRef] [PubMed]

15. Sharma, P.; Ambuj, B.J.; Rama, S.D.; Mohammad, P. Reactive oxygen species, oxidative damage, and antioxidative defense mechanism in plants under stressful conditions. *J. Bot.* **2012**, *2012*, 217037. [CrossRef]

16. Heyno, E.; Innocenti, G.; Lemaire, S.; Issakidis-Bourguet, E.; Krieger-Liszkay, A. Putative role of the malate valve enzyme NADP—Malate dehydrogenase in H_2O_2 signalling in *Arabidopsis*. *Philos. Trans. R. Soc. B* **2014**, *369*, 20130228. [CrossRef] [PubMed]

17. Srivastava, S.; Dubey, R.S. Manganese-excess induces oxidative stress, lowers the pool of antioxidants and elevates activities of key antioxidative enzymes in rice seedlings. *Plant Growth Regul.* **2011**, *19*, 1–16. [CrossRef]

18. Matos, A.C.B.; Borges, E.E.L.; Silva, L.J. Fisiologia da germinação de sementes de *Dalbergia nigra* (Vell.) Allemão ex Benth. sob diferentes temperaturas e tempos de exposição. *Rev. Árvore* **2015**, *39*, 115–125. [CrossRef]

19. Maguire, J.D. Speed of germination aid in selection and evaluation for seedling emergence and vigor. *Crop Sci.* **1962**, *2*, 176–177. [CrossRef]

20. Woodstock, L.W. Physiological and biochemical tests for seed vigor. *Seed Sci. Technol.* **1973**, *1*, 127–157.

21. Mohammadi, M.; Karr, A.L. Superoxide anion generation in effective and ineffective soybean root nodules. *J. Plant Physiol.* **2001**, *158*, 1023–1029. [CrossRef]

22. Misra, H.P.; Fridovich, I. The generation of superoxide radical during the autoxidation of ferredoxins. *J. Biol. Chem.* **1971**, *246*, 6886–6890. [PubMed]

23. Boveris, A. Determination of the production of superoxide radicals and hydrogen peroxide in mitochondria. *Methods Enzymol.* **1984**, *105*, 429–435. [CrossRef] [PubMed]

24. Kuo, M.C.; Kao, C.H. Aluminum effects on lipid peroxidation and antioxidative enzyme activities in rice leaves. *Biol. Plant.* **2003**, *46*, 149–152. [CrossRef]

25. Gay, C.; Gebicki, J.M. A critical evaluation of the effect of sorbitol on the ferric-xylenol orange hydroperoxide assay. *Anal. Biochem.* **2000**, *284*, 217–220. [CrossRef] [PubMed]

26. Araujo, J.M.A. Oxidação de lipídios. In *Química de Alimentos—Teoria e Prática*, 2nd ed.; Editora da Universidade Federal de Viçosa (UFV): Viçosa, Brazil, 1995; p. 22.

27. Lehner, A.; Mamadou, N.; Poels, P.; Côme, D.; Bailly, C.; Corbineau, F. Changes in soluble carbohydrates, lipid peroxidation and antioxidant enzyme activities in the embryo during ageing in wheat grains. *J. Cereal Sci.* **2008**, *47*, 555–565. [CrossRef]

28. Hodges, P.W.; Butler, J.E.; Mckenzie, D.; Gandevia, S.C. Contraction of the human diaphragm during postural adjustments. *J. Physiol.* **1997**, *505*, 539–548. [CrossRef] [PubMed]

29. Del Longo, O.T.; Gonzales, C.A.; Pastori, G.M.; Trippi, V.S. Antioxidant defenses under hyperoxygenic and hyperosmotic conditions in leaves of two lines of maize with differential sensitivity to drought. *Plant Cell Physiol.* **1993**, *34*, 1023–1028. [CrossRef]

30. Giannopolitis, C.N.; Ries, S.K. Superoxide dismutases: I. Occurrence in higher plants. *Physiol. Plant.* **1977**, *59*, 309–314. [CrossRef]

31. Beuchamp, C.; Fridovick, I. Superoxide dismutase improved as says and as say applicable to acrylamide gels. *Anal. Biochem.* **1971**, *44*, 276–287. [CrossRef]

32. Ramalheiro, J.P.S.C. Contribuição Para a Caracterização Bioquímica do Estado de Maturação de Azeitonas de Diferentes Variedades. Masters Dissertation, Universidade Técnica de Lisboa, Lisbon, Portugal, 2009. Available online: http://hdl.handle.net/10400.5/1940 (accessed on 10 February 2017).

33. Nakano, Y.; Asada, K. Hydrogen peroxide is scavenged by ascorbate-specific peroxidase in spinach chloroplasts. *Plant Cell Physiol.* **1981**, *22*, 867–880. [CrossRef]

34. Hodges, D.M.; Andrews, C.J.; Johnson, D.A.; Hamilton, R.I. Antioxidant enzyme responses to chilling stress in differentially sensitive inbred maize lines. *J. Exp. Bot.* **1997**, *48*, 1105–1113. [CrossRef]

35. Anderson, M.D.; Prasad, T.K.; Stewart, C.R. Changes in isozyme profiles of catalase, peroxidase, and glutathione reductase during acclimation to chilling in mesocotylus of maize seedlings. *Plant Physiol.* **1995**, *109*, 1247–1257. [CrossRef] [PubMed]

36. Kar, M.; Mishra, D. Catalase, peroxidase, and polyphenoloxidase activities during rice leaf senescence. *Plant Physiol.* **1976**, *57*, 315–319. [CrossRef] [PubMed]

37. Chance, B.; Maehley, A.C. Assay of catalases and peroxidases. *Methods Enzymol.* **1955**, *2*, 764–775. [CrossRef]

38. Bradford, M.M. A rapid and sensitive method for the quantification of microgram quantities of proteins utilizing the principle of protein-dye binding. *Anal. Biochem.* **1976**, *72*, 248–254. [CrossRef]

39. SAS Institute. *SAS/STAT: User's Guide*; Version 9.2; SAS Institute: Cary, NC, USA, 2009; 7869p.

40. Mukaka, M.M. Statistics Corner: A guide to appropriate use of Correlation coefficient in medical research. *Malawi Med. J.* **2012**, *24*, 69–71. [PubMed]

41. Carvalho, N.M.; Nakagawa, J. *Sementes: Ciência, Tecnologia e Produção*, 4th ed.; FUNEP: Jaboticabal, Brazil, 2000; 588p.

42. Albrecht, J.M.F.; Alburqueque, M.C.L.F.; Silva, V.S.M. Influência da temperatura e do tipo de substrato na germinação de sementes de cerejeira. *Rev. Bras. Semente* **1986**, *8*, 49–55. [CrossRef]

43. Godoi, S.; Takaki, M. Efeito da temperatura e a participação de fitocromo no controle da germinação de sementes de embaúba. *Rev. Bras. Sementes* **2005**, *27*, 87–90. [CrossRef]

44. Ronchi, C.P.; Almeida, W.L.; Souza, D.S.; Souza, J.M.S.J.; Guerra, A.M.N.M.; Pimenta, P.H.C. Morphophysiological plasticity of plagiotropic branches in response to change in the coffee plant spacing within rows. *Semina* **2016**, *37*, 3820–3834. [CrossRef]

45. Borges, E.E.L.; Ataíde, G.M.; Matos, A.C.B. Micropilar and embryonic events during hydration of *Melanoxylon brauna* Schott seeds. *J. Seed Sci.* **2015**, *37*, 192–201. [CrossRef]

46. Kärkönen, A.; Kuchitsu, K. Reactive oxygen species in cell wall metabolism and development in plants. *Phytochemistry* **2015**, *112*, 22–32. [CrossRef] [PubMed]

47. Prodanovic, O.; Prodanovic, R.; Pristov, J.B.; Mitrovic, A.; Radotic, K. Effect of cadmium stress on antioxidative enzymes during the germination of Serbian spruce (*Picea omorika* (Panc) Purkyne). *Afr. J. Biotechnol.* **2012**, *11*, 11377–11385. [CrossRef]

48. Cakmak, I.; Pfeiffer, W.; Mcclafferty, B. Review: Biofortification of durum wheat with inc and Iron. *Cereal Chem.* **2010**, *87*, 10–20. [CrossRef]

49. Kumar, S.; Gupta, D.; Nayyar, H. Comparative response of maize and rice genotypes to heat stress: Status of oxidative stress and antioxidants. *Acta Physiol. Plant.* **2012**, *34*, 75–86. [CrossRef]

50. Meloni, D.A.; Oliva, M.A.; Martinez, C.A.; Cambraia, J. Photosynthesis and activity of superoxide dismutase, peroxidase and glutathione reductase in cotton under salt stress. *Environ. Exp. Bot.* **2003**, *49*, 69–76. [CrossRef]

51. Mittler, R. Oxidative stress, antioxidants and stress tolerance. *Trends Plant Sci.* **2002**, *7*, 405–410. [CrossRef]

52. Deuner, C.; Maia, M.S.; Deuner, S.; Almeida, A.S.; Meneghello, G.E. Viabilidade e atividade antioxidante de sementes de genótipo de feijão-miúdo submetidos ao estresse salino. *Rev. Bras. Sementes* **2011**, *33*, 711–720. [CrossRef]

53. Asada, K. Ascorbate peroxidase—A hydrogen peroxide-scavenging enzyme in plants. *Physiol. Plant.* **1992**, *82*, 235–241. [CrossRef]

54. Sun, W.-H.; Duan, M.; Li, F.; Shu, D.-F.; Yang, S.; Weng, Q.-W. Overexpression of tomato tAPX gene in tobacco improves tolerance to high or low temperature stress. *Biol. Plant.* **2010**, *54*, 614–620. [CrossRef]

55. Fenton, H.J.H. Oxidation of tartaric acid in presence of iron. *J. Chem. Soc.* **1894**, *65*, 899–911. [CrossRef]

56. Bienert, G.P.; Schjoerring, J.K.; Jahn, T.P. Membrane transport of hydrogen peroxide. *Biochim. Biophys. Acta* **2006**, *1758*, 994–1003. [CrossRef] [PubMed]

57. Birk, J.; Meyer, M.; Aller, I.; Hansen, H.G.; Odermatt, A.; Dick, T.P.; Meyer, A.J.; Appenzeller-Herzog, C. Endoplasmic reticulum: Reduced and oxidized glutathione revisited. *J. Cell Sci.* **2013**, *126*, 1604–1617. [CrossRef] [PubMed]

forests

MDPI

Article

Taxon-Independent and Taxon-Dependent Responses to Drought in Seedlings from *Quercus robur* L., *Q. petraea* (Matt.) Liebl. and Their Morphological Intermediates

Kristine Vander Mijnsbrugge [1,*], Arion Turcsán [1,2,3], Jorne Maes [4], Nils Duchêne [4],
Steven Meeus [4], Beatrijs Van der Aa [1], Kathy Steppe [5] and Marijke Steenackers [1]

[1] Department of Forest Genetic Resources, Research Institute for Nature and Forest, 9500 Geraardsbergen,
Belgium; raup25@gmail.com (A.T.); beatrijs.vanderaa@inbo.be (B.V.d.A.);
marijke.steenackers@inbo.be (M.S.)
[2] Department of Biometrics and Agricultural Informatics, Szent István University, 1118 Budapest, Hungary
[3] Department of Forest Reproductive Material and Plantation Management, Institute of Silviculture and
Forest Protection, West-Hungarian University, 9400 Sopron, Hungary
[4] Department of Agro- and Biotechnology, School of Technology, Odisee University College, 9100 Sint-Niklaas,
Belgium; jorne.maes@gmail.com (J.M.); nils.duchene@hotmail.com (N.D.); steven.meeus@odisee.be (S.M.)
[5] Laboratory of Plant Ecology, Faculty of Bioscience Engineering, Ghent University, 9000 Ghent, Belgium;
kathy.steppe@UGent.be
* Correspondence: kristine.vandermijnsbrugge@inbo.be; Tel.: +32-478-28-21-37

Received: 31 August 2017; Accepted: 24 October 2017; Published: 27 October 2017

Abstract: The increasing severity and frequency of summer droughts at mid-latitudes in Europe may impact forest regeneration. We investigated whether the sympatric species *Quercus robur* L., *Q. petraea* (Matt.) Liebl., and their morphological intermediates respond differentially to water deficit. Acorns were sourced from a naturally mixed population. Half of the potted seedlings were subjected to two successive drought periods during the first growing season, each followed by a plentiful re-watering. The surviving drought-exposed seedlings subsisted independent of the taxon of the mother tree. The phenological responses were also taxon-independent. However, drought-exposed plants showed a retarded height growth in the year following the treatment which was taxon-dependent. Offspring from *Q. robur* and from trees with leaves resembling *Q. robur* leaves and infructescences resembling *Q. petraea* infructescences showed a stronger decrease in height growth compared to the offspring from *Q. petraea* and from trees with leaves resembling *Q. petraea* leaves and infructescences resembling *Q. robur* infructescences. Diameter growth in the year following the drought treatment showed a weak taxon-dependent response. Together, our results may suggest that the composition of oak species and their hybrids in natural oak forests could be altered upon prolonged periods of precipitation deficit.

Keywords: sessile oak; pedunculate oak; hybridization; survival; leaf senescence; growth

1. Introduction

Climate change may alter temperature and precipitation regimes across Europe, which may result in longer and more severe summer droughts that in turn will challenge forest vitality [1,2]. Increases in tree mortality have been documented in temperate forests due to rising temperatures and water limitation [3], and multiple recurrent drought events were found to be more damaging than a single drought [4,5]. Generally speaking, the seedling stage of a forest tree is the most vulnerable phase in its life cycle. Therefore, comprehension of stress responses in seedlings is fundamental for predicting forest regeneration [6,7].

Among the prominent broad-leaved European tree species, oaks are well-known to be tolerant to drought. Therefore, *Quercus robur* L. and *Q. petraea* (Matt.) Liebl. are proposed as candidate tree species to replace more drought-sensitive species such as beech (*Fagus sylvatica* L.) or spruce (*Picea abies* (L.) Karst.) in warm and dry sites in Europe [8]. Responses to water deficit in oaks have been described in several studies. Water limitations induce a reduced above-ground growth pattern, a diminished biomass production, and a shift towards below-ground root growth [9–13]. Dry growing conditions induce an earlier stop of height growth [12,14] and an earlier cessation of secondary (radial) growth [15]. In the spring that follows a growing season with a drought treatment and an earlier autumnal growth stop, an advanced bud burst was noticed [16]. Re-watering after a period of water deficit increased the chance of an extra shoot as compensation growth [17]. A delay in autumnal senescence, with a delayed bud burst in the subsequent spring, was observed upon re-watering after a drought period in late summer, and is interpreted as a compensation time for physiological repair before entering the next developmental phase of senescence [18]. To date, these studies focus on pure oak species and do not account for interspecific hybrids that are known to occur regularly.

Q. robur and *Q. petraea* are two native mid-successional European oak species, and are among the most frequent tree species in Central and Western Europe. Both oaks occur widely across most of Europe, reaching northwards to southern Scandinavia, southwards to the northern part of the Mediterranean region, westwards to Ireland, and eastwards to southern European Russia. *Q. robur* has a more extended distribution, reaching more eastwards into continental central Russia. Both are of great ecological interest as habitat and food source for a great variety of insects, mammals, birds, fungi, lichens, and moss species. At the same time, they imply a considerable economic value for forest enterprises and the wood processing industry [19]. Sympatric in large geographical areas, these interfertile species are well-known to hybridise in natural conditions, giving rise to the hybrid taxon *Q. x rosacea* Bechst [20–22]. It was already suggested in 1950 that the number of genes by which species of oaks differ from each other is considerably smaller than is the case for many related and interfertile species in other plant genera [23]. Both species still deviate in their ecological requirements. *Q. robur* flourishes in nutrient-rich, humid sites, often in lowlands, whereas *Q. petraea* is more dominantly present in drier and warmer sites that are less nutrient-rich, often at mid-altitudes [24]. For instance, in Germany only 25% of 1200 studied natural oak stands consisted of only one of the two species, while the other 75% contained both species in variable proportions, with the latter sites being characterised by mosaics of dry and humid microenvironments and the occurrence of intermediate morphological forms [24]. *Q. petraea* is considered the more drought-tolerant species, and detailed studies have corroborated this. For instance, a study on excised branches showed that *Q. robur* was more vulnerable than *Q. petraea* to water-stress-induced cavitation [25] and *Q. petraea* displayed a higher water use efficiency compared to *Q. robur* [26]. In addition, *Q. petraea* is less adapted to anoxia than *Q. robur* [27]. The behaviour of hybrids in this respect is unknown.

In general, *Q. robur* and *Q. petraea* are characterized by distinct differences in leaf morphology, although overlap exists between the two species when individual traits are considered [28–30]. The two species do not differ in alleles but in allele frequencies, as detected by molecular marker studies [30]. A joint analysis of morphology and genetic information is rare, often applying one approach as confirmation for the other, with genetic taxon assignment coinciding well with morphological classification [30–32]. In all these cases, leaf morphological traits are applied. *Q. robur* is characterized by a short leaf petiole, secondary veins running to the sinuses between leaf lobes, and a leaf lamina base being typically lobated [33]. *Q. petraea* leaves have longer petioles, no secondary veins that end in the sinuses, and a more symmetrical leaf shape. In comparison to other leaf traits, the largest part of the variation attributable to species distinction between *Q. robur* and *Q. petraea* was found for the trait leaf petiole length, indicating the importance of this trait for species determination [29]. Still, infructescence traits can enhance the morphological separation of the two species [20]. Intermediate individuals can be described based on a statistical multivariate analysis of leaf morphological traits, denoting individuals that are situated in between two peaks of a bimodal distribution along a synthetic axis where the peaks represent *Q. robur* and *Q. petraea* [28,29]. Intermediate individuals have also been defined based on a combination of leaf and infructescence traits when leaves resemble one species and infructescence traits resemble the other [20].

As the interfertile *Q. robur* and *Q. petraea* differ in their drought tolerance and as drought tolerance in their hybrids is unknown, we hypothesized in this study that *Q. robur*, *Q. petraea*, and their morphological intermediates respond differentially upon experimental drought, with *Q. petraea* being more drought tolerant and morphologically intermediate individuals displaying intermediate responses to water limitation. We question whether these taxa display variable responses (i) for survival after two successive drought periods in the first year; (ii) for leaf senescence in the first two years; (iii) for bud burst in the second year; (iv) for height growth in the second year; and (v) diameter growth in the second year.

2. Materials and Methods

2.1. Source Material

The sourced oak population in the northern part of Belgium (51°0′57.8556″ N, 5°31′57.0384″ E) is autochthonous [34]. A cpDNA analysis revealed a uniform haplotype. This haplotype fitted in the reconstructed postglacial migration routes [35]. The trees form a small relict of abandoned coppice wood growing on inland sand dunes within a former heath land where no tradition existed among the relatively poor farmers of introducing foreign provenances. Acorns were collected on 26 September 2013 from 18 mother trees in an area of 150 m × 150 m. At this time, acorns had ripened but were still hanging in the trees. As the oaks are growing widely spaced on the sand dunes, they have low reaching branches and the acorns were picked by hand, excluding any mixing of the acorns among mother trees. The taxon of the mother trees was identified in the field [20]. *Q. robur* showed a leaf stalk smaller than 1 cm and an infructescence stalk larger than 2 cm (six mother trees sampled and abbreviated as r). *Q. petraea* was characterised by a leaf stalk larger than 1 cm and an infructescence stalk smaller than 2 cm (six mother trees sampled and abbreviated as p). Any other combination of these two measures was considered to belong to the mother trees with intermediate morphological traits, further called intermediates. Two types of intermediates were present: the very rare trees with long leaf stalks (>1 cm) and long infructescence stalks (>2 cm), further called the long-stalked intermediates (one mother tree sampled and abbreviated as pr), and the more common trees with short leaf stalks (<1 cm) and short infructescence stalks (<2 cm), further called the short-stalked intermediates (five mother trees sampled and abbreviated as rp). The rarity of the long-stalked intermediates is also suggested by the results from controlled crosses. Interspecific crosses with *Q. robur* as a female parent showed higher acorn production rates, producing offspring with leaves resembling *Q. robur* and most probably resulting in short-stalked intermediates at fertile ages, compared to interspecific crosses with *Q. petraea* as a female parent generating offspring with leaves resembling *Q. petraea* and most probably yielding long-stalked intermediates at fertility [24].

2.2. Germination of the Acorns

In October 2013, the collected seeds were sown in forestry trays with one seed per cell, using standard nursery potting soil (organic matter 20%, pH 5.0–6.5, electrical conductivity (EC): 450 μS/cm, dry matter 25%, fertilization: 1.5 kg/m^3 powdered compound fertilizer NPK 12 + 14 + 24). During winter, the trays were watered manually, keeping the soil moist. The experiment took place in a greenhouse with automatic temperature regulation, keeping the greenhouse frost-free in wintertime, but without additional heating. An automatic internal grey shade cloth system operated in the greenhouse, protecting the plants from high levels of insolation. In total, 392 seeds germinated, 143 from *Q. petraea*, 109 from *Q. robur*, and 140 from the intermediates (Table 1). Germination success per sampled taxon was 90%, 96%, 78%, and 79% for *Q. petraea*, the long-stalked intermediate, the short-stalked intermediate and *Q. robur* respectively. All germinating plants were given water at regular times according to the visual needs as judged by the experienced greenhouse workers. Seedlings were transferred in April 2014 to one-litre pots (12 × 11 × 11 cm) using standard nursery potting soil. The seedlings were not additionally fertilised.

Table 1. Number of oak seedlings according to the treatment and the taxon of the mother tree (n_t) and number of seedlings that survived the first growing season (n_{su}), also expressed in % (n_{su}%).

Treatment	Taxon Mother Tree *	n_t	n_{su} (n_{su}%)
control	p	69	68 (99)
	pr	13	13 (100)
	rp	55	55 (100)
	r	58	58 (100)
drought	p	74	50 (68)
	pr	14	11 (79)
	rp	58	38 (66)
	r	51	36 (71)

* p: *Q. petraea*, pr: long-stalked intermediate, rp: short-stalked intermediate, r: *Q. robur*.

2.3. Drought Treatment, Measurements, and Scoring

For the summer months, climate scenarios for Flanders indicate a decrease in average precipitation together with an increase in extreme short rainfall events [36]. Our experimental set-up mimicked the summer scenario by withholding any watering to potted seedlings (heat waves) interrupted by plentiful re-watering (short heavy rainfall). The pots were divided in two groups: a control and a treatment group. In both groups, the offspring of the 18 mother trees were individually mingled at random (completely randomised). Two successive drought periods were imposed on the oak seedlings during the first growing season. In 2014 on DOY (day of the year) 134 and DOY 217, respectively, the two groups of plants were soaked overnight in a basin with the water level up to two cm above the bottom of the pots to reach a fully-water-saturated condition. Watering was withheld from the drought-treated group up to DOY 182 and DOY 290, respectively, whereas the control group was further watered. All plants were re-watered on DOY 183 and DOY 291, respectively, by soaking the two groups of plants in the basins in the same way. The first drought period lasted 48 days and the second 73 days. At the end of the first drought period, only 1% of the seedlings in the treated group showed wilting and/or curling of the leaves, and no seedlings had died among the treated plants nor in the control group (Figure S1). The second drought period lasted until a larger number of plants (43%) showed wilting and/or curling of the leaves and started dying off (Figure S2). This period was considered as having had the strongest effect on the seedlings. When a drought period is mentioned in this paper, it concerns the second drought period of 2014. It was not a part of the experimental design to unravel the effect of the first drought period on the responses of the plants upon the second drought period. Still, results should be interpreted in light of a putative contribution of the first drought period. After the second re-watering, all plants—including both control plants and treated plants—were kept well-watered. They were well-watered during the whole following growing year by soaking all plants on a regular basis in the basins for several hours and subsequent draining. In this way, plants were brought to field capacity on a regular basis using a relatively easy-to-apply method of watering the individual pots in an equal manner.

During the two drought periods, all the pots were weighed nearly weekly to measure the water loss (Figures S1 and S2). On the same days, the wilting and/or curling of the leaves was observed (Figures S1 and S2). The initial weight at the beginning of the treatment period was measured after the pots had been drained of excess water. We described the soil water reserve with a relative value which is related to the term relative extractable water (REW). REW is calculated for a given day (j) using the formula REWj = (Rj − Rmin)/RU. Rj is the soil water content on a given day (j), calculated using the rooting depth. Rmin represents the minimum soil water content observed at the permanent wilting point, also expressed at the same depth. RU represents the total amount of extractable soil water in the rooted zone. It is the difference between the soil water reserve at field capacity and at the permanent wilting point. Water availability in the soil for plants to access will obviously depend on the size of the available reserve; i.e., the depth of rooting and the physical soil properties. In our experiment, all pots had the same size and the same volume and type of soil. Therefore, we modified the formula of REW for a given day (j) to an adjusted REWj = (weight DOY j − mean weight DOY 290)/(mean weight DOY 218 − mean weight DOY 290). Mean weight DOY 290 was the mean weight of the pots on DOY 290 for which the plant had died off at the end of the second drought period and did not recover anymore afterwards. This represented an approximation of the permanent wilting point. Mean weight DOY 218 was the mean weight of all the pots on DOY 218 and approximated field capacity. The calculation did not take into account possible weight gain by plant growth. This approximation is easily applicable to a larger amount of potted plants.

As acorns were collected within one natural stand with co-occurring *Q. robur*, *Q. petraea*, and morphologically intermediate forms; it cannot be excluded that hybrid individuals were sampled among the descendants of *Q. robur* and *Q. petraea*. A morphological evaluation was made of a well-developed leaf per plant in the first growing season. Morphological analyses of the leaves in the juvenile phase of oaks are known to have a diagnostic value, albeit weak [37]. Four scores were given per seedling, as indicated in Table S1. The principal component analysis was run on these scores to control the identification of the mother trees in the field.

During the second drought period, a significant amount of plants died off. Therefore, survival was monitored as a separate binary variable. The height of all the seedlings and the diameter of the plants at 1 cm above the soil level was measured at the end of the first and second growing seasons. Height and diameter growth during the second growing season were calculated by subtracting the height and diameter at the end of 2014 from the height and diameter at the end of 2015, respectively. Two phenophases were scored on all plants: leaf senescence in autumn 2014 and 2015, and bud burst in spring 2015. Leaf senescence was scored following an 8-level scoring protocol (Table S2) on DOY 335 in 2014 and on DOY 292, 313, and 330 in 2015. All the leaves of a seedling were observed together, and a visual mean of colour change was made. Bud burst and leaf unfolding in the apical bud was scored according to a 6-level scoring protocol (Table S2) on DOY 128 in 2015.

2.4. Data Analysis

All the statistical analyses were performed in the open source software R 3.1.2 [38]. Five response variables were modelled using generalised linear mixed models: survival, leaf senescence, bud burst, height growth, and diameter growth. Survival was examined using logistic regression (generalised linear mixed models) in the package lme4 [39], whereas the phenological response variables were modelled using cumulative logistic regression in the package ordinal [40]. We ordered the ordinal response variables bud burst and leaf senescence in decreasing order: going from unfolded leaves to buds in winter rest (from 6 to 1) and from shed leaves to dark green leaves (from 8 to 1). This way, the probability to have reached maximally for instance bud burst score 4 equalled the probability to have reached scores 6, 5, or 4. This included all plants with an apical bud from which leaves are protruding but not yet unfolding up to leaves fully unfolded (Table S2). This was interpreted as the probability of having reached at least bud burst score 4. Height and diameter growth in the second growing year—thus in the year following a growing season with a drought treatment—were processed with linear mixed models [39].

In the fixed part of the models, several covariates were examined for significant explanatory power: the taxon of the mother tree, the adjusted relative extractable water of the pots at the end of the second drought period, and the plant height at the end of the first growing season. The interaction between taxon of the mother tree and the adjusted relative extractable water at the end of the second drought period indicated whether the taxa experienced water limitation differentially. The phenological model leaf senescence got two additional covariates (day of observation and year of observation), as for this response variable repeated observations per plant were available. In all the models, the mother plant from which acorns were collected was in the random part (random intercept). For the phenological model leaf senescence, an additional unique plant identity variable was added in the random part of the models (random intercept) to account for the repeated measurements on the same plants. We simplified all full models to allow an easier and better interpretation. Using drop1 (a likelihood ratio test), the fixed part of all five models was reduced up to only significant terms. With a significant interaction term, the corresponding covariates (main effects) remained in the model. When taxon appeared not significant in a model, it was discarded from the model, indicating that the response variable was independent from the taxon of the mother tree.

Here we show the full models. The chance (p) of survival was calculated following a logistic regression:

$$\log(p/(1 - p)) = \alpha + \beta_T T + \beta_H H + \beta_A A + \beta_{TA} TA + \beta_{HA} HA$$

with α as the estimated intercept and the β's as the estimated parameters of the fitted model. T is the taxon of the mother tree (r: *Q. robur*, p: *Q. petraea*, pr: long-stalked intermediate, rp: short-stalked intermediate); A is the adjusted relative extractable water of the pots at the end of the second drought period, accounting for the water deficient condition; and H is the plant height at the end of the first growing season.

For bud burst, the chance (p) to have reached at least a given phenological score level on the day of observation was calculated following a cumulative logistic regression:

$$\log(p/(1 - p)) = \alpha_T - \beta_T T - \beta_H H - \beta_A A - \beta_{TA} TA - \beta_{HA} HA$$

with α_T as an estimated threshold value for the passing on from one level of the phenological variable to the next.

For leaf senescence, the chance (p) to have reached at least a given phenological score level on the days of observation (both in 2014 and 2015) was calculated following a cumulative logistic regression:

$$\log(p/(1 - p)) = \alpha_T - \beta_D D - \beta_Y Y - \beta_T T - \beta_H H - \beta_A A - \beta_{YA} YA - \beta_{TA} TA - \beta_{HA} HA$$

D is the day of observation and Y is the year of observation (factor variable with two levels, 2014 and 2015).

Both diameter and height growth in the second growing year (I) were analysed with linear mixed models. Apart from the plants that had died off totally ($n = 63$), also the plants that had died off above ground ($n = 10$) were excluded from the dataset.

$$I = \alpha + \beta_T T + \beta_H H + \beta_A A + \beta_{TA} TA + \beta_{HA} HA$$

The full and reduced models, using drop1, were run with the "maximum likelihood" method until only significant terms remained in the fixed part. Model statistics were taken from the final reduced model run with "restricted maximum likelihood".

Confidence intervals (95%) were calculated based on the estimates of the parameters and their variance-covariance matrices in the final models.

To quantify the relative variability of the individual mother tree in comparison to the taxon of the mother tree, all final models were run with the taxon of the mother tree in the random part. In this way, the variance attributable to the mother trees could be compared to the variance attributable to the taxa to which these mother trees belonged to. In the linear models (response variables of height and diameter growth in second growing year), the residual variance indicated the variance of the individual seedlings and therefore allowed the comparison of the variance of the mother taxa and of the individual mother trees (nested within mother taxa) with the variance among the seedlings (nested within the individual mother trees). Logistic regression and cumulative logistic regression models did not have an error term, and therefore residual variance was not calculated.

3. Results

3.1. Morphological Evaluation of Seedlings

A principal component analysis was applied on the four morphological leaf traits of the seedlings to control the identification of the mother trees in the field. The first principal component discriminates the parental species *Q. robur* and *Q. petraea* mainly based on the length of the leaf petiole (PL) and the presence or absence of leaf ears at the lamina base (LE) (Figures S3 and S4; Table S3). Offspring of the two types of intermediates (short-stalked and long-stalked individuals) show a tendency to deviate from the parental species. The second and third PC axes mainly account for the number of intercalary veins and display less discriminating power between the different sampled oak taxa. The majority of the seedlings displayed a petiole length (PL) and a leaf lamina base (LE) that matched the expected petiole length according to the mother tree (Table S4). The small amount of seedlings that did not coincide could concern first-generation hybrids, as sourced mother trees were open-pollinated.

3.2. Survival

At the end of the first drought period, only 1% of the seedlings in the treated group showed wilting and/or curling of the leaves, and no seedlings had died among the treated plants nor in the control group (Figure S1). The second drought period lasted until a larger number of plants (43%) showed wilting and/or curling of the leaves (Figure S2) and started dying off (Table 1). The taxon of the mother tree was not significant in the survival model. Survival depended quite evidently on the weight loss of the pots at the end of the second drought period. The height of the seedlings at the end of the first growing season was an influencing trait, depending on the adjusted relative extractable water (significant interaction term; Table 2). The taller the seedlings, the greater the probability to die off in the drought-exposed group of plants (Figure 1 and Table 2).

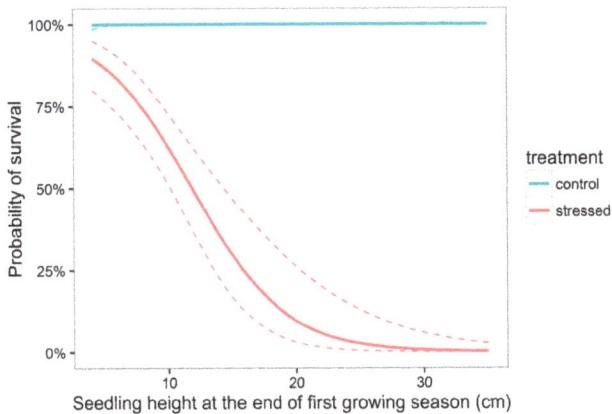

Figure 1. Modelled probability of survival depending on the plant height at the end of the first growing season. The mean adjusted relative extractable water of the pots in the control and treated group of plants was applied to calculate the probabilities; 95% confidence intervals are shown with dashed lines.

Table 2. Model statistics for the general linear mixed model of the binary response variable survival.

Covariate	Estimate	St. er.	z Value	p Value
Intercept	3.26	0.59	5.52	**<0.001 *****
H	−0.28	0.06	−4.98	**<0.001 *****
A	−3.66	1.95	−1.88	0.061
H:A	2.49	0.53	4.69	**<0.001 *****

A: adjusted relative extractable water (continuous variable), H: height of the plant (continuous variable). Significant results are in bold: *** $p < 0.001$.

3.3. Leaf Senescence

In the phenological model describing leaf senescence, the taxon of the mother tree was not significant (with or without interaction term with adjusted relative extractable water), whereas the height of the seedlings was significant without interaction term with adjusted relative extractable water (Table 3). Leaf senescence appeared earlier in the taller seedlings, independent of the drought treatment (thus, both in the control group and in the drought-exposed seedlings). The interaction term between year of observation and the adjusted relative extractable water was significant in the model, indicating that the senescence response in 2014 differed from 2015 depending on the amount of weight loss in the pots during the drought period in 2014 (Table 3). A more severe drought period, as expressed by a low adjusted relative extractable water, retarded the decolouration of the leaves in 2014 and advanced the decolouration in 2015 (Figure 2).

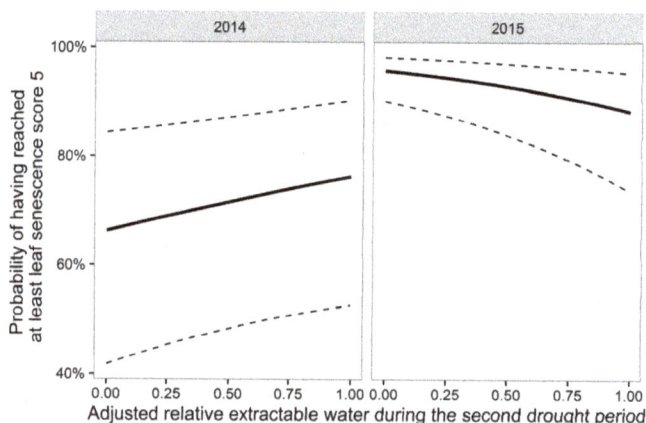

Figure 2. Modelled probability of having reached at least leaf senescence score 5 (yellowing leaves with brown parts) depending on the adjusted relative extractable water of the pots during the second drought period in the first growing season. To calculate the probabilities, the mean plant height of 9.2 cm in 2014 and of 17.6 cm in 2015 was applied, together with $DOY_{2014} = 340$ and $DOY_{2015} = 325$; 95% confidence intervals are shown with dashed lines.

Table 3. Model statistics for the general linear mixed models of the ordinal phenological response variables leaf senescence and bud burst.

Covariate	Leaf Senescence (2014 and 2015)				Bud Burst (2015)			
	Estimate	St. er.	z Value	p Value	Estimate	St. er.	z Value	p Value
D	−0.19	0.008	−25.20	**<0.001 *****				
Y	4.80	0.295	16.25	**<0.001 *****				
A	1.09	0.252	4.33	**<0.001 *****	−2.24	0.30	−7.40	**<0.001 *****
H	−0.07	0.013	−5.14	**<0.001 *****				
Y:A	−1.58	0.331	−4.78	**<0.001 *****				

D: day of observation (continuous variable), Y: year of observation (factor variable with 2015 as standard level), A: adjusted relative extractable water (continuous variable), H: height of the plant (continuous variable). Significant results are in bold: *** $p < 0.001$.

3.4. Bud Burst

Modelling the phenological variable bud burst revealed that the taxon of the mother tree was not significant. The only significant influence was the adjusted relative extractable water of the pots during the second drought treatment in the first growing season (2014), indicating that the drought-exposed group of plants burst buds later compared to the control group (Figure 3 and Table 3). Next to the taxon of the mother tree, plant height displayed no significant explanatory power (with or without interaction term with adjusted relative extractable water).

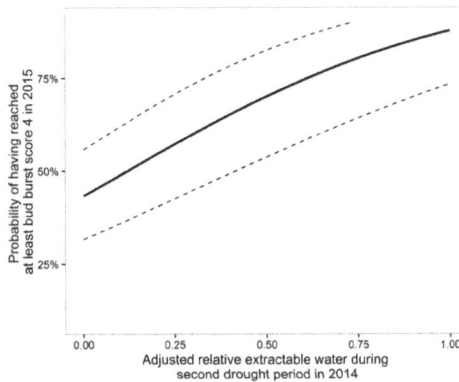

Figure 3. Modelled probability of having reached at least bud burst score 4 (leaves protruding from the apical bud) depending on the adjusted relative extractable water of the pots during the second drought period in the first growing season; 95% confidence intervals are shown with dashed lines.

3.5. Height and Diameter Growth in the Second Growing Year

Total plant height and stem diameter growth in the two growing seasons are shown in Figures S5 and S6. In the height growth model, the covariable taxon of the mother tree appeared significant in interaction with adjusted relative extractable water of the pots during the second drought treatment of 2014 (Table 4), indicating that the reduction in height growth in 2015 in comparison to the control group (and thus due to the water deficit in 2014) depended on the taxon of the mother tree. Seedlings from *Q. robur* and short-stalked intermediates showed a stronger decrease in height growth due to the drought treatment than the seedlings from *Q. petraea* and the long-stalked intermediate (Figure 4). In addition to the taxon of the mother tree, the height at the end of the growing season in 2014 was significant in the height model, independent of the weight loss of the pots during the second drought period in 2014 (Figure 4 and Table 4), showing that the taller plants displayed a larger height growth in the second growing season. The taxon of the mother tree was analogously significant in the model of the diameter growth in the second growing season, also in interaction with the weight loss of the pots (Figure 5 and Table 4) but independent of the height of the seedlings at the end of the growing season in 2014.

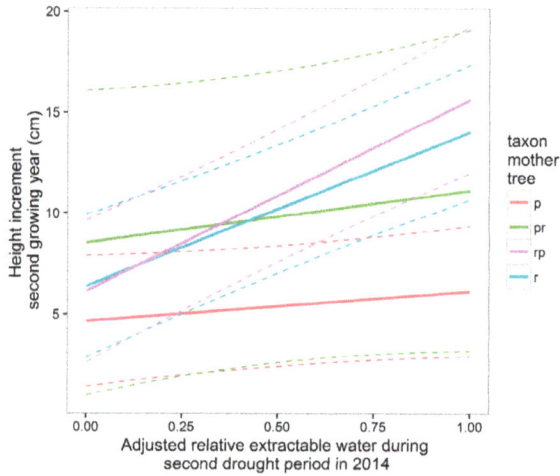

Figure 4. Modelled height growth in the second growing season (2015) depending on the adjusted relative extractable water of the pots during the second drought period in the first growing season and on the taxon of the mother tree. The modelled height is shown for seedlings reaching up to 8 cm (median height) at the end of 2014. p: *Q. petraea*; pr: long-stalked intermediate; rp: short-stalked intermediate; r: *Q. robur* (95% confidence intervals are shown with dashed lines).

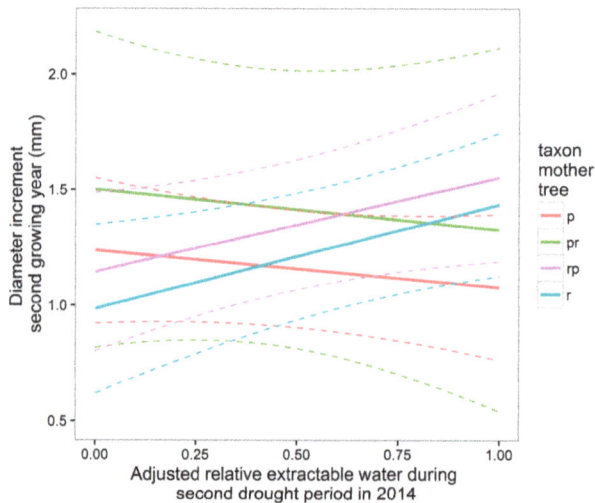

Figure 5. Modelled diameter growth in the second growing season (2015) depending on the adjusted relative extractable water of the pots during the second drought period in the first growing season and on the taxon of the mother tree. p: *Q. petraea*; pr: long-stalked intermediate; rp: short-stalked intermediate; r: *Q. robur* (95% confidence intervals are shown with dashed lines).

Table 4. Model statistics for the linear mixed models of the continuous response variables height and diameter growth in the second growing year.

Covariate	Height Growth in 2015					Diameter Growth in 2015				
	Estimate	St. er.	DF	*t* Value	*p* Value	Estimate	St. er.	DF	*t* Value	*p* Value
Intercept	5.86	1.72	296	3.40	<0.001 ***	123.7	16.1	297	7.71	<0.001 ***
A	1.48	1.25	296	1.18	0.238	−15.9	19.1	297	−0.83	0.40
Tpr	3.87	4.19	14	0.92	0.371	26.4	38.4	14	0.69	0.50
Tr	1.73	2.46	14	0.70	0.494	−25.2	24.6	14	−1.03	0.32
Trp	1.47	2.44	14	0.60	0.557	−9.3	23.7	14	−0.39	0.70
H	−0.15	0.08	296	−1.92	0.056					
A:Tpr	1.13	3.10	296	0.37	0.716	−1.45	47.5	297	−0.03	0.98
A:Tr	6.17	1.84	296	3.35	<0.001 ***	61.1	27.9	297	2.19	0.029 *
A:Trp	8.03	1.90	296	4.23	<0.001 ***	56.8	29.0	297	1.96	0.051

A: adjusted relative extractable water (continuous variable), T: taxon of the mother tree with pr: long-stalked intermediate, r: *Q. robur*, rp: short-stalked intermediate (factor variable with *Q. petraea* as standard level to which pr, r, and rp are compared to), H: height of the plant (continuous variable). Significant results are in bold: *** $p < 0.001$; * $p < 0.05$.

3.6. Variance Components

To assess the influence of the individual mother trees in the analysis, variance components were calculated (Table 5). Taxon of the mother tree was not significant in the models with response variables survival, bud burst, and leaf senescence. This was reflected in the low levels of variance attributed to the taxon of the mother tree in comparison to the individual mother tree (individual mother tree nested within taxon of the mother tree). The models for height and diameter growth in the second growing season allowed an additional comparison with variance attributable to the individual seedlings (the residual variance). In both models, the largest part of the variance was attributable to variability among individual seedlings, being more pronounced in the diameter model. For the height model, the variance component attributable to the variation among individual mother trees was roughly double the variance component of the taxon of the mother tree and roughly half of the variance attributable to the variation among the seedlings. In the diameter model, this variance component of the individual mother trees was roughly 12 times the variance component of the taxon of the mother trees, and about 1/9th of the variance attributable to variation among individual seedlings.

Table 5. Variance components calculated in the different models for the taxon of the mother tree (Mother taxon), the individual mother tree (Mother ID), and the residual variance. Only linear regression models have a calculation of the residual variance.

Response Variable	Variance Component		
	Mother Taxon	Mother ID	Residual
Survival [1]	<0.0001	0.23	-
Bud burst [2]	<0.0001	0.52	-
Senescence [2]	<0.0001	0.55	-
Height [3]	5.63	12.50	25.36
Diameter [3]	0.55	6.72	59.72

[1] Logistic regression model; [2] Cumulative logistic regression model; [3] Linear regression model.

4. Discussion

4.1. Taxon-Independent Responses to Drought

Q. petraea is regarded as more drought-tolerant than *Q. robur* [25–27]. Still, in our experiment survival of the seedlings after a period of water limitation appeared independent of the taxon of the mother tree when taking into account the height of the seedlings (taller seedlings have a higher number of leaves, leading to a higher evaporation). This indicates that under life-threatening stress caused by drought, equally sized seedlings of both species and their intermediate morphological forms likely die off at the same rate. This observation is not in accordance with Vivin, Aussenac, and Levy [41], describing a higher mortality rate among *Q. robur* upon water deficit. There is no indication of the provenances used in this study. Possibly, the observed deviating responses between the two species may (partly) be due to different provenances rather than due to the different taxa. Different groups of neutral Simple Sequence Repeats (SSR) have been identified that either discriminate between species or between provenances, indicating that provenance is not merely a type of taxon [42]. In our case, the mother trees from the different taxa belong to the same natural population. The delayed leaf senescence upon re-watering after a severe water deficit, the delayed subsequent bud burst, and the advanced leaf senescence in the following growing year were found both in the here-described experiment with offspring of *Q. petraea*, *Q. robur*, and morphological intermediates, as well as for different provenances of *Q. petraea* [18,43]. For the different taxa sourced in the same provenance (here-described experiment), no dependency on the taxon of the mother tree could be detected, whereas phenological responses in a similar experiment with different provenances of *Q. petraea* were dependent on the provenance [18]. These dissimilar responses for the provenance and taxon can likely be related to recent genomic findings. Provenance-specific and taxon-specific loci have been discovered in the genomes of *Q. robur* and *Q. petraea*, with "species discriminant" loci representing genome regions affected by directional selection maintaining species' identity, and "provenance-specific" loci representing genome regions with high interspecific gene flow and common adaptive patterns (e.g., phenological responses) to local growth environment [42]. The delayed leaf senescence is suggested to be caused by a compensation time upon re-watering after a severe drought in which physiological repair mechanisms act before the seedlings enter the next developmental phenological phase [18]. The observed reduction in height growth in the second growing season likely caused an earlier growth stop, resulting in an advanced leaf senescence in this year.

4.2. Taxon-Dependent Responses to Drought

Drought stress causes a reduction in tree stem growth. Büsgen et al. [44] describe that the summer drought of short duration in Germany may seriously reduce the height growth of spruce and other tree species and, in addition, that repeated summer droughts may diminish stem growth for decades, leading to enormous losses of increment. Drought-exposed saplings of *Q. robur* and *Q. petraea* display a reduced secondary growth and adjust their xylem wood-anatomical structure to improve resistance and repairing abilities after cavitation [45]. This supports the hypothesis that carbon allocation attributes lowest priority to stem growth under stress [46]. This allows woody plants to redirect assimilates and energy otherwise used for shoot growth to maintain respiration, to stimulate root growth, or to favour other protective adjustments. Additionally, increased energy storage in the xylem parenchyma cells is believed to be induced by drought to repair embolised vessels as soon as the conditions improve [45]. In our experiment, both *Q. robur* and *Q. petraea* displayed a reduced height growth in the year succeeding the period of water deficit in comparison to the control group. This carry-over effect on shoot length can simply be explained by an insufficient resource storage for bud formation and growth [47]. Interestingly, compared to the control plants, *Q. petraea* showed a relatively lesser diminution of height growth in the year succeeding the period of drought than *Q. robur* (significant interaction terms between taxon of the mother tree and adjusted relative extractable water during drought period in the height model, Table 4), indicating a more stable growth pattern for *Q. petraea* when confronted with water-limiting growth conditions. This observation is consistent with the fact that *Q. petraea* withstands more xeric growth conditions compared to *Q. robur* [25–27]. However, it is not in agreement with the findings of Kuster et al. [16], where no difference is reported in shoot elongation, radial stem growth, and shoot biomass production between the two species in a three-year drought experiment. Possibly, the different provenances sourced for both species in this experiment may have faded the influence of taxon. In addition, our observed height growth response that differentiated between *Q. robur* and *Q. petraea* could be stronger, taking into account that F1 hybrids may be present in the collected acorns on *Q. robur* and *Q. petraea* mother trees. These may have attenuated this result.

The offspring of the short-stalked intermediate mother trees tended to a similar reduction in height growth in 2015 to offspring of *Q. robur* when compared to the control groups that did not experience water limitation in the preceding growing season. The reduction in height growth of 2015 depended on the adjusted relative extractable water during the drought period, and is visualised in Figure 4 with similar slopes between offspring of *Q. robur* and offspring of the short-stalked intermediates. On the other hand, the offspring of the long-stalked intermediate mother tree (only one mother tree sampled) behaved similar to the offspring of *Q. petraea* (also similar slopes between the offspring of *Q. petraea* and the long-stalked intermediate in Figure 4). This can be partly explained by matroclinal inheritance, at least in the juvenile phase, as has already been observed for juvenile leaf morphology in the offspring of controlled crosses between *Q. robur* and *Q. petraea*, with interspecific hybrids displaying leaf morphological features of the mother tree [24]. As we sampled morphological intermediates in a natural population, we have no indication of the number of generations (the number of back crosses) that may have passed since the original (natural) cross between *Q. robur* and *Q. petraea*. In this sense, our results may suggest that the more *Q. robur* is introgressed in the short-stalked intermediates (leaf morphology resembling *Q. robur*) or the more *Q. petraea* is introgressed in the long-stalked intermediates (leaf morphology resembling *Q. petraea*), the more the height growth response of these hybrids upon water deficit may mirror the respective original maternal parental response.

For the diameter growth in 2015, there are relative weak significant interaction terms between the taxa of the mother trees and the adjusted relative extractable water during the drought period in 2014. This means that the offspring from *Q. robur* and the short-stalked intermediates differ from *Q. petraea* offspring, with *p*-values of 0.029 and 0.051, respectively (Table 4). As with height growth, this indicates that the offspring of short-stalked intermediates tend to have a similar response as the offspring of *Q. robur*, and the offspring of the long-stalked intermediate behave similar to the offspring of *Q. petraea*. Strangely, a tendency can be observed for an increased diameter growth in 2015, measured at 1 cm above soil level, upon higher pre-drought (lower adjusted relative extractable water in 2014; Figure 5) in the offspring from *Q. petraea* and the long-stalked intermediate. This is possibly due to an enhanced allocation to root growth, which has been observed in oak upon water deficit [10]. A slightly larger diameter increment at the stem base could therefore be considered as a side-effect of the re-allocation of resources due to the pre-drought period.

The variance component analysis allowed an assessment of the relative importance of the taxon of the mother tree in comparison to the individual mother trees belonging to each taxon (which can be interpreted as the maternal effect) and the individual seedlings belonging to each mother tree. When compared to the height growth in the second growing season, the diameter growth showed more relative variance among the individual seedlings. This suggests that height growth, in comparison to diameter growth, is relatively more influenced by both taxon of the mother tree and the individual mother tree, and relatively less variation resides among the different seedlings.

5. Conclusions

Together, we showed that in pedigrees of a naturally mixed population of *Q. robur* and *Q. petraea* in Belgium, the survival rate after a severe drought and phenological responses of the persisting seedlings were independent of the taxon of the mother trees, whereas a taxon-effect was detected for height and diameter growth in the year succeeding the growing season with water limitation. The frequency of extreme climate events will increase [36], and plant responses may differ depending on the timing of these events within the year [48,49]. The impact of drought on perennial herbaceous plant growth and biomass production were found to be the least apparent in spring, possibly due to lower leaf area of the plants and a seasonally differing water potential in the leaves that may not fully reflect the water potential of the soil [49]. In our experiment, the first early summer drought event did not severely impact the plants as visually judged. Still, it possibly aggravated the responses to the second longer lasting drought period that occurred later in the year [4,5]. This growth response upon water limitation may influence the growth dynamics of seedling competition in natural conditions. It is generally known among plant species that in a population where plants compete with each other, larger individuals grow faster than smaller individuals and that initial height differences in trees can build over time as larger individuals pre-empt available light and suppress the growth of smaller neighbours [50–53]. Complex stand structures and diverse community assemblages can develop from the legacies of small differences in initial size and growth rates between individuals and species [54]. The rate of hybridisation between *Q. robur* and *Q. petraea* may augment in the future due to the predicted climate change, which may decrease species clustering and density [55]. As *Q. petraea* is characterised by stronger post-pollination hybridisation barriers than *Q. robur* [22,56], it can be expected that in the forest stands in Belgium (and by extension in Europe) which harbour both species, an evolution may occur towards larger amounts of short-stalked intermediate forms. If our observed taxon-specific responses to water deficit in controlled conditions can be extrapolated to field conditions, it can be hypothesised that the composition of oak species and their hybrids in the natural oak forests may alter upon prolonged periods of precipitation deficit, promoting *Q. petraea* and the long-stalked intermediates. Clearly, longer-term experiments are needed here.

Supplementary Materials: The following are available online at www.mdpi.com/1999-4907/8/11/407/s1. Table S1: Morphological descriptors of fully developed oak leaves on first year seedlings discriminating *Q. robur* from *Q. petraea*. Table S2: Description of the score levels of the two phenological response variables leaf senescence and bud burst in oak seedlings. Table S3: Loadings of leaf morphological traits in PCA of the analysed oak seedlings. Abbreviation of traits are in Table S1. Table S4: Number of seedlings (in %) in the different score levels of the two morphological leaf traits petiole length (PL) and leaf ear (LE). Total number of seedlings are in Table 1. Descriptions of score levels are in Table S1. p: *Q. petraea*. pr: long-stalked intermediate. rp: short-stalked intermediate. r: *Q. robur*. Figure S1: Average and standard deviation of the weights of the pots in the control and the drought treated group of plants during the first drought period in the first growing year. Percentage of seedlings in the treated group of plants showing wilting and/or curling of the leaves is indicated. p: *Q. petraea*, pr: long-stalked intermediate, rp: short-stalked intermediate, r: *Q. robur*. Figure S2: Average and standard deviation of the weights of the pots in the control and the drought treated group of plants during the second drought period in the first growing year. Percentage of seedlings in the treated group of plants showing wilting and/or curling of the leaves is indicated. p: *Q. petraea*, pr: long-stalked intermediate, rp: short-stalked intermediate, r: *Q. robur*. Figure S3: Biplots of a Principal Components Analysis on leaf morphological traits of oak seedlings. Seedlings are coloured according to the field identification of the mother trees that was based on both leaf and infructescence traits. p: *Q. petraea*, pr: long-stalked intermediate, rp: short-stalked intermediate, r: *Q. robur*. Abbreviation of traits are in Table S1. Figure S4: Principal components on leaf morphological traits of oak seedlings in relation to the field identification of the mother trees that was based on both leaf and infructescence traits. p: *Q. petraea*, pr: long-stalked intermediate, rp: short-stalked intermediate, r: *Q. robur*. Figure S5: Height growth in the two growing seasons 2014 and 2015 for both the control and the stressed group of plants. p: *Q. petraea*, pr: long-stalked intermediate, rp: short-stalked intermediate, r: *Q. robur*. Figure S6: Diameter growth in the two growing seasons 2014 and 2015 for both the control and the stressed group of plants. p: *Q. petraea*, pr: long-stalked intermediate, rp: short-stalked intermediate, r: *Q. robur*. Table S5: Basic data.

Acknowledgments: In the first place, we like to thank Stefaan Moreels for the elaborate and meticulous nursery work and indispensable help during the measurements and observations. We thank Ghislain Mees, Marc Missoorten and Eddy Hubrechts from the Agency for Nature and Forest for facilitating the seed collections. We also thank the COST FP1106 network Studying Tree Responses to Extreme Events: A Synthesis (STReESS) for the short-term scientific mission granted to AT, and the ERASMUS exchange program for funding. We are grateful to Pieter Verschelde for help with the Principal Components Analysis.

Author Contributions: K.V.M., A.T., M.S., K.S., B.V.d.A. and S.M. conceived of and supervised the whole study, while A.T., K.V.M., J.M. and N.D. collected the seedlings and conducted the measurements and observations on the seedlings. A.T., J.M. and N.D. were responsible for the weight loss and height/diameter measurements. A.T., J.M. and N.D. observed the leaf senescence. K.V.M. observed the bud burst. K.V.M., A.T., J.M., B.V.d.A. and N.D. performed the statistical analyses. All authors (K.V.M., A.T., J.M., N.D., S.M., M.S., K.S., B.V.d.A.) contributed substantially to the manuscript preparation.

Conflicts of Interest: The authors declare no conflict of interest.

References

1. Lindner, M.; Fitzgerald, J.B.; Zimmermann, N.E.; Reyer, C.; Delzon, S.; van der Maaten, E.; Schelhaas, M.J.; Lasch, P.; Eggers, J.; van der Maaten-Theunissen, M.; et al. Climate change and european forests: What do we know, what are the uncertainties, and what are the implications for forest management? *J. Environ. Manag.* **2014**, *146*, 69–83. [CrossRef] [PubMed]

2. Intergovernmental Panel on Climate Change (IPCC). *Climate Change 2014*; Synthesis Report; Contribution of Working Groups I, II and III to the Fifth Assessment Report of the Intergovernmental Panel on Climate Change; Intergovernmental Panel on Climate Change (IPCC): Geneva, Switzerland, 2014.

3. Anderegg, W.R.L.; Berry, J.A.; Field, C.B. Linking definitions, mechanisms, and modeling of drought-induced tree death. *Trends Plant Sci.* **2012**, *17*, 693–700. [CrossRef] [PubMed]

4. Scheffer, M.; Carpenter, S.; Foley, J.A.; Folke, C.; Walker, B. Catastrophic shifts in ecosystems. *Nature* **2001**, *413*, 591–596. [CrossRef] [PubMed]

5. Benigno, S.M.; Dixon, K.W.; Stevens, J.C. Seedling mortality during biphasic drought in sandy mediterranean soils. *Funct. Plant Biol.* **2014**, *41*, 1239–1248. [CrossRef]

6. Niinemets, U. Responses of forest trees to single and multiple environmental stresses from seedlings to mature plants: Past stress history, stress interactions, tolerance and acclimation. *For. Ecol. Manag.* **2010**, *260*, 1623–1639. [CrossRef]

7. Psidova, E.; Ditmarova, L.; Jamnicka, G.; Kurjak, D.; Majerova, J.; Czajkowski, T.; Bolte, A. Photosynthetic response of beech seedlings of different origin to water deficit. *Photosynthetica* **2015**, *53*, 187–194. [CrossRef]

8. Leuschner, C.; Backes, K.; Hertel, D.; Schipka, F.; Schmitt, U.; Terborg, O.; Runge, M. Drought responses at leaf, stem and fine root levels of competitive *Fagus sylvatica* L. and *Quercus petraea* (Matt.) Liebl. trees in dry and wet years. *For. Ecol. Manag.* **2001**, *149*, 33–46. [CrossRef]

9. Broadmeadow, M.S.J.; Jackson, S.B. Growth responses of *Quercus petraea, Fraxinus excelsior* and *Pinus sylvestris* to elevated carbon dioxide, ozone and water supply. *New Phytol.* **2000**, *146*, 437–451. [CrossRef]

10. Thomas, F.M.; Gausling, T. Morphological and physiological responses of oak seedlings (*Quercus petraea* and *Q. robur*) to moderate drought. *Ann. For. Sci.* **2000**, *57*, 325–333. [CrossRef]

11. Arend, M.; Kuster, T.; Gunthardt-Goerg, M.S.; Dobbertin, M. Provenance-specific growth responses to drought and air warming in three european oak species (*Quercus robur, Q. petraea* and *Q. pubescens*). *Tree Physiol.* **2011**, *31*, 287–297. [CrossRef] [PubMed]

12. Spiess, N.; Oufir, M.; Matusikova, I.; Stierschneider, M.; Kopecky, D.; Homolka, A.; Burg, K.; Fluch, S.; Hausman, J.F.; Wilhelm, E. Ecophysiological and transcriptomic responses of oak (*Quercus robur*) to long-term drought exposure and rewatering. *Environ. Exp. Bot.* **2012**, *77*, 117–126. [CrossRef]

13. Kuster, T.M.; Arend, M.; Gunthardt-Goerg, M.S.; Schulin, R. Root growth of different oak provenances in two soils under drought stress and air warming conditions. *Plant Soil* **2013**, *369*, 61–71. [CrossRef]

14. Jensen, J.S.; Hansen, J.K. Genetic variation in responses to different soil water treatments in *Quercus robur* L. *Scand. J. For. Res.* **2010**, *25*, 400–411. [CrossRef]

15. Pflug, E.E.; Siegwolf, R.; Buchmann, N.; Dobbertin, M.; Kuster, T.M.; Gunthardt-Goerg, M.S.; Arend, M. Growth cessation uncouples isotopic signals in leaves and tree rings of drought-exposed oak trees. *Tree Physiol.* **2015**, *35*, 1095–1105. [CrossRef] [PubMed]

16. Kuster, T.M.; Dobbertin, M.; Gunthardt-Goerg, M.S.; Schaub, M.; Arend, M. A phenological timetable of oak growth under experimental drought and air warming. *PLoS ONE* **2014**, *9*, e89724. [CrossRef] [PubMed]

17. Turcsán, A.; Steppe, K.; Sárközi, E.; Erdélyi, É.; Missoorten, M.; Mees, G.; Vander Mijnsbrugge, K. Early summer drought stress during the first growing year stimulates extra shoot growth in oak seedlings (*Quercus petraea*). *Front. Plant Sci.* **2016**, *7*. [CrossRef] [PubMed]

18. Vander Mijnsbrugge, K.; Turcsan, A.; Maes, J.; Duchene, N.; Meeus, S.; Steppe, K.; Steenackers, M. Repeated summer drought and re-watering during the first growing year of oak (*Quercus petraea*) delay autumn senescence and bud burst in the following spring. *Front. Plant Sci.* **2016**, *7*. [CrossRef] [PubMed]

19. Ducousso, A.; Bordacs, S. *Euforgen Technical Guidelines for Genetic Conservation and Use for Pedunculate and Sessile Oaks (Quercus robur and Q. petraea)*; International Plant Genetic Resources Institute: Rome, Italy, 2004.

20. Vander Mijnsbrugge, K.; De Cleene, L.; Beeckman, H. A combination of fruit and leaf morphology enables taxonomic classification of the complex *Q. robur* L.—*Q. x rosacea* Bechst.—*Q. petraea* (Matt.) Liebl. in autochthonous stands in flanders. *Silvae Genet.* **2011**, *60*, 139–148. [CrossRef]

21. Bacilieri, R.; Ducousso, A.; Petit, R.J.; Kremer, A. Mating system and asymmetric hybridization in a mixed stand of european oaks. *Evolution* **1996**, *50*, 900–908. [CrossRef] [PubMed]

22. Lepais, O.; Petit, R.J.; Guichoux, E.; Lavabre, J.E.; Alberto, F.; Kremer, A.; Gerber, S. Species relative abundance and direction of introgression in oaks. *Mol. Ecol.* **2009**, *18*, 2228–2242. [CrossRef] [PubMed]

23. Stebbins, G.L. *Variation and Evolution in Plants*; Columbia University Press: New York, NY, USA, 1950.

24. Kleinschmit, J.R.G.; Bacilieri, R.; Kremer, A.; Roloff, A. Comparison of morphological and genetic traits of pedunculate oak (*Q. robur* L.) and sessile oak (*Q. petraea* (Matt) Liebl). *Silvae Genet.* **1995**, *44*, 256–269.

25. Cochard, H.; Breda, N.; Granier, A.; Aussenac, G. Vulnerability to air-embolism of 3 european oak species (*Quercus petraea* (Matt) Liebl, *Quercus pubescens* Willd, *Quercus robur* L.). *Ann. Sci. For.* **1992**, *49*, 225–233. [CrossRef]

26. Ponton, S.; Dupouey, J.L.; Breda, N.; Feuillat, F.; Bodenes, C.; Dreyer, E. Carbon isotope discrimination and wood anatomy variations in mixed stands of *Quercus robur* and *Quercus petraea*. *Plant Cell Environ.* **2001**, *24*, 861–868. [CrossRef]

27. Le Provost, G.; Sulmon, C.; Frigerio, J.M.; Bodenes, C.; Kremer, A.; Plomion, C. Role of waterlogging-responsive genes in shaping interspecific differentiation between two sympatric oak species. *Tree Physiol.* **2012**, *32*, 119–134. [CrossRef] [PubMed]

28. Bacilieri, R.; Ducousso, A.; Kremer, A. Genetic, morphological, ecological and phenological differentiation between *Quercus petraea* (Matt.) Liebl. and *Quercus robur* L. in a mixed stand of northwest of France. *Sylvae Genet.* **1995**, *44*, 1–10.

29. Saintagne, C.; Bodénès, C.; Barreneche, T.; Pot, D.; Plomion, C.; Kremer, A. Distribution of genomic regions differentiating oak species assessed by QTL detection. *Heredity* **2004**, *92*, 20–30. [CrossRef] [PubMed]

30. Rellstab, C.; Bühler, A.; Graf, R.; Folly, C.; Gugerli, F. Using joint multivariate analyses of leaf morphology and molecular-genetic markers for taxon identification in three hybridizing European white oak species (*Quercus* spp.). *Ann. For. Sci.* **2016**, *73*, 669–679. [CrossRef]

31. Curtu, A.; Gailing, O.; Finkeldey, R. Patterns of contemporary hybridization inferred from paternity analysis in a four-oak-species forest. *BMC Evolut. Biol.* **2007**, *9*. [CrossRef] [PubMed]

32. Gugerli, F.; Walser, J.; Dounavi, K.; Holderegger, R.; Finkeldey, R. Coincidence of small-scale spatial discontinuities in leaf morphology and nuclear microsatellite variation of *Quercus petraea* and *Q. robur* in a mixed forest. *Ann. Bot.* **2007**, *99*, 713–722. [CrossRef] [PubMed]

33. Kremer, A.; Dupouey, J.L.; Deans, J.D.; Cottrell, J.; Csaikl, U.; Finkeldey, R.; Espinel, S.; Jensen, J.; Kleinschmit, J.; Van Dam, B.; et al. Leaf morphological differentiation between *Quercus robur* and *Quercus petraea* is stable across western European mixed oak stands. *Ann. For. Sci.* **2002**, *59*, 777–787. [CrossRef]

34. Vander Mijnsbrugge, K.; Cox, K.; Van Slycken, J. Conservation approaches for autochthonous woody plants in flanders. *Silvae Genet.* **2005**, *54*, 197–206. [CrossRef]

35. Vander Mijnsbrugge, K.; Coart, E.; Beeckman, H.; Van Slycken, J. Conservation measures for autochthonous oaks in flanders. *For. Genet.* **2003**, *10*, 207–217.

36. Brouwers, J.; Peeters, B.; Van Steertegem, M.; van Lipzig, N.; Wouters, H.; Beullens, J.; Demuzere, M.; Willems, P.; De Ridder, K.; Maiheu, B.; et al. *Mira Climate Report 2015: About Observed and Future Climate Changes in Flanders and Belgium*; Flanders Environment Agency: Aalst, Belgium, 2015.

37. Boratynski, A.; Marcysiak, K.; Lewandowska, A.; Jasinska, A.; Iszkulo, G.; Burczyk, J. Differences in leaf morphology between *Quercus petraea* and *Q. robur* adult and young individuals. *Silva Fenn.* **2008**, *42*, 115–124. [CrossRef]

38. R Core Team. *R: A Language and Environment for Statistical Computing*; R Foundation for Statistical Computing: Vienna, Austria, 2014.

39. Bates, D.; Machler, M.; Bolker, B.M.; Walker, S.C. Fitting linear mixed-effects models using lme4. *J. Stat. Softw.* **2015**, *67*, 1–48. [CrossRef]

40. Christensen, R.H.B. Ordinal: Regression Models for Ordinal Data. R Package Version 2013.10–31. 2013. Available online: http://www.Cran.R-project.Org/package=ordinal/ (accessed on 3 July 2015).

41. Vivin, P.; Aussenac, G.; Levy, G. Differences in drought resistance among 3 deciduous oak species grown in large boxes. *Ann. Sci. For.* **1993**, *50*, 221–233. [CrossRef]

42. Neophytou, C.; Aravanopoulos, F.A.; Fink, S.; Dounavi, A. Detecting interspecific and geographic differentiation patterns in two interfertile oak species (*Quercus petraea* (Matt.) Liebl. and *Q. robur* L.) using small sets of microsatellite markers. *For. Ecol. Manag.* **2010**, *259*, 2026–2035. [CrossRef]

43. Vander Mijnsbrugge, K.; Turcsan, A.; Maes, J.; Duchene, N.; Meeus, S.; Steppe, K.; Steenackers, M. Research Institute for Nature and Forest, Geraardsbergen, Belgium. Unpublished work. 2016.

44. Büsgen, M.; Münch, E.; Thomson, T. *The Structure and Life of Forest Trees*; Chapman and Hall: London, UK, 1929.

45. Fonti, P.; Heller, O.; Cherubini, P.; Rigling, A.; Arend, M. Wood anatomical responses of oak saplings exposed to air warming and soil drought. *Plant Biol.* **2013**, *15*, 210–219. [CrossRef] [PubMed]

46. Waring, R.H. Characteristics of trees predisposed to die. *Bioscience* **1987**, *37*, 569–574. [CrossRef]

47. Dobbertin, M. Tree growth as indicator of tree vitality and of tree reaction to environmental stress: A review. *Eur. J. For. Res.* **2005**, *124*, 319–333. [CrossRef]

48. De Boeck, H.; Dreesen, F.; Janssens, I.; Nijs, I. Climatic characteristics of heat waves and their simulation in plant experiments. *Glob. Chang. Biol.* **2010**, *16*, 1992–2000. [CrossRef]

49. De Boeck, H.; Dreesen, F.; Janssens, I.; Nijs, I. Whole-system responses of experimental plant communities to climate extremes imposed in different seasons. *New Phytol.* **2011**, *189*, 806–817. [CrossRef] [PubMed]

50. D'Amato, A.W.; Puettmann, K.J. The relative dominance hypothesis explains interaction dynamics in mixed species *Alnus rubra/Pseudotsuga menziesii* stands. *J. Ecol.* **2004**, *92*, 450–463.

51. Schwinning, S.; Weiner, J. Mechanisms determining the degree of size asymmetry in competition among plants. *Oecologia* **1998**, *113*, 447–455. [CrossRef] [PubMed]

52. Cannell, M.G.R.; Rothery, P.; Ford, E.D. Competition within stands of *Picea sitchensis* and *Pinus contorta*. *Ann. Bot. (Lond.)* **1984**, *53*, 349–362. [CrossRef]

53. Van Couwenberghe, R.; Gegout, J.C.; Lacombe, E.; Collet, C. Light and competition gradients fail to explain the coexistence of shade-tolerant *Fagus sylvatica* and shade-intermediate *Quercus petraea* seedlings. *Ann. Bot. (Lond.)* **2013**, *112*, 1421–1430. [CrossRef] [PubMed]

54. Boyden, S.B.; Reich, P.B.; Puettmann, K.J.; Baker, T.R. Effects of density and ontogeny on size and growth ranks of three competing tree species. *J. Ecol.* **2009**, *97*, 277–288. [CrossRef]

55. Lagache, L.; Klein, E.K.; Guichoux, E.; Petit, R.J. Fine-scale environmental control of hybridization in oaks. *Mol. Ecol.* **2013**, *22*, 423–436. [CrossRef] [PubMed]

56. Petit, R.J.; Bodenes, C.; Ducousso, A.; Roussel, G.; Kremer, A. Hybridization as a mechanism of invasion in oaks. *New Phytol.* **2004**, *161*, 151–164. [CrossRef]

forests

MDPI

Article

Influence of Scarification on the Germination Capacity of Acorns Harvested from Uneven-Aged Stands of Pedunculate Oak (*Quercus robur* L.)

Zdzisław Kaliniewicz [1,*] and Paweł Tylek [2]

[1] Faculty of Technical Sciences, University of Warmia and Mazury in Olsztyn, ul. Oczapowskiego 11, 10-719 Olsztyn, Poland

[2] Faculty of Forestry, University of Agriculture in Cracow, Al. 29 Listopada 46, 31-425 Kraków, Poland; rltylek@cyf-kr.edu.pl

* Correspondence: zdzislaw.kaliniewicz@uwm.edu.pl; Tel.: +48-089-523-3934

Received: 8 February 2018; Accepted: 23 February 2018; Published: 26 February 2018

Abstract: Scarification involves the partial removal of the seed coat on the side of the hilum, opposite the radicle, to speed up germination in acorns. The aim of this study was to determine the influence of scarification on the germination capacity of pedunculate oak acorns, selected and prepared for sowing. The diameter, length and mass of acorns were measured before and after scarification in four batches of acorns harvested from uneven-aged trees (76, 91, 131 and 161 years). The measured parameters were used to determine the correlations between acorn dimensions and mass, and to calculate the dimensional scarification index and the mass scarification index in acorns. Individual complete and scarified acorns from every batch were germinated on sand and peat substrate for 28 days. The analyzed acorns were characterized by average size and mass. Scarification decreased acorn mass by around 22% and acorn length by around 31% on average. Scarification and the elimination of infected acorns increased germination capacity from around 64% to around 81% on average. Acorns can be divided into size groups before scarification to obtain seed material with varied germination capacity. Larger acorns with higher germination capacity can be used for sowing in container nurseries, whereas smaller acorns with lower germination capacity can be sown in open-field nurseries.

Keywords: *Quercus robur* L.; seed size; scarification index; germination

---

1. Introduction

Pedunculate oak (*Quercus robur* L.) is a tree species measuring up to 40 m in height and up to 3 m in diameter at breast height. It is the main, dominant or co-dominant species in mixed-species forests, in particular in fresh mixed broadleaved forests, moist mixed broadleaved forests, fresh broadleaved forests, moist broadleaved forests, riparian forests, moist upland forests and upland forests [1]. Pedunculate oak is widely distributed throughout the European continent, excluding northern Europe and parts of Mediterranean Europe [1–5]. The species thrives in fertile and moist habitats, on loamy and sandy loam soils with high humus content and a moderately acidic or neutral pH. On nutrient-poor soils, oaks have an irregular growth pattern, they are smaller, produce twisted trunks and resemble shrubs [1,3].

Pedunculate oaks produce flowers and fruit at 40–50 years of age or even later (60–80 years) when they are grown in dense stands. Acorns or oak nuts ripen in September or October. The common name of *Quercus robur* is derived from the fact that the species produces several acorns per peduncle. The peduncle measures 5 to 12 cm in length. They are ellipsoidal in shape, and they are enclosed by woody cupules to one-third of their height. The hilum is located at the base of the acorn, and it is covered by the cupules. Fresh and rehydrated acorns have green and, subsequently, olive-brown stripes which disappear with moisture loss. Oak trees shed acorns in October, and empty or worm-riddled acorns are usually discarded first [2]. For this reason, acorns should be harvested only after the first batch of nuts has been shed. Fresh acorns are characterized by high moisture content and high susceptibility to fungal infections. Therefore, harvested acorns should be quickly transported to a processing facility in open boxes, baskets or bags made of loose fabric of plastic mesh to enable ventilation and prevent overheating [2].

In the processing plant, acorns are cleaned, sorted, immersed in water, subjected to heat treatment, dried, dressed with fungicides and prepared for cold storage. Acorns are immersed in water to remove weakly developed, damaged, almost empty and empty nuts. They are heated to eliminate fungal spores, in particular *Ciboria batschiana* which is responsible for black rot and mummification of acorns [6,7]. Acorns are immersed in water heated to a temperature of 41 °C for 2.5 h. The moisture content of acorns should not drop below 40% by dry weight during processing. Processing temperature has to be rigorously controlled because overheating decreases the germination capacity. Acorns with moisture content higher than 45% can be dried. Acorns can be stored in non-tight containers at a temperature of around −3 °C for up to two years without loss of germination capacity [2,7].

Pedunculate oak acorns do not enter winter dormancy. However, germination is strongly suppressed, and the seed coat prevents water and air from penetrating the acorn. Germination can be enhanced through scarification, namely the partial excision of the seed coat on the side of the hilum, opposite the radicle [2,8]. In other plant species, the seed coat is also excised to promote germination. The seed coat can be punctured, scarified with sharp sand, excised or removed chemically with concentrated acid. Scarification is recommended in around 7% of tree species, including in the Persian turpentine tree [9], honey locust [10], common myrtle [11], black velvet tamarind [12], black locust [13], lebbeck tree [14], African locust bean [15], Judas tree [16], noni [17], afzelia and African teak [18].

According to Suszka et al. [2], pedunculate oak acorns are scarified by reducing their length by one-third to one-quarter, usually with the use of shears or a grinding disc. Scarification exposes the cotyledon and enables visual evaluation of acorn health. Mummified acorns are eliminated at this stage [8,19]. Researchers are currently designing a robot system that will eliminate manual sorting, increase scarification efficiency and maximize the percentage of healthy acorns in the sorted batch [8,20–24]. Automated scarification will be a highly accurate process, and the removed portion of the acorn will be minimized to guarantee the highest germination capacity.

The aim of this study was to determine the influence of acorn scarification on the germination capacity of pedunculate oak acorns, selected and prepared for sowing.

2. Materials and Methods

2.1. Sample Preparation

The experiment was performed on pedunculate oak acorns harvested manually in uneven-aged tree stands (76, 91, 131 and 161 years) in seed zone Dbs 20, a fresh mixed broadleaved forest in Szczytno municipality in north-eastern Poland. Acorns were harvested with the use of collection nets between 10 and 14 October 2016. Each batch of harvested acorns was stored separately in non-heated and well ventilated premises. Every day, acorns were shoveled into piles not exceeding 10 cm in height. When the relative moisture content of acorns reached around 42%, acorns were subjected to heat treatment by immersion in water with a temperature of 41 °C for 2.5 h. After the treatment, acorns were surface dried, and samples of around 2 kg each were collected from every batch and refrigerated at a temperature of around 5 °C. The remaining acorns were placed in plastic kegs and freeze stored at a temperature of −3 °C. Two samples of 96 acorns each were selected from the refrigerated acorns by the survey sampling method [25]. The size of each sample corresponded to the number of cells in seeding containers.

2.2. Determination of Physical Properties

The length L, diameter D (Figure 1) and mass m of every acorn were determined, and acorns from one sample in each batch were scarified by reducing their length by one-quarter to one-third with the use of shears. Acorn health was evaluated visually, and only acorns without visible symptoms of pathological changes were used in the experiment. Rejected acorns were randomly replaced with new acorns whose geometric properties and mass were determined. The length L_s and mass m_s of scarified acorns were measured.

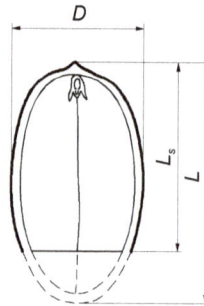

Figure 1. Acorn dimensions: D—diameter, L, L_s—length before and after scarification.

The length and diameter of acorns were measured with a caliper to the nearest 0.02 mm. Acorn diameter was determined as the average of two measurements performed at the widest point, perpendicular to the longitudinal axis. Acorn mass was determined with the Hornady 1500GR Bench Scale (Hornady®, Grand Island, NE, USA) to the nearest 0.01 g.

The following parameters were determined in each acorn:

- arithmetic mean diameter D_a and the geometric mean diameter D_g [26]:

$$D_a = \frac{2D + L}{3} \tag{1}$$

$$D_g = \left(D^2 \times L\right)^{1/3} \tag{2}$$

- specific mass m_D [27]:

$$m_D = \frac{m}{D_g} \tag{3}$$

- shape factors K_1 and K_2 [26,28]:

$$K_1 = \frac{D}{L} \tag{4}$$

$$K_2 = \frac{D_g}{L} \tag{5}$$

- and in scarified acorns—the dimensional scarification index S_L and the mass scarification index S_m:

$$S_L = \frac{L - L_s}{L} \tag{6}$$

$$S_m = \frac{m - m_s}{m} \tag{7}$$

2.3. Comparative Germination of Complete and Scarified Acorns

Individual acorns, whose physical parameters had been determined previously, were placed in the cell of plastic containers measuring $51 \times 33 \times 8$ cm. Each container was composed of 96 square cells measuring 4×4 cm. The cells were filled with sand and peat substrate (1:1) with approximate moisture content of 55%, which was compacted by twice dropping the container on the floor from a height of approximately 10 cm. Excess substrate was removed with a flat wooden slat positioned obliquely across the container, in two perpendicular motions. Acorns were pushed into the substrate with the radicle up and the upper portion of each acorn 2–3 mm below the edge of the cell, according to the method described by Tylkowski and Bujarska-Borkowska [29]. Acorns were covered with a layer of the peat substrate, and excess substrate was removed as described previously. The containers with the seeded acorns were stored indoors at a temperature of around 20 °C and were exposed to artificial light for 8 h daily. The germination test was carried out for 28 full days (from 14 November to 12 December 2016). The upper surface of the cells and the substrate were sprayed with tap water (electric conductance—0.25 mS/cm) once a day between 6 p.m. and 7 p.m. Acorns that were pushed up by the root at least 10 mm above the upper edge of the cell were regarded as germinated. Germination capacity was determined as the percentage of germinated acorns in the total number of tested acorns [2].

2.4. Statistical Analysis

The physical parameters of acorns were analyzed statistically in the Statistica PL program (version 12.5, StatSoft Polska Sp. z o.o., Crakow, Poland) at a significance level of $\alpha = 0.05$. Differences between the measured parameters were determined by one-way ANOVA, and differences in the physical parameters of complete and scarified acorns or germinated and non-germinated acorns were determined by the Student's *t*-test for independent samples. The normality of each group was verified by the Shapiro–Wilk *W*-test, and the homogeneity of variance was assessed with Levene's test. Where the null hypothesis of equal population means was rejected, the significance of differences was determined by Duncan's test, and homogenous groups were identified [30].

3. Results

3.1. Experimental Material

The physical parameters of acorns from the analyzed batches (harvested from uneven-aged tree stands) are presented in Table 1. The average values of the measured parameters were determined in the following ranges: length—28.10–28.82 mm, diameter—16.25–16.54 mm, mass—4.35–4.87 g, arithmetic mean diameter—20.21–20.63 mm, geometric mean diameter—19.49–19.88 mm, specific mass—0.22–0.24 g mm^{-1}, shape factor K_1—0.58, shape factor K_2—0.69–0.70. Acorns from the evaluated batches differed most significantly in length and mass. The analyzed acorns did not differ significantly in diameter, arithmetic and geometric mean diameter, specific mass and shape factors K_1 and K_2. The results of the analysis revealed that the largest acorns were harvested from a 76-year-old tree stand, and the smallest acorns were harvested from a 91-year-old tree stand.

3.2. Germination Capacity of Acorns

The germination capacity (Figure 2) of complete acorns was estimated in the range of 61% (batch O-91) to 66% (batches O-76 and O-161), which places the evaluated material in quality class I (germination capacity of 61–100%). Scarification and the removal of infected acorns increased germination capacity from around 77% (batch O-161) to around 86% (batch O-76), i.e., by around 16 percentage points on average. It should be noted that germination capacity was not significantly influenced by the age of the parent tree stand.

Table 1. Statistical distribution of the physical properties (mean value ± standard deviation) of acorns and significant differences between batches.

Property/Indicator	Acorn Batch			
	O-76	O-91	O-131	O-161
Length (mm)	28.82 ± 2.20 [b]	28.10 ± 2.36 [a]	28.74 ± 2.41 [ab]	28.49 ± 2.88 [ab]
Diameter (mm)	16.54 ± 1.62 [a]	16.27 ± 1.52 [a]	16.53 ± 1.53 [a]	16.25 ± 1.56 [a]
Mass (g)	4.87 ± 1.16 [b]	4.35 ± 0.98 [a]	4.61 ± 1.22 [ab]	4.43 ± 1.12 [a]
Arithm. mean diameter (mm)	20.63 ± 1.50 [a]	20.21 ± 1.39 [a]	20.60 ± 1.53 [a]	20.33 ± 1.57 [a]
Geom. mean diameter (mm)	19.88 ± 1.55 [a]	19.49 ± 1.41 [a]	19.86 ± 1.53 [a]	19.56 ± 1.55 [a]
Specific mass (g mm^{-1})	0.24 ± 0.04 [a]	0.22 ± 0.04 [a]	0.23 ± 0.04 [a]	0.22 ± 0.04 [a]
Shape factor K_1 (-)	0.58 ± 0.06 [a]	0.58 ± 0.07 [a]	0.58 ± 0.06 [a]	0.58 ± 0.07 [a]
Shape factor K_2 (-)	0.69 ± 0.05 [a]	0.70 ± 0.05 [a]	0.69 ± 0.05 [a]	0.69 ± 0.06 [a]

[a,b]—superscript letters denote significant differences between the corresponding properties (indicators).

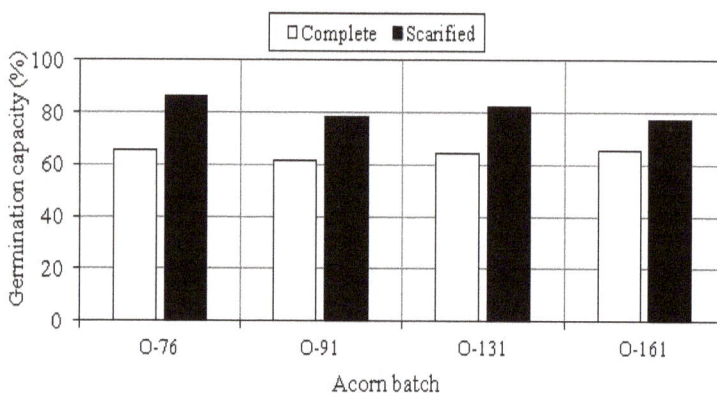

Figure 2. Germination capacity of complete and scarified pedunculate oak acorns.

The results of the Student's *t*-test for independent samples revealed that germinated and non-germinated complete acorns from four batches (Figure 3) differed mainly in average length and arithmetic mean diameter. In three batches (excluding O-91), significant differences were also observed in diameter, mass, geometric mean diameter and specific mass. Unlike in the remaining batches, germinated acorns in batch O-91 had a somewhat different shape than non-germinated acorns. These acorns were slimmer, and their average shape factor values were lower than those determined in non-germinated acorns. An analysis of the physical parameters of the evaluated acorns revealed that up to 2% of the shortest acorns can be removed from each batch without a loss of germinating complete acorns. The above will increase germination capacity by around 1.5 percentage points on average. The greatest improvement could be achieved in batch O-161 where the removal of around 15% of the shortest acorns would increase germination capacity from around 65% to around 76%. However, the above would lead to the loss of around 3% of viable acorns.

The results of the Student's t-test for independent samples revealed that germinated and non-germinated acorns from four batches of scarified material (Figure 4) differed in length, diameter, mass, arithmetic and geometric mean diameter, and specific mass. The above parameters were higher in germinated than in non-germinated acorns. In most cases (excluding batch O-91), no significant differences in shape were observed in germinated or non-germinated acorns. An analysis of the measured physical parameters revealed that the elimination of non-germinating acorns always leads to a certain loss of viable acorns.

3.3. Evaluation of Scarification Treatment

The statistical distribution of scarification index values is presented in Table 2. Scarification reduced acorn length by 15% to 42% (31% on average) and decreased acorn mass by 12% to 35% (22% on average). Batch O-76 differed significantly from the remaining acorn batches in terms of the dimensional scarification index. Batch O-91 differed significantly from the remaining acorn batches in terms of the mass scarification index. The coefficient of variation of the above parameters ranged from 9.54% to 19.47%.

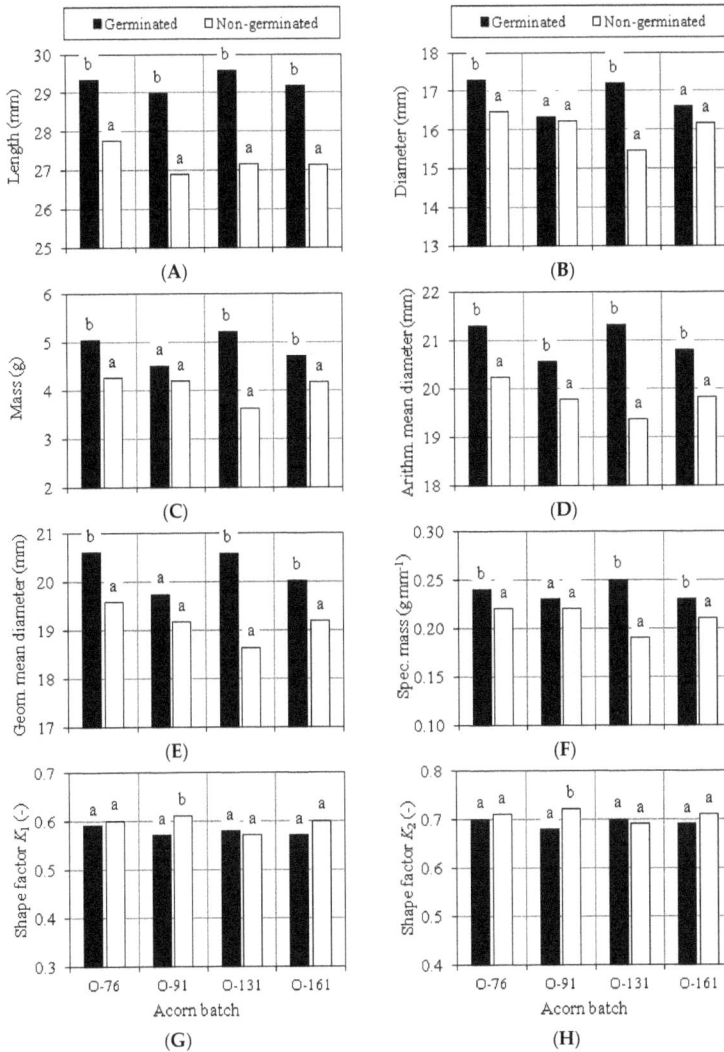

Figure 3. Significance of differences in the length (**A**), diameter (**B**), mass (**C**), arithmetic mean diameter (**D**), geometric mean diameter (**E**), specific mass (**F**), shape factor K_1 (**G**) and shape factor K_2 (**H**) of germinated and non-germinated complete acorns; a, b—different letters denote statistically significant differences.

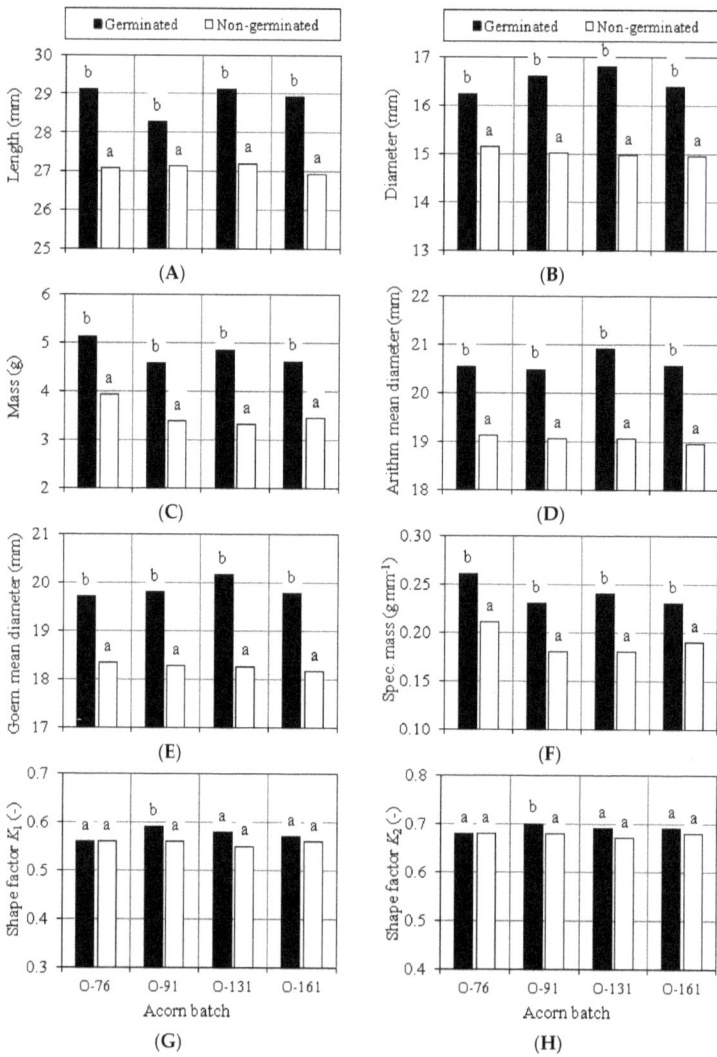

Figure 4. Significance of differences in the length (**A**), diameter (**B**), mass (**C**), arithmetic mean diameter (**D**), geometric mean diameter (**E**), specific mass (**F**), shape factor K_1 (**G**) and shape factor K_2 (**H**) of germinated and non-germinated acorns subjected to scarification; a, b—different letters denote statistically significant differences.

Table 2. Statistical distribution and significant differences in the scarification index of acorns from four batches.

Scarification Index	Acorn Batch	Value of Trait			Standard Deviation of Trait	Coefficient of Trait Variability (%)
		Minimum	Maximum	Average		
Dimensional S_L	O-76	0.15	0.39	0.28 [a]	0.046	16.18
	O-91	0.19	0.40	0.32 [b]	0.041	12.69
	O-131	0.24	0.42	0.32 [b]	0.031	9.77
	O-161	0.22	0.38	0.31 [b]	0.030	9.54
Mass S_m	O-76	0.12	0.35	0.22 [a]	0.043	19.47
	O-91	0.13	0.33	0.24 [b]	0.043	18.16
	O-131	0.13	0.33	0.22 [a]	0.034	15.64
	O-161	0.14	0.33	0.22 [a]	0.042	19.23

[a,b]—superscript letters denote significant differences between the corresponding properties.

The scarification index of acorns that germinated and acorns that did not germinate during the 28-day germination test is analyzed in Figure 5. No significant differences in the dimensional scarification index were found in either group in all batches, which indicates that the degree of scarification did not influence germination. In batches O-91 and O-161, minor (but statistically significant) differences were observed in the mass scarification index of germinated and non-germinated acorns, where non-germinated acorns lost more mass than germinated acorns.

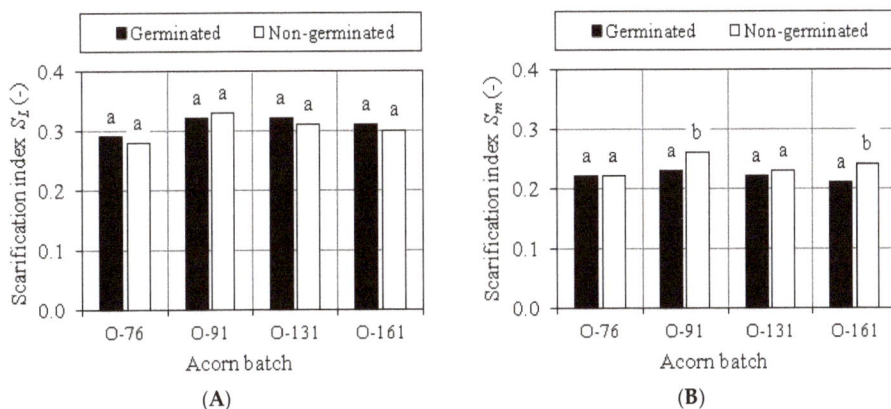

Figure 5. Significance of differences in the dimensional (**A**) and mass (**B**) scarification index of germinated and non-germinated acorns: a, b—different letters denote statistically significant differences.

3.4. Germination Capacity of Scarified Acorns

The germination capacity of scarified acorns divided into three size groups based on their diameter is presented in Figure 6. The germination capacity of the smallest acorns ranged from around 33% (batch O-131) to around 73% (batch O-76). The largest acorns were characterized by the highest germination capacity in the estimated range of 89% (batch O-161) to 100% (batch O-91).

Similar relationships were noted when acorns were divided into two size groups based on their diameter (Figure 7). Germination capacity ranged from around 59% (batch O-91) to around 83% (batch O-76) in acorns measuring up to 16 mm in diameter, and it exceeded 90% in acorns with a diameter larger than 16 mm.

In most cases, the above size groups did not differ significantly in their scarification index (Figure 8). Differences in the values of the dimensional scarification index were noted only in batch O-76, and differences in the values of the mass scarification index were found only in batch O-161.

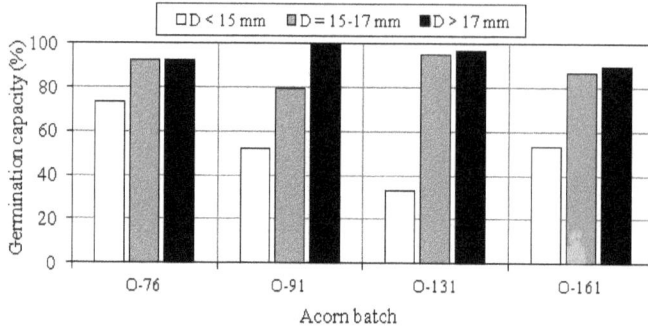

Figure 6. Germination capacity of pedunculate oak acorns divided into three size groups.

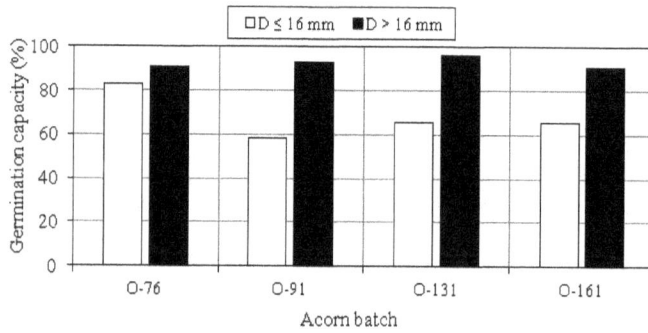

Figure 7. Germination capacity of pedunculate oak acorns divided into two size groups.

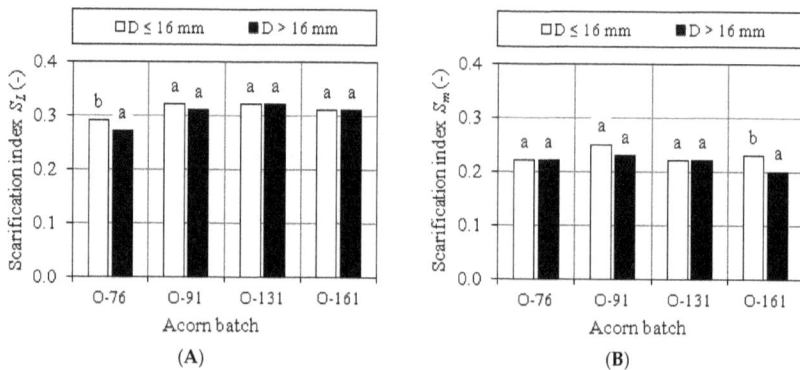

Figure 8. Significance of differences in the dimensional (**A**) and mass (**B**) scarification index of acorns measuring up to 16 mm and more than 16 mm in diameter: a, b—different letters denote statistically significant differences.

The germination capacity of the analyzed size groups relative to the values of the dimensional scarification index is presented in Table 3. These groups differed significantly in their germination capacity which ranged from 0% to even 100%. The scarification index and germination capacity were not directly correlated within the analyzed range of values of the dimensional scarification index. In the group of acorns measuring up to 16 mm in diameter, germination capacity exceeded 90% in acorns with a scarification index of 0.31 to 0.35 (batch O-76) and in acorns with a scarification index higher than 0.35 (batches O-131 and O-161). Acorns measuring more than 16 mm in diameter were characterized by significantly higher germination capacity which did not drop below 84% regardless of the value of the scarification index.

Table 3. Germination capacity of acorns from different size groups relative to their dimensional scarification index.

Acorn Batch	Size Group	Germination Capacity of Acorns with Dimensional Scarification Index S_L:			
		≤ 0.25	0.26–0.30	0.31–0.35	>0.35
	$D \leq 16$ mm	60.0	83.3	92.7	75.0
O-76	$D > 16$ mm	100.0	84.2	92.2	-
	Total	88.2	83.6	92.3	75.0
	$D \leq 16$ mm	0	63.6	57.9	60.0
O-91	$D > 16$ mm	100.0	88.9	95.8	88.9
	Total	80.0	79.3	79.1	73.7
	$D \leq 16$ mm	-	50.0	70.4	100.0
O-131	$D > 16$ mm	100.0	94.4	96.3	100.0
	Total	100.0	76.7	81.8	100.0
	$D \leq 16$ mm	33.3	64.7	64.0	100.0
O-161	$D > 16$ mm	-	81.0	100.0	100.0
	Total	33.3	73.2	78.6	100.0

Similar results were noted in an analysis of the mass scarification index (Table 4). The germination capacity of different size groups ranged from 0% to 100%, and germination capacity was not directly correlated with the scarification index. In most acorns measuring more than 16 mm in diameter (excluding acorns from batch O-161 with a scarification index of 0.26 to 0.30), germination capacity exceeded 80%, and it reached 100% in 11 out of 19 cases. In acorns measuring up to 16 mm in diameter, germination capacity was highest when the scarification index was below 0.15 (batches O-91 and O-131), 0.16–0.20 (batch O-131) and above 0.30 (batch O-76).

Table 4. Germination capacity of acorns from different size groups relative to their mass scarification index.

Acorn Batch	Size Group	Germination Capacity of Acorns with Mass Scarification Index S_m:				
		≤ 0.15	0.16–0.20	0.21–0.25	0.26–0.30	>0.30
	$D \leq 16$ mm	66.6	82.3	82.6	88.9	100.0
O-76	$D > 16$ mm	100.0	100.0	81.0	100.0	100.0
	Total	83.3	90.3	81.8	91.7	100.0
	$D \leq 16$ mm	100.0	66.7	66.7	42.9	50.0
O-91	$D > 16$ mm	100.0	92.3	96.4	90.0	66.7
	Total	100.0	87.5	84.8	62.5	57.1
	$D \leq 16$ mm	0	93.3	40.0	63.6	-
O-131	$D > 16$ mm	100.0	100.0	93.1	100.0	100.0
	Total	50.0	97.1	75.0	73.3	100.0
	$D \leq 16$ mm	100.0	50.0	79.2	57.1	0
O-161	$D > 16$ mm	100.0	100.0	81.3	75.0	-
	Total	100.0	82.8	80.0	61.1	0

4. Discussion

An analysis of the physical parameters of acorns harvested from uneven-aged tree stands revealed that the largest acorns were harvested from 76-year-old trees and the smallest acorns were harvested from 91-year-old trees. Despite significant differences in length and mass, the acorns from the above batches were characterized by similar diameter and shape. Acorns were harvested from tree stands in the same geographical location; therefore, differences in acorn size can probably be attributed to genetic variations which significantly influence the physical properties of seeds [31–33]. In the current study, the age of the parent tree stand (76 to 161 years) did not exert a significant influence on the physical parameters of acorns. Different results were reported by Kaliniewicz et al. [34] in Scotts pine where the physical dimensions and mass of seeds decreased with the age of parent trees. Similar trends were noted by Suszka et al. [2] based on long-term observations of tree stands rather than a comparison of the physical properties of acorns harvested from uneven-aged tree stands where genetic variations could play a key role. The significant influence of tree age on the physical parameters of seeds was also noted in a study of Norway spruce, but the nature of the observed changes was difficult to describe due to the disrupting influence of genetic factors [35]. In the present experiment, the dimensions and mass of pedunculate oak acorns were within the range of values reported by Suszka et al. [2], Nikolić and Orlović [36] and Tylkowski and Bujarska-Borkowska [29]. The evaluated acorns were somewhat smaller than those harvested in southern Poland [8,37] and in Serbia [38]. Seed size and mass generally decrease in northern regions of the globe [39–41].

In terms of germination capacity, acorns from uneven-aged tree stands were within the lower range of values in quality class I (61.5–65.6%). Thermal treatment was effective in preventing fungal diseases and acorn mummification, but failed to eliminate already infected and partially damaged acorns from the processed batches. According to Tylek [37] and Tylek et al. [42], small and large acorns are equally susceptible to fungal infections; therefore, they cannot be effectively separated based on their geometric parameters or shape. The results of the present study indicate that germination capacity can be somewhat improved (by around 1.5 percentage points) by eliminating around 2% of the shortest acorns from each batch. Scarification was a more effective treatment which increased germination capacity from around 64% to around 81%. Similar results were reported by Giertych and Suszka [19]. During scarification, the seed coat and cotyledons are partially removed, which improves water penetration and aeration, thus accelerating germination. Symptoms of disease are also more visible in scarified acorns which can be removed from the batch. Unlike the seeds of other forest trees [43], pedunculate oak acorns cannot be sorted effectively based on physical parameters; therefore, the optical parameters of acorn cross-sections could be used as an innovative selection trait. Optical parameters cannot be reliably evaluated by the naked eye, which is why an automated scarification device with a vision system has been developed [8,24] to identify early symptoms of disease, eliminate damaged acorns and increase germination capacity by up to 10% relative to manually processed material. However, evaluations of acorn health can be compromised by two types of errors. Firstly, acorns with normally developed cotyledons are often classified as healthy despite the presence of necrotic changes in the radicle, which are not visible to the evaluator. Secondly, acorns with damaged cotyledons can be classified as unfit for sowing even when the radicle is healthy and potentially capable of germinating. Nonetheless, the germination capacity of acorns is influenced mainly by the severity of pathological changes and effective removal of non-viable acorns. Some batches contain up to several dozen percent of damaged acorns [6].

Seed batches for sowing should contain both small and large acorns to preserve the genetic diversity of the future generations [2]. Gradual removal of small acorns can lead to the elimination of acorns produced by old trees, which are best adapted to a given habitat, local soil and weather conditions. For this reason, the quality of seed material can be more effectively improved through scarification than through the elimination of the shortest acorns—a procedure that induces only a minor increase in germination capacity (around 1.5 percentage points in the analyzed case).

According to Tylkowski and Bujarska-Borkowski [29], pedunculate oak acorns should be sorted based on size before planting. Acorn mass is positively correlated with seedling size [44–49]; therefore, similarly sized acorns should be planted separately to promote even emergence of seedlings. The results of the present study also demonstrate that acorns should be sorted into size groups before scarification and sowing. Germination capacity decreased with a decrease in acorn diameter, which implies that the seeding rate of acorns from different size groups should be adjusted accordingly to obtain the required number of seedlings. Larger acorns with a higher germination capacity (>90%) should be used mainly in container nurseries, whereas smaller acorns should be sown in open-field nurseries. The seeding rate should be determined based on the germination capacity of acorns. Acorns are easy to separate with the use of conventional sorting devices, and mesh sieves with longitudinal openings are particularly recommended for separating acorns into size groups based on their diameter.

Acorns with partially excised seed cover and cotyledons germinate faster [2,8,19,50]. This is a particularly important consideration in container nurseries where the growth cycle is relatively short and where polyethylene tents are used several times during the growing season. In the current study, the variations in the values of the dimensional scarification index (0.15 to 0.42) and the mass scarification index (0.12 to 0.35) did not influence the germination capacity of differently sized acorns. Shi et al. [51] reported the best results where acorns were reduced in length by one-third to one-half. In the cited study, scarification increased fertilizer absorption by oak acorns and seedlings grown in a nursery. According to Giertych and Suszka [19] and Tadeusiewicz et al. [8], the reduction in acorn mass during scarification should not exceed 20%. More extensive scarification increases the accuracy of health assessments, but it also compromises seedling growth. The presence of intact nutrient reserves in acorns promotes embryonic development, increases seedling resistance to adverse environmental factors and improves the morphological parameters of developing plants [19,50,52].

5. Conclusions

The results of this study indicate that the age of parent pedunculate oak trees (76 to 161 years) generally does not influence the physical parameters or the germination capacity of acorns. The germination capacity of complete acorns ranged from 61.5% to 65.6%. Up to 2% of the shortest acorns can be removed from the processed batch without the loss of germinating acorns.

Scarification and the elimination of acorns with symptoms of disease are the most effective methods of improving the quality of pedunculate oak acorns for sowing. The above treatments increased the germination capacity of acorns from around 64% to around 81%. Germination capacity was not correlated with the dimensional scarification index (15% to 42%) or the mass scarification index (12% to 35%).

The germination capacity of scarified acorns was correlated with their diameter. Acorns should be sorted into size groups before scarification and sowing to promote even seedling emergence. The germination capacity of acorns with the largest diameter (e.g., above 16 mm) exceeds 90%, and these acorns are recommended for sowing in container nurseries. Acorns with the smallest diameter (up to 16 mm) are characterized by lower germination capacity (around 58% to 83%), and they are more suited for sowing in open-field nurseries where the seeding rate should be determined based on the germination capacity of a given acorn batch.

Acknowledgments: The authors would like to thank Anna Prusik, a student of the University of Warmia and Mazury in Olsztyn majoring in Agricultural and Forest Engineering, for determining the physical parameters of acorns and conducting acorn germination tests as part of her Master's degree research.

Author Contributions: Z.K. conceived and designed the experiment; Z.K. and P.T. analyzed the data; Z.K. contributed materials/analysis tools; Z.K. wrote the paper.

Conflicts of Interest: The authors declare no conflict of interest.

References

1. Jaworski, A. *Hodowla lasu. Tom III. Charakterystyka hodowlana drzew i krzewów leśnych (Silviculture. Volume 3. Breeding Characteristics of Forest Trees and Shrubs)*; PWRiL: Warszawa, Poland, 2011; pp. 242–266. ISBN 9788309010760. (In Polish)

2. Suszka, B.; Muller, C.; Bonnet-Masimber, M. *Nasiona leśnych drzew liściastych od zbioru do siewu (Seeds of Deciduous Forest Trees—From Harvest to Sowing)*; Wydawnictwo Naukowe PWN: Warszawa-Poznań, Poland, 2000; pp. 1–274. ISBN 8301133430. (In Polish)

3. Rodriguez-Campos, A.; Diaz-Maroto, I.J.; Barcala-Perez, E.; Vila-Lameiro, P. Comparison of the autoecology of *Quercus robur* L. and *Q. petraea* (Mattuschka) Liebl. Stands in the Northwest of the Iberian Peninsula. *Ann. For. Res.* **2010**, *53*, 7–25.

4. Brus, D.J.; Hengeveld, G.M.; Walvoort, D.J.J.; Goedhart, P.W.; Heidema, A.H.; Nabuurs, G.J.; Gunia, K. Statistical mapping of tree species over Europe. *Eur. J. For. Res.* **2012**, *131*, 145–157. [CrossRef]

5. Peguero-Pina, J.J.; Sisó, S.; Sancho-Knapik, D.; Díaz-Espejo, A.; Flexas, J.; Galmés, J.; Gil-Pelegrín, E. Leaf morphological and physiological adaptations of a deciduous oak (*Quercus faginea* Lam.) to the Mediterranean climate: A comparison with a closely related temperate species (*Quercus robur* L.). *Tree Phys.* **2016**, *36*, 287–299. [CrossRef] [PubMed]

6. Knudsen, I.M.B.; Thomsen, K.A.; Jensen, B.; Poulsen, K.M. Effect of hot water treatment, biocontrol agents, disinfectants and a fungicide on storability of English oak acorns and control of the pathogen, Ciboria batschiana. *For. Pathol.* **2004**, *1*, 47–64. [CrossRef]

7. Schröder, T.; Kehr, R.; Procházková, Z.; Sutherland, J.R. Practical methods for estimating the infection rate of *Quercus robur* acorn seedlots by *Ciboria batschiana*. *For. Pathol.* **2004**, *34*, 187–196. [CrossRef]

8. Tadeusiewicz, R.; Tylek, P.; Adamczyk, F.; Kiełbasa, P.; Jabłoński, M.; Pawlik, P.; Piłat, A.; Walczyk, J.; Szczepaniak, J.; Juliszewski, T.; et al. Automation of the acorn scarification process as a contribution to sustainable forest management. Case study: Common oak. *Sustainability* **2017**, *9*, 2276. [CrossRef]

9. Chebouti-Meziou, N.; Merabet, A.; Chebouti, Y.; Bissaad, F.Z.; Behidj-Benyounes, N.; Doumandji, S. Effect of cold and scarification on seeds germination of *Pistacia atlantica* L. for rapid multiplication. *Pak. J. Bot.* **2014**, *46*, 441–446.

10. Asl, M.B.; Sharivivash, R.; Rahbari, A. Effect of different treatments on seed germination of honey locust (*Gleditschia triacanthos*). *Mod. Appl. Sci.* **2011**, *5*, 200–204.

11. Ciccarelli, D.; Andreucci, A.C.; Pagni, A.M.; Garbari, F. The role of the elaiosome in the germination of seeds of *Myrtus communis* L. (Myrtaceae). *Atti Soc. Tosc. Sci. Nat.* **2004**, *111*, 143–146.

12. Nwaoguala, C.N.C.; Osaigbovo, A.U. Enhancing seedling production of black velvet tamarind (*Dialium guineense* Willd). *J. Appl. Nat. Sci.* **2009**, *1*, 36–40.

13. Tylkowski, T.; Grupa, R. Effectiveness of pre-sowing scarification methods of black locust seeds. *Sylwan* **2010**, *154*, 33–40.

14. Missanjo, E.; Maya, C.; Kapira, D.; Banda, H.; Kamanga-Thole, G. Effect of seed size and pretreatment methods on germination of *Albizia lebbeck*. *ISRN Bot.* **2013**. [CrossRef]

15. Aliero, B.L. Effects of sulphuric acid, mechanical scarification and wet heat treatments on germination of seeds of African locust bean tree, *Parkia biglobosa*. *Afric. J. Biotechnol.* **2004**, *3*, 179–181. [CrossRef]

16. Pipinis, E.; Milios, E.; Smiris, P.; Gioumousidis, C. Effect of acid scarification and cold moist stratification on the germination of *Cercis siliquastrum* L. seeds. *Turk. J. Agric. For.* **2011**, *35*, 259–264. [CrossRef]

17. Nelson, S. Noni seed handling and seedling production. *Fruit. Nut.* **2005**, *10*, 1–4.

18. Botsheleng, B.; Mathowa, T.; Mojeremane, W. Effects of pre-treatments methods on the germination of pod mahogany (*Afzelia quanzensis*) and mukusi (*Baikiaea plurijuga*) seeds. *Int. J. Innov. Res. Sci.* **2014**, *3*, 8108–8113.

19. Giertych, M.J.; Suszka, J. Consequences of cutting off distal ends of cotyledons of *Quercus robur* acorns before sowing. *Ann. For. Sci.* **2011**, *68*, 433–442. [CrossRef]

20. Jabłoński, M.; Tylek, P.; Walczyk, J.; Tadeusiewicz, R.; Piłat, A. Colour-based binary discrimination of scarified *Quercus robur* acorn under varying illumination. *Sensors* **2016**, *16*, 1319. [CrossRef] [PubMed]

21. Grabska-Chrząstowska, J.; Kwiecień, J.; Drożdż, M.; Bubliński, Z.; Tadeusiewicz, R.; Szczepaniak, J.; Walczyk, J.; Tylek, P. Comparison of selected classification methods in automated oak seed sorting. *J. Res. Appl. Agric. Eng.* **2017**, *62*, 31–33.

22. Pawlik, P.; Jabłoński, M.; Bubliński, Z.; Tadeusiewicz, R.; Walczyk, J.; Tylek, P.; Juliszewski, T.; Adamczyk, F. Use of Harris detector for determination of orientation of acorns in the process of automated scarification. *J. Res. Appl. Agric. Eng.* **2017**, *62*, 163–165.

23. Przybyło, J.; Jabłoński, M.; Pociecha, D.; Tadeusiewicz, R.; Piłat, A.; Walczyk, J.; Kiełbasa, P.; Szczepaniak, J.; Adamczyk, F. Application of model-based design in algorithms' prototyping for experimental acorn scarification rig. *J. Res. Appl. Agric. Eng.* **2017**, *62*, 166–170.

24. Tadeusiewicz, R.; Tylek, P.; Adamczyk, F.; Kiełbasa, P.; Jabłoński, M.; Bubliński, Z.; Grabska-Chrząstowska, J.; Kaliniewicz, Z.; Walczyk, J.; Szczepaniak, J.; et al. Assessment of selected parameters of the automatic scarification device as an example of a device for sustainable forest management. *Sustainability* **2017**, *9*, 2370. [CrossRef]

25. Greń, J. *Statystyka matematyczna. Modele i zadania (Mathematical Statistics. Models and Tasks)*; PWN: Warszawa, Poland, 1984; pp. 237–280. ISBN 8301036990. (In Polish)

26. Mohsenin, N.N. *Physical Properties of Plant and Animal Materials*; Gordon and Breach Science Public: New York, NY, USA, 1986; pp. 1–891. ISBN 9780677213705.

27. Kaliniewicz, Z.; Tylek, P.; Markowski, P.; Anders, A.; Rawa, T.; Jóźwiak, K.; Fura, S. Correlations between the germination capacity and selected physical properties of Scots pine (Pinus sylvestris L.) seeds. *Balt. For.* **2013**, *19*, 201–211.

28. Grochowicz, J. *Maszyny do czyszczenia i sortowania nasion (Seed Cleaning and Sorting Machines)*; Akademia Rolnicza: Lublin, Poland, 1994; pp. 25–28. ISBN 839016129X. (In Polish)

29. Tylkowski, T.; Bujarska-Borkowska, B. Effect of acorn size and sowing depth on *Quercus robur* and *Q. petraea* seedling emergence and height. *Sylwan* **2011**, *155*, 159–170. (In Polish)

30. Rabiej, M. *Statystyka z programem Statistica (Statisctics in Statistica Software)*; Helion: Gliwice, Poland, 2012; pp. 1–344. ISBN 9788324641109. (In Polish)

31. García-Mozo, H.; Gómez-Casero, M.T.; Domínguez, E.; Galán, C. Influence of pollen emission and weather-related factors on variations in holm-oak (*Quercus ilex* subsp. *ballota*) acorn production. *Environ. Exp. Bot.* **2007**, *61*, 35–40. [CrossRef]

32. Koenig, W.D.; Knops, J.M.H.; Carmen, W.J.; Sage, D. No trade-off between seed size and number in the valley oak *Quercus lobata*. *Am. Nat.* **2009**, *173*, 682–688. [CrossRef] [PubMed]

33. Alejano, R.; Vázquez-Piqué, J.; Carevic, F.; Fernández, M. Do ecological and silvicultural factors influence acorn mass in Holm Oak (southwestern Spain)? *Agroforest Syst.* **2011**, *83*, 25–39. [CrossRef]

34. Kaliniewicz, Z.; Rawa, T.; Tylek, P.; Markowski, P.; Anders, A.; Fura, S. The effect of the age of Scots pine (*Pinus sylvestris* L.) stands on the physical properties of seeds and the operating parameters of cleaning machines. *Tech. Sci.* **2013**, *16*, 63–72.

35. Kaliniewicz, Z.; Markowski, P.; Anders, A.; Tylek, P.; Krzysiak, Z.; Fura, S. Influence of the age of parent stand on selected physical properties of Norway spruce seeds. *Sylwan* **2017**, *161*, 548–557. (In Polish)

36. Nikolić, N.P.; Orlović, S. Genotypic variability of morphological characteristics of English oak (*Quercus robur* L.) acorn. *Zb. Matice Srp. Prir. Nauk.* **2002**, *102*, 53–58. [CrossRef]

37. Tylek, P. Size and shape as separation properties of pedunculate oak seeds (*Quercus robur* L.). *Acta Agrophys.* **2012**, *19*, 673–687. (In Polish)

38. Rakić, S.; Povrenović, D.; Tešević, V.; Simić, M.; Maletić, R. Oak acorn, polyphenols and antioxidant activity in functional food. *J. Food Eng.* **2006**, *74*, 416–423. [CrossRef]

39. Aizen, M.A.; Patterson, W.A. Acorn size and geographical range in the North American oaks (*Quercus* L.). *J. Biogeogr.* **1990**, *17*, 327–332. [CrossRef]

40. Oleksyn, J.; Reich, P.B.; Tjoelker, M.G.; Chalupka, W. Biogeographic differences in shoot elongation pattern among European Scots pine populations. *For. Ecol. Manag.* **2001**, *148*, 207–220. [CrossRef]

41. Moles, A.T.; Westoby, M. Latitude, seed predation and seed mass. *J. Biogeogr.* **2003**, *30*, 105–128. [CrossRef]

42. Tylek, P.; Kaliniewicz, Z.; Kiełbasa, P.; Zagrobelny, T. Mass and density as separation criteria of pedunculate oak (*Quercus robur* L.) seeds. *Electron. J. Pol. Agric. Univ. Ser. For.* **2015**, *18*. Available online: http://www.ejpau.media.pl/volume18/issue4/art-05.html (accessed on 5 February 2018).

43. Kaliniewicz, Z.; Tylek, P.; Anders, A.; Markowski, P.; Rawa, T.; Ołdakowski, M.; Wąsowski, L. An analysis of the physical properties of seeds of selected deciduous tree species. *Balt. For.* **2016**, *22*, 169–174.

44. Kleinschmit, J. Intraspecific variation of growth and adaptive traits in European oak species. *Ann. For. Sci.* **1993**, *50*, 166–185. [CrossRef]

45. Kormanik, P.P.; Sung, S.S.; Kormanik, T.L.; Schlarbaum, S.E.; Zarnoch, S.J. Effect of acorn size on development of northern red oak 1-0 seedlings. *Can. J. For. Res.* **1998**, *28*, 1805–1813. [CrossRef]

46. Quero, J.L.; Villar, R.; Marañón, T.; Zamora, R.; Poorter, L. Seed-mass effects in four Mediterranean *Quercus* species (Fagaceae) growing in contrasting light environments. *Am. J. Bot.* **2007**, *94*, 1795–1803. [CrossRef] [PubMed]

47. Tilki, F. Influence of acorn size and storage duration on moisture content, germination and survival of *Quercus petraea* (Mattuschka). *J. Environ. Biol.* **2010**, *31*, 325–328. [PubMed]

48. Yi, X.; Zhang, J.; Wang, Z. Large and small acorns contribute equally to early-stage oak seedlings: A multiple species study. *Eur. J. For. Res.* **2015**, *134*, 1019–1026. [CrossRef]

49. Llanderal-Mendoza, J.; Gugger, P.F.; Oyama, K.; Uribe-Salas, D.; González-Rodríguez, A. Climatic determinants of acorn size and germination percentage of Quercus rucosa (Fagaceae) along a latitudinal gradient in Mexico. *Bot. Sci.* **2017**, *95*, 37–45. [CrossRef]

50. García-Cebrián, F.; Esteso-Martínez, J.; Gil-Pelegrín, E. Influence of cotyledon removal on early seedling growth in *Quercus robur* L. *Ann. For. Sci.* **2003**, *60*, 69–73. [CrossRef]

51. Shi, W.; Bloomberg, M.; Li, G.; Su, S.; Jia, L. Combined effects of cotyledon excision and nursery fertilization on root growth, nutrient status and outplanting performance of *Quercus variabilis* container seedlings. *PLoS ONE* **2017**, *12*, e0177002. [CrossRef] [PubMed]

52. Hou, X.; Yi, X.; Yang, Y.; Liu, W. Acorn germination and seedling survival of Q. *variabilis*: Effects of cotyledon excision. *Ann. For. Sci.* **2010**, *67*. [CrossRef]

forests

MDPI

Article

First-Year Vitality of Reforestation Plantings in Response to Herbivore Exclusion on Reclaimed Appalachian Surface-Mined Land

Zachary J. Hackworth *, John M. Lhotka, John J. Cox, Christopher D. Barton and Matthew T. Springer

Department of Forestry and Natural Resources, University of Kentucky, 105 T.P. Cooper Building, Lexington, KY 40546, USA; john.lhotka@uky.edu (J.M.L.); jjcox@uky.edu (J.J.C.); barton@uky.edu (C.D.B.); mattspringer@uky.edu (M.T.S.)
* Correspondence: zachary.hackworth@uky.edu; Tel.: +1-859-257-7596

Received: 30 March 2018; Accepted: 19 April 2018; Published: 21 April 2018

Abstract: Conventional Appalachian surface-mine reclamation techniques repress natural forest regeneration, and tree plantings are often necessary for reforestation. Reclaimed Appalachian surface mines harbor a suite of mammal herbivores that forage on recently planted seedlings. Anecdotal reports across Appalachia have implicated herbivory in the hindrance and failure of reforestation efforts, yet empirical evaluation of herbivory impacts on planted seedling vitality in this region remains relatively uninitiated. First growing-season survival, height growth, and mammal herbivory damage of black locust (*Robinia pseudoacacia* L.), shortleaf pine (*Pinus echinata* Mill.), and white oak (*Quercus alba* L.) are presented in response to varying intensities of herbivore exclusion. Seedling survival was generally high, and height growth was positive for all species. The highest herbivory incidence of all tree species was observed in treatments offering no herbivore exclusion. While seedling protectors lowered herbivory incidence compared with no exclusion, full exclusion treatments resulted in the greatest reduction of herbivore damage. Although herbivory from rabbits, small mammals, and domestic animals was observed, cervids (deer and elk) were responsible for 95.8% of all damaged seedlings. This study indicates that cervids forage heavily on planted seedlings during the first growing-season, but exclusion is effective at reducing herbivory.

Keywords: mine reclamation; browse; black locust; shortleaf pine; white oak; elk; white-tailed deer; rabbit; small mammal

1. Introduction

Surface mining for coal has negatively impacted forest resources across Appalachia, including the loss of over 1.1 million ha of forests [1] and the fragmentation of at least an additional 1 million ha [2,3]. Federal regulations of the Surface Mining Control and Reclamation Act of 1977 (SMCRA) led to reclamation methods that, while intended to limit soil destabilization and water-quality impairment, resulted in compacted post-mining landscapes that greatly hinder forest regeneration. Compacted mine soils inhibit water infiltration, increase the frequency of ponding, and suppress root spreading [4–6], which diminishes water and nutrient absorption and root anchoring ability critical for vertical stability with tree maturation [7]. Post-mining vegetation communities in Appalachia are typically composed of planted invasive, exotic woody and herbaceous species (e.g., autumn-olive (*Elaeagnus umbellata* Thunb.), sericea lespedeza (*Lespedeza cuneata* (Dum. Cours.) G. Don), and multi-flora rose (*Rosa multiflora* Thunb.)) that rapidly colonize disturbed areas and outcompete native pioneer species [8–10]. Additionally, intensive vegetation control in popular agricultural post-mining land-uses, such as hayland pasture and crop production, can preclude forest succession and reforestation efforts.

Motivated by the exigencies of mine reforestation under conventional reclamation standards, a multi-disciplinary group of investigators initiated a large-scale study of techniques that would improve the favorability of post-mining landscapes for reforestation [11]. The Forestry Reclamation Approach (FRA) advocates a broad five-step method for mine reforestation that includes site preparation to create adequate rooting media and the use of proper tree planting techniques [11,12]. Heavy machinery is typically used to reduce pre-existing competing vegetation and alleviate soil compaction to create proper rooting media for planted seedlings. Restoration of native forests on reclaimed mined lands is reliant upon artificial regeneration. Distance to native seed sources, absence of soil seed bank, and abundant seed availability from non-native invasive species often hinder natural regeneration and necessitate tree planting to commence forest growth. However, after planting, seedlings are subject to a variety of factors that can decrease survival, growth, and subsequent forest maturation, of which herbivory can be among the most impactful.

Herbivory can greatly influence vegetation communities. Individual plant factors, such as species, life stage, nutrient quality, and defensive chemical potency [13–16], contribute to the extent of herbivore damage to plant communities. Community-level impacts, including floral dynamics [17,18], herbivory timing and intensity [19,20], and trophic interactions [21,22], also dictate the influence of herbivory. The loss of apex predators in the eastern U.S. has aided in the overabundance of primary consumers, specifically white-tailed deer (*Odocoileus virginianus* Zimmermann), a species noted for its impact on plant composition and structure in eastern U.S. ecosystems, including the biodiverse mixed-mesophytic forests of Appalachia [23]. Vulnerable plants, such as American ginseng (*Panax quinquefolius* L.) and several understory forbs, have experienced sharp declines in numbers and population viability as a result of increased deer browsing [14,24,25]. Areas with high deer densities commonly experience regenerating forests with compositions reflective of differences in plant species palatability and defensive mechanisms to reduce browsing; less palatable and more defensive plant species become more common in these areas, which dictates compositional and structural changes manifested with forest aging [26].

Herbivory can be particularly detrimental to newly established tree plantations. Artificial regeneration is often selected to alter pre-existing cohort species compositions, to reforest (or afforest) a non-forested area, and/or to accelerate the rate of regeneration. Therefore, plantation failure can prove both ecologically and financially costly, especially to highly denatured surface-mined lands where tree planting is vital to successful reforestation. Recently, herbivore damage of reforested seedlings has been implicated in the widespread damage to several FRA plantings across Appalachia [27]. However, aside from anecdotal claims and isolated information in a few published studies [28,29], a formal investigation of herbivory impacts on mine reforestation remains lacking. We present the first empirical study of herbivory damage to tree seedlings planted under the FRA on reclaimed Appalachian mined lands. We examined survival, height growth, and relative cause-specific herbivory of black locust (*Robinia pseudoacacia* L.), shortleaf pine (*Pinus echinata* Mill.), and white oak (*Quercus alba* L.) seedlings in response to herbivore exclusion.

2. Materials and Methods

2.1. Plot Design and Data Collection

We selected four ~0.4-ha sites across a complex of surface-mined tracts owned by the University of Kentucky in Breathitt County, KY, USA (Figure 1). Following FRA site preparation recommendations [12], each of the sites was bulldozed to reduce pre-existing vegetation (primarily invasive, exotic species), and compacted soils were ripped with a ripping shank mounted behind a Caterpillar D-11 bulldozer. Each site was partitioned into three, 36-m square plots, and ~108 1-0 bare-root seedlings of each of black locust, shortleaf pine, and white oak were planted randomly in rows on a 2-m spacing within each plot (4 sites × 3 plots/site = 12 plots). Seedlings were purchased from the Kentucky Division of Forestry nursery and were planted by experienced reforestation contractors in March 2017.

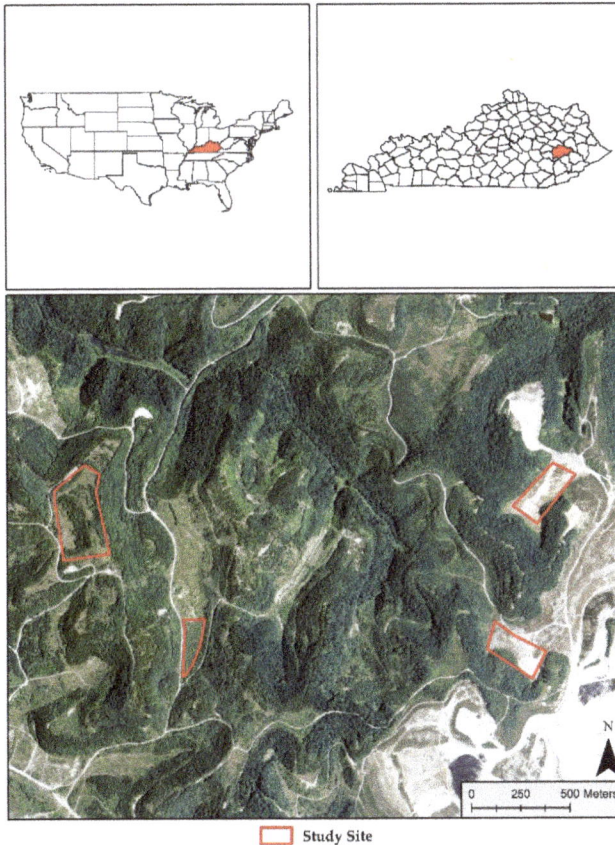

Figure 1. Study location prior to site preparation, Breathitt County, KY, USA. Exotic shrubs and conifers were common in the two western plot locations, and vegetation was relatively absent in the eastern plots.

Similar to many legacy mined lands across Appalachia, the study site harbors a number of herbivores capable of damaging planted seedlings, including elk (*Cervus canadensis* Erxleben), white-tailed deer, rabbits (*Sylvilagus* spp.), and small mammals. Small mammal communities across study sites were predominantly composed of white-footed mice (*Peromyscus leucopus* Rafinesque) [Z. Hackworth, unpublished data], which is consistent with prior work on adjacent mined lands in eastern Kentucky [30]. The study site also harbors a semi-feral horse (*Equus ferus caballus* L.) population and, occasionally, domestic cattle (*Bos taurus* L.) that have escaped from neighboring properties. Since domestic animal occupancy of abandoned mined lands is common throughout Appalachia and confirmed in our study area, damage caused by this group was included in the analysis. We were only interested in seedling damage mediated by mammal herbivores and did not examine herbivory from other taxa (e.g., insects).

A randomized complete block experimental design was used, whereby each plot within a site was randomly prescribed one of three herbivore exclusion treatments: no exclusion, seedling protectors, or full exclusion. The no exclusion treatment served as the control within a site replicate and offered unobstructed access to all herbivores. Within plots assigned protector treatments, an 8.5-cm × 46-cm (diameter x height) plastic diamond-mesh seedling protector (Forestry Suppliers, Jackson, MS, USA) was installed around each seedling, the base of which was entrenched in the soil 2–3 cm below the surface, and was anchored with a bamboo stake. Protector plots were designed to exclude small mammals and rabbits, but allow ungulate herbivory. In full exclusion treatments, a 2.4-m fence constructed from treated wooden posts and 12.5-gauge woven wire (Kencove Farm Fence Supplies, Blairsville, PA, USA) was installed around the perimeter of the plot, and each seedling within the plot was surrounded by a seedling protector according to the protocol implemented for protector treatments. Full exclusion was designed to prohibit seedling access to all aboveground mammal herbivores of interest to this study.

Soil samples were collected from all experimental plots to determine variability in edaphic characteristics across the experiment. Each plot was halved longitudinally, and a sample aggregated from three random subsamples was collected from each half of the plot prior to planting. Soil samples were analyzed for the following soil parameters: pH, P, K, Ca, Mg, Zn, N, and exchangeable K, Ca, Mg, and Na. Soil pH was calculated in a 1:1 soil:water solution [31]. P, K, Ca, Mg, and Zn concentrations were extracted via Mehlich III [32]. Relative sand, silt, and clay percentages were calculated with the micropipette method [33]. Exchangeable nutrient concentrations were determined after ammonium acetate extraction with ICP [32]. Total N (%) was evaluated with a LECO CHN-2000 Analyzer (LECO Corporation, St. Joseph, MI, USA). Cation exchange capacity was assessed by the ammonium acetate method at pH 3 [34]. Soil parameter differences among exclusion treatments were compared via a linear mixed-effect model with exclusion treatment as a fixed effect and site as a random effect. Significant differences were evaluated using a Type III ANOVA model. No significant differences among exclusion treatments were observed for any of the selected soil parameters (Table 1).

Table 1. Edaphic characteristics (Mean ± SE) across herbivore exclusion treatments on reclaimed mined lands in southeastern KY. No significant differences among exclusion treatments were detected for any of the soil parameters based upon individual Type III ANOVA models and a 0.05 significance level.

	Treatment		
Parameter	No Exclusion	Protector	Full Exclusion
Soil pH	5.60 ± 0.52	5.62 ± 0.75	6.08 ± 0.69
P (mg/kg)	5.94 ± 0.91	11.19 ± 6.06	6.00 ± 1.01
K (mg/kg)	71.06 ± 17.17	65.81 ± 14.32	63.94 ± 12.11
Ca (mg/kg)	580.13 ± 168.38	653.44 ± 183.04	640.44 ± 163.42
Mg (mg/kg)	251.19 ± 83.16	225.38 ± 75.81	269.06 ± 73.44
Zn (mg/kg)	3.65 ± 1.35	3.49 ± 1.49	3.49 ± 1.16
Total N (%)	0.07 ± 0.03	0.10 ± 0.06	0.08 ± 0.03
Sand (%)	52.18 ± 10.14	56.85 ± 8.48	57.25 ± 8.43
Silt (%)	33.38 ± 7.94	29.10 ± 6.31	39.33 ± 6.24
Clay (%)	14.44 ± 2.28	14.05 ± 2.37	13.42 ± 2.20
CEC * (meq/100 g)	7.09 ± 1.25	6.91 ± 1.40	6.65 ± 1.47
Exch † K (meq/100 g)	0.20 ± 0.05	0.18 ± 0.04	0.18 ± 0.04
Exch Ca (meq/100 g)	2.97 ± 1.00	3.11 ± 0.93	3.49 ± 1.01
Exch Mg (meq/100 g)	1.94 ± 0.74	1.69 ±0.64	2.17 ± 0.68
Exch Na (meq/100 g)	0.02 ± 0.003	0.02 ± 0.001	0.02 ± 0.004

* CEC indicates cation exchange capacity. † Exch indicates exchangeable.

First growing-season survival, height growth, and herbivore damage of seedlings were monitored via a series of seedling assessments. In May 2017, each seedling was assessed for survival, and the initial heights of all seedlings were measured. In October 2017, the end-of-growing-season survival of all seedlings was recorded, and the height of all seedlings was remeasured. Seedling heights were measured from the ground line to the tip of the apical bud of the tallest seedling branch. In February 2018, each seedling was evaluated for the presence of mammal herbivory to assess cumulative herbivory across all seasons. Herbivory indicators were categorized into four cause-specific groups: cervids, rabbits, small mammals, and domestic animals. Elk and deer produced nearly identical browse indicators, and, since one-year-old seedlings were below the "browse line" of both species, herbivory could not be distinguished between them and was, therefore, classified as "cervids". Cervid herbivory was typically identified by the damage or removal of shoot terminal buds which left a characteristically ragged edge due to the lack of upper incisors and biting of the bottom teeth against the upper lip pad. A clean, angular branch severance near the base of the seedling or complete seedling severance near the ground was attributed to rabbit herbivory. Basal bark gnawing was considered characteristic of small mammal herbivory.

Since herbicide was not employed during site preparation, regrowing competing vegetation could impact seedling survival and growth. However, fencing in full exclusion treatments may produce taller competing vegetation heights due to the exclusion of large herbivores. Height of competing vegetation was measured via ten random subsamples within each experimental plot in October 2018. Mean vegetation height of treatments without fencing (i.e., no exclusion and protector treatments) was 54.2 cm, and mean height of vegetation within full exclusion treatments was 75.9 cm, indicating that fencing promotes higher levels of competing vegetation compared with non-fenced treatments.

2.2. Statistical Analysis

First-year seedling survival, height growth, and herbivory damage were evaluated using a model with species and exclusion treatment as the main effects, a species x treatment interaction term, and site as a random effect. All analyses were performed in Program R 3.4 [35]: generalized linear models were fit using functions in the "lme4" package [36]; overall species and treatment effects were evaluated using a Type III ANOVA model within the "car" package [37]; and differences among levels in significant main effects were calculated with Tukey-corrected pairwise comparisons in the "lsmeans" package [38]. A 0.05 significance level was observed for all statistical tests.

Using survival data collected in the May and October 2017 assessments, first growing-season survival was calculated per the following formula:

$$\text{Survival} = \frac{\text{Seedlings alive}_{\text{October 2017}}}{\text{Seedlings counted}_{\text{May 2017}}}. \tag{1}$$

To elucidate plot-level survival differences among tree species and herbivore exclusion treatments, survival was tested as the response variable in a mixed-effect generalized linear model using the binomial distribution and logit link function.

With seedling height data collected in the May and October 2017 assessments, plot-level height growth of live seedlings was calculated per the following formula:

$$\text{Growth} = \text{Mean Height}_{\text{October 2017}} - \text{Mean Height}_{\text{May 2017}} \tag{2}$$

Height growth was analyzed using a linear mixed-effect model, and a natural logarithmic transformation of growth was used as the response variable to satisfy model assumptions.

With February 2018 seedling data, the proportion of herbivory-damaged trees was calculated as:

$$\text{Herbivory} = \frac{\text{Seedlings damaged}}{\text{Seedlings assessed}}. \tag{3}$$

Herbivory was first modeled via a mixed-effect generalized linear model using a binomial distribution and logit link function with species and exclusion treatment as the main effects, a species x treatment interaction, and site as a random effect. However, due to model non-convergence, the random effect was removed, and the model was refit with only fixed effects.

3. Results

3.1. Survival

Survival estimates from the first growing-season (May–October) are presented in Table 2. A significant interaction was observed for mean survival between tree species and exclusion treatment ($\chi^2 = 28.6$, $p < 0.001$). Black locust demonstrated higher mean survival in protector (80.3%) and full exclusion (81.7%) treatments compared with no exclusion treatments (73.1%). Shortleaf pine survival was low across all treatments: while mean survival was similar in no exclusion (37.8%) and protector (36.5%) treatments, and shortleaf pines in full exclusion plots exhibited lower survival (28.5%). White oak survival was higher in protector (80.5%) and full exclusion (80.5%) treatments compared with no exclusion treatments (68.2%). In no exclusion treatments, black locust and white oak survivals were higher than that of shortleaf pine. Similarly, in protector and full exclusion treatments, no significant differences were present in black locust and white oak survival; however, survivals of both species were higher than that of shortleaf pine.

Table 2. First growing-season seedling survival (%; Mean ± SE) among tree species and exclusion treatments on reclaimed mined lands in southeastern KY. Means with differing letters indicate significant differences among exclusion treatments within a species, and means with different symbols indicate significant differences among species within an exclusion treatment, as determined via Type III ANOVA and subsequent Tukey-corrected pairwise comparisons at a 0.05 significance level.

Species	Treatment		
	No Exclusion	Protector	Full Exclusion
Black Locust	73.1b * ± 10.6	80.3a * ± 6.0	81.7a * ± 9.9
Shortleaf Pine	37.8a † ± 10.0	36.5a † ± 9.7	28.5b † ± 8.7
White Oak	68.2b * ± 10.4	80.5a * ± 5.3	80.5a * ± 6.0

3.2. Height Growth

The Type III ANOVA model testing for differences in mean height between tree species and exclusion treatments provided little evidence for an interaction ($\chi^2 = 3.5$, $p = 0.463$). After removing the interaction term and refitting the model, tree species ($\chi^2 = 57.0$, $p < 0.001$) and exclusion treatment ($\chi^2 = 10.4$, $p = 0.005$) were found to be significant in predicting height growth. Among tree species, mean height growth of black locusts (30.3 cm) was significantly greater than that of shortleaf pine (11.9 cm) and white oak (8.6 cm); there was no difference in mean height growth between shortleaf pine and white oak (Figure 2). Protector treatments (mean = 20.3 cm) sustained significantly higher mean height growth compared with full exclusion treatments (mean = 13.5 cm; Figure 2). Significant differences were not present between protector and no exclusion treatments (mean = 17.0 cm) or between full exclusion and no exclusion treatments.

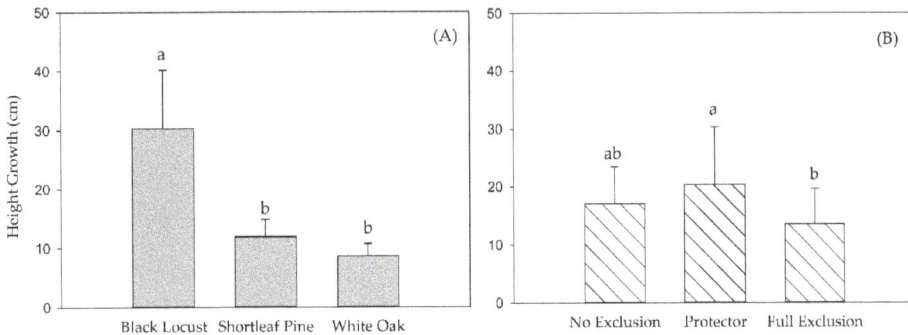

Figure 2. First growing-season height growth (Mean ± SE) among (**A**) tree species and (**B**) herbivore exclusion treatments on reclaimed Appalachian mined lands in southeastern KY. Different letters indicate significant differences among effect level means.

3.3. Herbivory

A significant interaction was present between tree species and exclusion treatment in modeling herbivory incidence ($\chi^2 = 105.5$, $p < 0.001$). Post-hoc pairwise comparisons demonstrated similar within-species trends for herbivore exclusion treatments: within each species, no exclusion treatments contained the highest herbivory percentages, followed by protector treatments, and full exclusion treatments (Table 3). Within no exclusion plots, black locust was damaged most frequently (85.1%); white oak herbivory was significantly lower (72.6%); and shortleaf pine was the least damaged of all species (34.1%). Black locust was the species damaged most often in protector treatments (73.8%); white oak seedlings were damaged less frequently (51.1%); and shortleaf pine damage was the least damaged of all species (2.9%). In full exclusion plots, herbivory was generally low: white oak was damaged most frequently (14.8%); black locust damage was lower (3.8%); and shortleaf pine herbivory in full exclusion plots was nearly absent (0.2%).

Table 3. First-year mammal herbivory incidence (%; Mean ± SE) among tree species and exclusion treatments on reclaimed mined lands in southeastern Kentucky. Means with differing letters indicate significant differences among exclusion treatments within a species, and means with different symbols indicate significant differences among species within an exclusion treatment, as determined via Type III ANOVA and subsequent Tukey-corrected pairwise comparisons at a 0.05 significance level.

	Treatment		
Species	**No Exclusion**	**Protector**	**Full Exclusion**
Black Locust	85.1a * ± 2.7	73.8b * ± 6.7	3.8c † ± 1.2
Shortleaf Pine	34.1a ‡ ± 7.0	2.9b ‡ ± 1.5	0.2c ‡ ± 0.2
White Oak	72.6a † ± 7.2	51.1b † ± 3.9	14.8c * ± 3.2

The cumulative herbivory rate for all seedlings in the study was 33.2%, of which cervids were responsible for 95.8%. Of all black locusts damaged in each of the exclusion treatments, cervid herbivory accounted for at least 93%, with minor contributions by rabbits (0.4–6.7%) and small mammals (1.7%; Table 4). Cervids mediated 74.7 and 50% of shortleaf pine damage in no exclusion treatments and protector treatments, respectively; rabbits were culpable in the damage of the remaining shortleaf pines in these treatments (25.3% and 50%, respectively). Rabbit herbivory comprised all damage to shortleaf pines in full exclusion treatments. Similar to black locust, cervids were responsible for at least 91% of all white oak herbivory in each exclusion treatment; rabbit contribution to white oak damage was also similar to that of black locust (1.8–6.7%). Small mammal herbivory was highest on white oaks in no exclusion treatments (3.7%). A single uprooted white oak (0.6%) in no exclusion treatments was attributable to herbivory by domestic animals (i.e., horse).

Table 4. Relative herbivore contribution (%) to herbivory incidence by tree species and exclusion treatment on reclaimed mined land in southeastern KY. Damage of a seedling by multiple taxa results in total contributions greater than 100%.

Treatment	Cervid	Rabbit	Small Mammal	Domestic Animal
Black locust				
No Exclusion	98.8	1.3	1.7	-
Protector	99.6	0.4	-	-
Full Exclusion	93.3	6.7	-	-
Shortleaf pine				
No Exclusion	74.7	25.3	-	-
Protector	50.0	50.0	-	-
Full Exclusion	-	100.0	-	-
White oak				
No Exclusion	97.5	1.8	3.7	0.6
Protector	97.9	2.1	-	-
Full Exclusion	91.7	6.7	1.7	-

4. Discussion

Tree species and herbivore exclusion treatment significantly influenced survival, height growth, and herbivory damage. Black locust and white oak survival increased with exclusion presence; however, there was no difference in survival between protector or full exclusion treatments. Conversely, while shortleaf pine survival was low across all treatments, survival was similar in no exclusion and protector treatments but significantly lower in full exclusion treatments. Black locust typically sustains moderate to high survival (53–100%) on mined sites in the first three to five years after planting [28,39], attributing to its favorability for mine reforestation. White oak survival in this study (68.2–80.5%) was also similar to that found by Emerson et al. (2009) [39] when planted within weathered gray and unweathered brown sandstone mine spoils (70–80%) and by Bell et al. (2017) [40] when planted in a mixed pine-hardwoods polyculture (50–80%). Shortleaf pine survivals observed in this study were at the lower extent of shortleaf pine survivals found by Bell et al. (2017; 29–58%) [40].

Herbivore exclusion has effectively increased the survival of natural regeneration and reforestation plantings in many systems, often due to a reduction in herbivory incidence and severity [41–43]. On reclaimed mined lands in eastern KY, tree shelters successfully increased the initial survival of direct-seeded chestnuts (*Castanea* spp.) [44]. Fencing is generally successful at increasing seedling survival through large-ungulate exclusion [45–48]; however, its use in Appalachian surface mine reforestation appears limited, as the present study is, to our knowledge, the first to evaluate the effectiveness of exclusion at reducing herbivore damage in this region. In this study, shortleaf pine survival in full exclusion treatments was significantly lower than that in other treatments, which is possibly due to higher levels of competing vegetation in full exclusion plots and lower initial heights of pine seedlings compared with black locust and white oak. Reduced survival rates as a result of fencing have been shown for black cherry (*Prunus serotina* Ehrh.) on reclaimed mines in Indiana [49].

Positive height growth was observed for all species in this study; however, black locust growth was significantly higher than that of shortleaf pine and white oak. Black locust is a pioneer species that naturally colonizes disturbed areas and can persist in environmentally harsh conditions due to its rapid initial growth rates [50,51] and ability to form symbiotic relationships with N_2-fixing bacteria [52], justifying its use for the reforestation of mined lands, landfills, and degraded areas that are often nutrient-depleted [39,53]. First-year growth of black locust in this study was much greater than that of black locusts planted on adjacent reclaimed mined sites in eastern KY (9.4 cm) and was even higher than that of fertilized black locusts (20.4 cm) [54]. Mean white oak growth in this study

(8.6 cm) was somewhat higher than that of white oaks planted in pine-hardwood polyculture in eastern KY (5.6 cm); height growth of northern red oak (*Quercus rubra* L.) and chestnut oak (*Quercus montana* Willd.) was also lower than that of white oak found in the present study [40]. Mean tree heights three years post-planting reported by Showalter et al. (2007) [55] in response to spoil type in Virginia appear to indicate growth rates similar to those in this study. Mean shortleaf pine heights reported by Bell et al. (2017; 10.5 cm) [40] were comparable to mean heights in this study. Similar first-year growth rates for shortleaf pine were also found by Kabrick et al. (2015) [56] for underplanted pines in the Missouri Ozark Highlands, indicating that shortleaf pine growth on reclaimed surface mines may approximate that of one-year-old pines regenerating under a closed-canopy forest.

Exclusion treatment significantly affected seedling height growth. Protector treatments cultivated the highest growth rates. Seedling protectors (or tree shelters/tubes) have increased the tree growth of a variety of deciduous and coniferous species [44,57–59], not only from a decreased impact of herbivory, but also in their effect on growing conditions. Protector construction can either improve or inhibit seedling growth rates [58,60–62]. Microclimate variables affecting growth rate (e.g., relative humidity, radiation absorption, CO_2 concentrations) vary with and within protector types [58,61]. Andrews et al. (2010) [62] demonstrated elevated hardwood growth rates in riparian forest corridors due to tree shelter use, attributed to woody debris retention around the protector and physical protection against flooding. Protectors selected for this study were manufactured of plastic interwoven in a diamond pattern with 2–3 cm openings. The protector's construction accommodated air flow between the atmosphere and the interior of the protector and limited shading effects to seedlings; therefore, the increased growth rate is, at best, marginally attributable to improved microclimate. Since soil analyses yielded no significant difference among treatments for the selected parameters, height growth responses of protector treatments are likely more associated with increased stem elongation as a result of protector presence and with competing vegetation dynamics. Growth in full exclusion treatments (also employing protectors) was significantly lower than that in protector treatments: competing vegetation was observed to be taller in full exclusion treatments compared with non-fenced treatments, potentially from decreased herbivory prevalence compared with outside of exclosures. Therefore, protector presence and reduced competing vegetation are likely responsible for the improved growth rate fostered by protectors.

Herbivory incidence in this study was driven by an interaction between tree species and exclusion treatment. All species in this study responded similarly to exclusion: herbivory was greatest in plots with no exclusion; protectors significantly lowered herbivory, but full exclusion treatments vastly reduced herbivory. Cervids were responsible for nearly 96% of all herbivory. Therefore, fencing was effective at limiting damage, but did not fully prohibit plot access to cervids. While no animals were observed within any fenced plots, beds and trails were observed within the plots, and deer and elk tracks, scat, and hair were found on fence perimeters on multiple occasions. Regardless, herbivory incidence was reduced as a result of fencing. Protectors also effectively reduced herbivory compared with no exclusion treatments. Although cervids damaged seedlings within protector treatments, the treatment effect is speculatively driven, in part, by relative seedling height: certain seedlings did not grow beyond the top of the protector in the first growing-season; thus, they were not available for browsing, demonstrating that smaller seedlings are protected against herbivory while gaining root mass and leaf area, which will aid in resilience to herbivory once the seedling has grown above the top of the protector.

In no exclusion treatments, a definitive herbivory preference was observed for black locust (85.1%) and white oak (72.6%). On adjacent mined lands in eastern Kentucky, black locusts in control plots sustained two-year browse rates of 76%, but as high as 91% black locust browse was observed after soil fertilization [28]. Due to elevated shoot N levels [28], black locust is foraged preferentially by ungulates [63]. While white oak was preferred significantly less than black locust, herbivory of this species was, nonetheless, considerable. On reclaimed mined land in Indiana, first-year deer browse rates of white oak in unexcluded plots was approximately 90% [49]. Mixed hardwoods, in general, appear to be heavily browsed during the first year: black cherry (90%), bur oak (*Quercus macrophylla* Michx.; 89%), and northern red oak (84%) were heavily damaged by deer in Indiana [49]. Likewise, Skousen et al. (2009) [64] reported "heavy browse" of white ash (*Fraxinus americana* L.) on mines in West Virginia. Negative height growth of chestnut oak and northern red oak was attributed by Bell et al. (2017) [40] to deer and elk browse; however, American sycamore (*Platanus occidentalis* L.) was relatively undamaged (<3%) in the second growing-season on mined land in eastern Kentucky [28]. Additionally, pines seem to be less preferred by herbivores compared with hardwoods. In this study, the shortleaf pine herbivory rate in no exclusion treatments was 34%. Cumulative browse of unexcluded underplanted eastern white pine (*Pinus strobus* L.) in northern forests was less than 43% [59]. Tree species selection for planting mixtures is an active area of research and one that will greatly benefit from mine reclamation efforts.

This study has revealed that herbivory on reclaimed Appalachian mined land is extensive and that techniques for control require further consideration. Although herbivory by rabbits, small mammals, and domestic animals was documented, cervids accounted for nearly all first-year herbivory damage. Deer populations have increased markedly across the eastern U.S. over the previous decades. Similarly, elk reintroduction has become a nearly widespread management goal of state wildlife agencies across Appalachia, with successful population establishment in five states (Kentucky, North Carolina, Pennsylvania, Tennessee, and West Virginia) [65]. Concomitant with deer population explosion, elk expansion will intensify herbivory pressure, especially of reforestation plantings on reclaimed surface mines, where most elk releases in Kentucky and neighboring states have occurred. Horse populations on reclaimed mines will likely continue to increase; although the results of this study indicate that horse impacts are minimal, this source of herbivory should continue to be monitored on a local scale, specifically in areas with high populations.

5. Conclusions

Seedling protectors successfully lowered herbivory incidence during the first growing-season following planting; however, full exclusion drastically reduced herbivory, yet fencing was not effective at fully excluding cervids. Exclusion treatments also generally increased seedling survival and height growth. While exclusion has been found to be effective at limiting herbivory damage, these treatments may likely prove economically or logistically unfeasible in some circumstances. The cost of fencing (material and labor) for this study was approximately $21,220 per ha, and protector material and installation costs were approximately $0.60 per seedling ($1,500 per ha at study planting densities). Although fencing effectively negated herbivory damage and increased first-year survival rates compared with no exclusion plots, managers must decide if the large initial investment in fencing is offset by the future value of the forest resources. Protectors are a more economical method of reducing herbivory and promoting height growth; however, once seedlings grow beyond the top the protector, cervids damage the upper shoots, which will, ultimately, hinder height growth and create poor growth form. Therefore, tree species less preferred by herbivores (i.e., cervids) should be identified for inclusion in planting mixes to reduce herbivory impacts to forest recruitment. Black locust and white oak were found to be highly preferred by cervids, but shortleaf pine was selected less frequently. These results indicate that hardwood regeneration on mined lands will likely prove difficult with current and projected future cervid population levels. Restoration of pine forests on Appalachian surface mines may be more successful given lower herbivory rates; however, low survival rates may preclude this effort. Follow-up seedling assessments in three to five years will provide additional results on herbivory impacts during the years when seedlings are most susceptible to herbivory damage.

Acknowledgments: Support for purchasing and planting of seedlings was provided by the KY Sustainable Forestry Initiative Sustainable Implementation Committee. Partial funding was also provided by the National Fish and Wildlife Foundation, Treecycler, Arbor Day Foundation, University of Kentucky College of Agriculture, Food, and Environment, and the University of Kentucky Appalachian Center. We thank all of the UK Robinson Center for Appalachian Resource Sustainability (RCARS) and Robinson Forest employees who helped with fence construction. We are grateful to J. Frederick, M. Anderson, D. Thomas, Z. Grigsby, W. Dixon, A. Davis, and W. Leuenberger for assistance with field measurements.

Author Contributions: All authors obtained funding, designed research methods, and wrote (or substantially edited) the manuscript. Z.J.H. collected and analyzed data.

Conflicts of Interest: The authors declare no conflict of interest.

References

1. National Mining Association. Available online: https://nma.org/wp-content/uploads/2017/11/Mine-Reclamation-2017-2.pdf (accessed on 18 March 2018).
2. Wickham, J.D.; Riitters, K.H.; Wade, T.G.; Coan, M.; Homer, C. The effect of Appalachian mountaintop mining on interior forest. *Landsc. Ecol.* **2007**, *22*, 179–187. [CrossRef]
3. Wickham, J.D.; Wood, P.B.; Nicholson, M.C.; Jenkins, W.; Druckenbrod, D.; Suter, G.W.; Strager, M.P.; Mazzarella, C.; Galloway, W.; Amos, J. The overlooked terrestrial impacts of mountaintop mining. *Bioscience* **2013**, *63*, 335–348. [CrossRef]
4. Thurman, N.C.; Sencindiver, J.C. Properties, classification, and interpretations of minesoils at two sites in West Virginia. *Soil Sci. Soc. Am. J.* **1986**, *50*, 181–185. [CrossRef]
5. Thompson, P.J.; Jansen, I.J.; Hooks, C.L. Penetrometer resistance and bulk density as parameters for predicting root system performance in mine soils. *Soil Sci. Soc. Am. J.* **1987**, *51*, 1288–1293. [CrossRef]
6. Chong, S.K.; Cowsart, P.T. Infiltration in reclaimed mined land ameliorated with deep tillage treatments. *Soil Tillage Res.* **1997**, *44*, 255–264. [CrossRef]
7. Conrad, P.W.; Sweigard, R.J.; Graves, D.H.; Ringe, J.M.; Pelkki, M.H. Impacts of spoil conditions on reforestation of surface mined land. *Min. Eng.* **2002**, *54*, 39–46.
8. Evans, D.M.; Zipper, C.E.; Burger, J.A.; Strahm, B.D.; Villamagna, A.M. Reforestation practice for enhancement of ecosystem services on a compacted surface mine: Path toward ecosystem recovery. *Ecol. Eng.* **2013**, *51*, 16–23. [CrossRef]
9. Bohrer, S.L.; Limb, R.F.; Daigh, A.L.; Volk, J.M.; Wick, A.F. Fine and coarse-scale patterns of vegetation diversity on reclaimed surface mine-land over a 40-year chronosequence. *Environ. Manag.* **2017**, *59*, 431–439. [CrossRef] [PubMed]
10. Oliphant, A.J.; Wynne, R.H.; Zipper, C.E.; Ford, W.M.; Donovan, P.F.; Li, J. Autumn olive (*Elaeagnus umbellata*) presence and proliferation on former surface coal mines in Eastern USA. *Biol. Invasions* **2017**, *19*, 179–195. [CrossRef]
11. Zipper, C.E.; Burger, J.A.; Skousen, J.G.; Angel, P.N.; Barton, C.D.; Davis, V.; Franklin, J.A. Restoring forests and associated ecosystem services on Appalachian coal surface mines. *Environ. Manag.* **2011**, *47*, 751–765. [CrossRef] [PubMed]
12. Burger, J.A.; Zipper, C.E.; Angel, P.N.; Hall, N.; Skousen, J.G.; Barton, C.D.; Eggerud, S. *Establishing Native Trees on Legacy Surface Mines, Forest Reclamation Advisory No. 11*; USDOI Office of Surface Mining: Washington, DC, USA, 2013.
13. Swihart, R.K.; Picone, P.M. Selection of mature growth stages of coniferous browse in temperate forests by white-tailed deer (*Odocoileus virginianus*). *Am. Midl. Nat.* **1998**, *139*, 269–274. [CrossRef]
14. Cleavitt, N.L.; Berry, E.J.; Hautaniemi, J.; Fahey, T.J. Life stages, demographic rates, and leaf damage for the round-leaved orchids, *Platanthera orbiculata* (Pursh.) Lindley and *P. macrophylla* (Goldie) PM Brown in a northern hardwood forest in New Hampshire, USA. *Botany* **2017**, *95*, 61–71. [CrossRef]
15. Dostaler, S.; Ouellet, J.P.; Therrien, J.F.; Cote, S.D. Are feeding preferences of white-tailed deer related to plant constituents? *J. Wildl. Manag.* **2011**, *75*, 913–918. [CrossRef]
16. Burney, O.T.; Jacobs, D.F. Ungulate herbivory of regenerating conifers in relation to foliar nutrition and terpenoid production. *For. Ecol. Manag.* **2011**, *262*, 1834–1845. [CrossRef]

17. Augustine, D.J.; McNaughton, S.J. Ungulate effects on the functional species composition of plant communities: Herbivore selectivity and plant tolerance. *J. Wildl. Manag.* **1998**, *62*, 1165–1183. [CrossRef]

18. Champagne, E.; Perroud, L.; Dumont, A.; Tremblay, J.P.; Cote, S.D. Neighbouring plants and perception of predation risk modulate winter browsing by white-tailed deer (*Odocoileus virginianus*). *Can. J. Zool.* **2018**, *96*, 117–125. [CrossRef]

19. Rodel, H.G.; Volkl, W.; Kilias, H. Winter browsing of brown hares: Evidence for diet breadth expansion. *Mamm. Biol.* **2004**, *69*, 410–419. [CrossRef]

20. Lehndal, L.; Agren, J. Latitudinal variation in resistance and tolerance to herbivory in the perennial herb *Lythrum salicaria* is related to intensity of herbivory and plant phenology. *J. Evol. Biol.* **2015**, *28*, 576–589. [CrossRef] [PubMed]

21. White, C.A.; Olmsted, C.E.; Kay, C.E. Aspen, elk, and fire in the Rocky Mountain national parks of North America. *Wildl. Soc. Bull.* **1998**, *26*, 449–462.

22. Ripple, W.J.; Beschta, R.L. Hardwood tree decline following large carnivore loss on the Great Plains, USA. *Front. Ecol. Environ.* **2007**, *5*, 241–246. [CrossRef]

23. Jenkins, L.H.; Murrary, B.D.; Jenkins, M.A.; Webster, C.R. Woody regeneration response to over a decade of deer population reductions in Indiana state parks. *J. Torrey Bot. Soc.* **2015**, *142*, 205–219. [CrossRef]

24. McGraw, J.B.; Furedi, M.A. Deer browsing and population viability of a forest understory plant. *Science* **2005**, *307*, 920–922. [CrossRef] [PubMed]

25. Leege, L.M.; Thompson, J.S.; Parris, D.J. The response of rare and common trilliums (*Trillium reliquum*, *T-cuneatum*, and *T-maculatum*) to deer herbivory and invasive honeysuckle removal. *Castanea* **2010**, *75*, 433–443. [CrossRef]

26. Bradshaw, L.; Waller, D.M. Impacts of white-tailed deer on regional patterns of forest tree recruitment. *For. Ecol. Manag.* **2016**, *375*, 1–11. [CrossRef]

27. Barton, C.D.; University of Kentucky, Lexington, KY, USA. Personal communication, 2016.

28. Brinks, J.S.; Lhotka, J.M.; Barton, C.D.; Warner, R.C.; Agouridis, C.T. Effects of fertilization and irrigation on American sycamore and black locust on a reclaimed surface mine in Appalachia. *For. Ecol. Manag.* **2011**, *261*, 640–648. [CrossRef]

29. Agouridis, C.; Barton, C.; Warner, R. Recreating a headwater stream system on a valley fill in the Appalachian coal field. In *Spoil to Soil: Mine Site Rehabilitation and Revegetation*; Bolan, N., Kirkham, M.B., Ok, Y.S., Eds.; Taylor and Francis: Boca Raton, FL, USA, 2018; pp. 147–174.

30. Larkin, J.L.; Maehr, D.S.; Krupa, J.J.; Cox, J.J.; Alexy, K.; Under, D.E.; Barton, C.D. Small mammal response to vegetation and spoil conditions on a reclaimed surface mine in eastern Kentucky. *Southeast. Nat.* **2008**, *7*, 401–412. [CrossRef]

31. Thomas, G. Soil pH and soil acidity. In *Methods of Soil Analysis Part 3—Chemical Methods*; Soil Science Society of America, American Society of Agronomy: Madison, WI, USA, 1996; pp. 475–490.

32. Soil and Plant Analysis Council. *Soil Analysis Handbook of Reference Methods*; CRC Press: Boca Raton, FL, USA, 2000.

33. Miller, W.; Miller, D. A micro-pipette method for soil mechanical analysis. *Commun. Soil Sci. Plant Anal.* **1987**, *18*, 1–15. [CrossRef]

34. Summer, M.E.; Miller, W.P. Cation exchange capacity and exchange coefficients. In *Methods of Soil Analysis. Part 3. Chemical Methods*; Sparks, D., Bartels, J.M., Eds.; Soil Science Society of America, American Society of Agronomy: Madison, WI, USA, 1996.

35. R Core Team. *R: A Language and Environment for Statistical Computing*; R Foundation for Statistical Computing: Vienna, Austria, 2017.

36. Bates, D.; Maechler, M.; Bolker, B.; Walker, S. Fitting linear mixed-effects models using lme4. *J. Stat. Softw.* **2015**, *67*, 1–48. [CrossRef]

37. Fox, J.; Weisberg, S. *An R Companion to Applied Regression*, 3rd ed.; Sage: Thousand Oaks, CA, USA, 2011.

38. Lenth, R.V. Least-squares means: The R package lsmeans. *J. Stat. Softw.* **2016**, *69*, 1–33. [CrossRef]

39. Emerson, P.; Skousen, J.; Ziemkiewicz, P. Survival and growth of hardwoods in brown versus gray sandstone on a surface mine in West Virginia. *J. Environ. Qual.* **2009**, *38*, 1821–1829. [CrossRef] [PubMed]

40. Bell, G.; Sena, K.L.; Barton, C.D.; French, M. Establishing pine monocultures and mixed pine-hardwood stands on reclaimed surface mined land in eastern Kentucky: Implications for forest resilience in a changing climate. *Forests* **2017**, *8*, 375. [CrossRef]

41. Taylor, M.; Haase, D.L.; Rose, R.L. Fall planting and tree shelters for reforestation in the east Washington Cascades. *West. J. Appl. For.* **2009**, *24*, 173–179.

42. Dick, K.; Alexander, H.D.; Moczygemba, J.D. Use of shelter tubes, grass-specific herbicide, and herbivore exclosures to reduce stressors and improve restoration of semiarid thornscrub forests. *Restor. Ecol.* **2016**, *24*, 785–793. [CrossRef]

43. Piiroinen, T.; Valtonen, A.; Roininen, H. The seed-to-seedling transition is limited by ground vegetation and vertebrate herbivores in a selectively logged rainforest. *For. Ecol. Manag.* **2017**, *384*, 137–146. [CrossRef]

44. Barton, C.; Miller, J.; Sena, K.; Angel, P.; French, M. Evaluating the use of tree shelters for direct seeding of *Castanea* on a surface mine in Appalachia. *Forests* **2015**, *6*, 3514–3527. [CrossRef]

45. Kelly, D.L. The regeneration of *Quercus petraea* (sessile oak) in southwest Ireland: A 25-year experimental study. *For. Ecol. Manag.* **2002**, *166*, 207–226. [CrossRef]

46. Drozdowski, S.; Bolibok, L.; Buraczyk, W.; Wisniowski, P. Effect of planting time and method of protection from deer on the growth of oak plantations on the former farmland. *Sylwan* **2011**, *155*, 610–621.

47. Schnurr, J.; Canham, C.D. Linkages among canopy tree neighbourhoods, small mammal herbivores and herbaceous communities in temperate forests. *J. Veg. Sci.* **2016**, *27*, 980–998. [CrossRef]

48. Miller, G.W.; Brose, P.H.; Gottschalk, K.W. Advanced oak seedling development as influenced by shelterwood treatments, competition control, deer fencing, and prescribed fire. *J. For.* **2017**, *115*, 179–189. [CrossRef]

49. Burney, O.T.; Jacobs, D.F. Species selection—A fundamental silvicultural tool to promote forest regeneration under high animal browsing pressure. *For. Ecol. Manag.* **2018**, *408*, 67–74. [CrossRef]

50. Boehm, C.; Quinkenstein, A.; Freese, D. Yield prediction of young black locust (*Robinia pseudoacacia* L.) plantations for woody biomass production using allometric relations. *Ann. For. Res.* **2011**, *54*, 215–227.

51. Kurokochi, H.; Toyama, K. Invasive tree species *Robinia pseudoacacia*: A potential biomass resource in Nagano Prefecture, Japan. *Small-Scale For.* **2015**, *14*, 205–215. [CrossRef]

52. Roberts, D.R.; Zimmerman, R.W.; Stringer, J.W.; Carpenter, S.B. The effects of combined nitrogen on growth, nodulation, and nitrogen fixation of black locust seedlings. *Can. J. For. Res.* **1983**, *13*, 1251–1254. [CrossRef]

53. Kim, K.D.; Lee, E.J. Potential tree species for use in the restoration of unsanitary landfills. *Environ. Manag.* **2005**, *36*, 1–14. [CrossRef] [PubMed]

54. Brinks, J.; Lhotka, J.; Barton, C. One-year response of American sycamore (*Platanus occidentalis* L.) and black locust (*Robinia pseudoacacia*) to granular fertilizer applications on a reclaimed surface mine in eastern Kentucky. *Proc. Cent. Hard. For. Conf.* **2011**, *17*, 306–313.

55. Showalter, J.M.; Burger, J.A.; Zipper, C.E.; Galbraith, J.M.; Donovan, P.F. Influence of mine soil properties on white oak seedling growth: A proposed mine soil classification model. *South. J. Appl. For.* **2007**, *31*, 99–107.

56. Kabrick, J.M.; Knapp, B.O.; Dey, D.C.; Larsen, D.R. Effect of initial seedling size, understory competition, and overstory density on the survival and growth of *Pinus echinata* seedlings underplanted in hardwood forests for restoration. *New For.* **2015**, *46*, 897–918. [CrossRef]

57. Engeman, R.M.; Anthony, R.M.; Krupa, H.W.; Evans, J. The effects of Vexar® seedling protectors on the growth and development of lodgepole pine roots. *Crop Prot.* **1997**, *16*, 57–61. [CrossRef]

58. Dubois, M.R.; Cappelka, A.H.; Robbins, E.; Somers, G.; Baker, K. Tree shelters and weed control: Effects on protection, survival and growth of cherrybark oak seedlings planted on a cutover site. *New For.* **2000**, *20*, 105–118. [CrossRef]

59. Ward, J.S.; Mervosh, T.L. Strategies to reduce browse damage on eastern white pine (*Pinus strobus*) in southern New England, USA. *For. Ecol. Manag.* **2008**, *255*, 1559–1567. [CrossRef]

60. Burger, D.W.; Svihra, P.; Harris, R. Treeshelter use in producing container-grown trees. *Hortscience* **1992**, *27*, 30–32.

61. Bellot, J.; Ortiz de Urbina, J.M.; Bonet, A.; Sanchez, J.R. The effects of treeshelters on the growth of *Quercus coccifera* L. seedlings in a semiarid environment. *Forestry* **2002**, *75*, 89–106. [CrossRef]

62. Andrews, D.M.; Barton, C.D.; Czapka, S.J.; Kolka, R.K.; Sweeney, B.W. Influence of tree shelters on seedling success in an afforested riparian zone. *New For.* **2010**, *39*, 157–167. [CrossRef]

63. Tripler, C.; Canham, C.; Inouye, R.; Schnurr, J. Soil nitrogen availability, plant luxury consumption, and herbivory by white-tailed deer. *Oecologia* **2002**, *133*, 517–524. [CrossRef] [PubMed]

64. Skousen, J.; Gorman, J.; Pena-Yewtukhim, E.; King, J.; Stewart, J.; Emerson, P.; DeLong, C. Hardwood tree survival in heavy ground cover on reclaimed land in West Virginia: Mowing and Ripping Effects. *J. Environ. Qual.* **2009**, *38*, 1400–1409. [CrossRef] [PubMed]

65. Cox, J.J. Tales of a repatriated megaherbivore: Challenges and opportunities in the management of reintroduced elk in Appalachia. *Proc. Cent. Hard. For. Conf.* **2011**, *17*, 632–642.

forests

MDPI

Article

Stocktype and Vegetative Competition Influences on *Pseudotsuga menziesii* and *Larix occidentalis* Seedling Establishment

Jeremiah R. Pinto [1], Bridget A. McNassar [2], Olga A. Kildisheva [3] and Anthony S. Davis [4,*]

[1] USDA Forest Service, Rocky Mountain Research Station, 1221 South Main Street, Moscow, ID 83843, USA; jpinto@fs.fed.us

[2] Oxbow Farm & Conservation Center, 10819 Carnation-Duvall Rd NE, Carnation, WA 98014, USA; bridget@oxbow.org

[3] School of Plant Biology, University of Western Australia, 35 Stirling Hwy, Crawley, WA 6009, Australia; olga.kildisheva@gmail.com

[4] College of Forestry, Oregon State University, 109 Richardson Hall, Corvallis, OR 97331, USA

* Correspondence: anthony.davis@oregonstate.edu; Tel.: +1-541-737-1585

Received: 31 March 2018; Accepted: 24 April 2018; Published: 26 April 2018

Abstract: Douglas fir (*Pseudotsuga menziesii* var. *glauca* (Mayr) Franco), and western larch (*Larix occidentalis* Nutt.) are species of ecological and commercial importance that occur throughout the Western United States. Effective reforestation of these species relies on successful seedling establishment, which is affected by planting stock quality, stocktype size, and site preparation techniques. This study examined the effects of container volume (80, 130, 200, and 250 cm^3) and vegetative competition on seedling survival and physiological and morphological responses for two years, post-outplanting. Glyphosate application (GS) and grass planting (HC) were used to achieve low and high levels of competition. For all measured attributes, the container volume × vegetative competition was not significant. Mortality was strongly influenced by competition, with higher mortality observed for Douglas fir and western larch planted in HC plots one (28% and 98%) and two (61% and 99%) years following outplanting. When competition was controlled, seedlings of both species exhibited greater net photosynthesis (>9 $\mu mol\ m^{-2}\ s^{-1}$), greater predawn water potential (>−0.35 MPa), and lower mortality (2–3%) following one year in the field, indicating establishment success. The 80 cm^3 stocktype remained significantly smaller and exhibited lower growth rates for the duration of the study, while all other stocktypes were statistically similar. Our results demonstrate the importance of controlling vegetative competition regardless of stocktype, especially for western larch, and suggest that benefits to post-planting seedling physiology and growth in relation to container size plateau beyond 130 cm^3 among the investigated stocktypes.

Keywords: container parameters; nursery culture; western larch; Douglas fir; herbicide

1. Introduction

In order to ensure success, reforestation efforts must meet diverse objectives, which can involve myriad species, fit within a variety of economic models, and account for different site characteristics. Thus, tree seedlings used in these efforts should be produced with specific parameters in mind (genetics, morphology, and physiology, for example) [1–3]. Successful seedling stocking on a site within the required planting window may have economic, ecological, and legal implications—calling for high-quality, specifically-cultured plant material. Managers continually need updated, science-based information to justify on-the-ground decision-making; however, determining which traits lead to optimal seedling performance for a specific site and reforestation objectives continues to be a challenge.

Seedling phenotypic traits are a result of genetics and the environment. Nurseries have the capacity to alter these traits through modifications in nursery culture, including container size. Container-grown seedlings provide a wealth of stocktype choices in terms of dimensions (i.e., depth, diameter, volume) and container composition—factors which have been found to affect seedling phenotype [4–6]. Container selection in seedling production also has economic implications. Smaller-volume containers require fewer inputs per plant, maximizing nursery growing space while minimizing media and fertilizer; conversely, larger volume containers require more inputs, but produce larger seedlings [7–12]. The choice and difference between the two, however, can offer competitive advantages for survival and growth in the field. For larger containers, i.e., larger seedlings, such advantages stem from the ability to outcompete existing vegetation on the site as a result of higher nutrient reserves, increased photosynthetic capacity, and enhanced water use efficiency [3,13–15]. Villar-Salvador et al. [16] also argue that larger seedlings do better in environments that experience seasonal drought. Many studies have shown that larger stocktype seedlings initially remain larger and grow at a more rapid rate following outplanting [8,17–19]; however, results are often species- and site-specific and are subject to change following the first growing season [17,20,21].

After outplanting, seedling establishment during the first growing season depends upon initiation and maintenance of a positive feedback loop with its new environment. This loop requires uptake of soil water to support increased stomatal conductance and photosynthesis. The newly-assimilated carbon allows for enhanced root growth, thus continually increasing seedling access to soil water [22]. Competing vegetation is often an obstacle in the effective establishment of this loop; additionally, regional climate regimes can also be of influence. In the Inland Northwest region of the United States, where pronounced summer moisture limitations are common [23], plants must establish adequate root systems before the onset of drought. This is especially imperative for seedlings outplanted in the spring, which have a limited window for root growth to occur. In such conditions, stocktypes with large or deep root systems can be advantageous for overcoming drought stress [24]. Several studies have examined the effects of stocktype size and site treatments aimed at controlling competing vegetation on seedling performance [14,15,25,26]. Much of this work points to stocktype differences in the ability of seedlings to establish root contact with soil water, either through enhancing functional capacity to promote root growth in order to reach deeper soil moisture reserves or through reducing vegetative competition to increase available soil moisture. This is manifested by improved performance (higher survival, growth, and carbon assimilation capacity) among larger seedlings relative to smaller stocktypes under high competition [14,15,26]. The benefits of larger stocktypes, however, are less clear in the absence of vegetative competition or ecosystems without water limitations [14,25].

Despite the abundance of stocktype studies to date, only a small group of them examine the morpho-physiological performance of container stocktypes under different levels of vegetative cover during summer drought [14,25,26]. Unfortunately, these studies, among others, contain some level of confounding, thereby significantly limiting the transferability of findings. Much of this methodology shows that, during nursery cultivation or genetic selection, biases are introduced, thus optimizing growth of a single container size (or stocktype) and creating sub-optimal conditions for all other sizes. Addressing this issue, Pinto et al. [27] outline key considerations for conducting stocktype studies that minimize confounding. With these considerations, our study aims to examine the role of container size on seedling performance, while reducing confounding commonly associated with nursery culture, and create distinct drought differences between vegetative competition treatments at the site level. In order to expand the utility of the study, we use only stocktypes considered operationally feasible and focus on species of interest specific to the Western United States. We hypothesized that (1) larger stocktypes would have an advantage in accessing soil water due to larger root systems and would outperform smaller stocktypes during the summer drought, and (2) that this difference would be greater under high vegetative competition and low soil moisture. Evaluations of seedling performance included survival, growth, net photosynthesis, and plant water potential.

2. Materials and Methods

2.1. Nursery Culture

Open-pollinated, orchard-grown Douglas fir (*Pseudotsuga menziesii* var. *glauca* (Mayr) Franco) (Potlatch Lot ID#: DF-CL-Z7, 1000–1200 m elevation) and western larch (*Larix occidentalis* Nutt.) (Potlatch Lot ID#: WL-09-75 Improved BC, 1100–1350 elevation) seed was obtained in March 2012. Seed was cold, moist stratified for 30 days at 1.7 °C, and sown five per cell into 10 Styroblock® trays (Beaver Plastics, Edmonton, AB, Canada). Recognizing the operational range of seedling containers used in the region, two different tray sizes (Styroblock® 415C and 515A tray models) were used to obtain four container volume treatments which had approximate starting cell volumes of 130 and 250 cm^3. Two experimental container volumes were created by cutting off a 50 cm^3 portion of 415C and 515A models, resulting in cell volumes of 80 and 200 cm^3 to complement the original 130 and 250 cm^3 cells (Table 1).

Table 1. Container specifications used to produce Douglas fir and western larch seedlings. Two types of containers were used to create four container treatment volumes.

Container	ID Code Cells/mL	Metric Number	Cell Depth (cm)	Cell Volume (cm^3)	Cell Diameter (cm)	Cells Per Container	Cells Per m^2
Styroblock®	91/130	415C	11.0 [a] / 15.1	80 [a] / 130	3.9	91	430
Styroblock®	60/250	515A	12.4 [a] / 15.1	200 [a] / 250	5.1	60	284

[a] Containers were modified to a shorter depth to achieve a smaller volume without changing density.

To account for differences in growing space by container size, sowing was staggered by one week for each container size, beginning with the 250 cm^3 containers in late April for Douglas fir and mid-May for western larch. Growing medium consisted of sphagnum peat moss:vermiculite:aged fine bark (2:1:1, *v:v:v*; Sun Gro Horticulture, Bellevue, WA, USA) with a bulk density of 0.14 g cm^{-3}. Following sowing, media surface and seeds were covered with a medium-sized forestry nursery grit (Target Products Ltd., Burnaby, BC, Canada). Seedlings were grown in a greenhouse at the University of Idaho Pitkin Forest Nursery, Moscow, Idaho (46.7255° N, 116.9563° W) from April to November 2011. Culturing temperatures averaged 15 and 26 °C (minimum and maximum, respectively) over this time period. Unless otherwise noted, all methods were identical for both species.

Trays were arranged on nursery benches in a randomized complete block design (RCBD), with the arrangement repeated for each species. Blocking was determined by proximity to a heating tube underneath the nursery benches, with two Styroblocks® of the same cell volume serving as a single block. Each of the four tested container volumes consisted of five blocks and individual seedlings were considered to be experimental units (3020 total seedlings per species). Within each block, container locations were re-randomized every two weeks.

Irrigation and fertigation timing was individually tailored to each block. The saturated weight of one Styroblock® in each pair was determined at the onset of nursery culture and re-weighed daily to determine gravimetric water content [28]. Throughout the growing season, seedlings were fertigated as they reached gravimetric targets (see below), usually 2–3 times per week.

Immediately following sowing, containers were misted using an overhead boom twice daily for two weeks and thinned to 2–3 trees per cell after this period. After the initial two weeks, seedlings were thinned to one tree per cell and received applications of Peters® Conifer Starter™ (10:20:30 [N:P$_2$O$_5$:K$_2$O], The Scotts Company, Marysville, OH, USA) at 42 mg N L^{-1}, for the following three weeks, as blocks reached 85% of their saturated weight. During the rapid growth phase, seedlings received applications of Wilbur Ellis® Pro-Grower™ (20:7:19, Wilbur Ellis, Walnut Creek, CA, USA) at 60 mg N L^{-1} for western larch and 100 mg N L^{-1} for Douglas fir, and calcium nitrate (CN; 15:0:0) as containers dried to 80–85% of their saturated weight. Alternating applications of fertilizer during this stage was supplemented with Peters® S.T.E.M.™ (Soluble Trace Element Mix) micronutrient mix. As seedlings within container volume groups reached approximately two-thirds of their target heights (15 cm), excess nutrients were leached using a water flush, and seedlings were moved to the hardening phase fertilizer regime. During this phase, seedlings received Wilbur Ellis® Pro-Finisher™ (4:25:35) at 24 mg N L^{-1}, applied when blocks dried down to 65 and eventually 55% to initiate budset.

2.2. Nursery Phase Sampling

Height and root-collar diameter (RCD) measurements were obtained for a random subsample of 20 seedlings per Styroblock® tray every other week beginning 10 weeks after sowing, through the end of the season. These measurements were used to determine fertilization phase timing.

Assessment of morphological characteristics for each seedling at the end of the nursery culture (December 2011) included height and RCD and the respective growth increments for each variable. In addition, a random subset of 25 seedlings from each container volume × block combination was subject to destructive sampling ($n = 500$). After carefully washing roots to remove media, root volume (RV) was determined using the water displacement method described by Burdett [29]. The sampled seedlings were oven dried at 60 °C for 72 h, after which root and shoot dry masses were measured. The root-to-shoot (R:S) biomass ratios were calculated for all destructively-sampled seedlings. Remaining seedlings were lifted, placed into plastic bags inside wax boxes, and stored at −1.4 °C (±0.5) for five months, in line with standard nursery protocol.

2.3. Outplanting

A total of 1600 seedlings were outplanted between 25 and 31 May 2012 on a 15-hectare cut site in the East Hatter Creek Unit of the University of Idaho Experimental Forest (46.8445° N, 116.7960° W, 860 m a.s.l.). The site had a south-facing aspect, ranged in slope from 5–20%, and was logged in August 2011, with approximately 100 trees remaining to meet leave-tree obligations of the Idaho Forest Practices Act. The residual slash was broadcast burned in October 2011. Tree species present prior to harvest included western larch and Douglas fir, as well as ponderosa pine (*Pinus ponderosa* Dougl. Ex Laws. var. *ponderosa*). Soils were classified in the Santa series of Alfisols, described as moderately well-drained and moderately deep, formed in deep loess with a small amount of volcanic ash in the upper horizons [30].

For each species, a split-plot, randomized complete block design was used. Each block was the whole-plot, which consisted of two 21 × 10 m areas that were randomly assigned one of the two vegetative competition treatments (glyphosate-sprayed or high competition). In the glyphosate-sprayed (GS) treatments, competing vegetation was removed with two applications (7 and 29 June 2012) of glyphosate (41%, Glystar® Plus, Albaugh, Inc., Ankeny, IA, USA) using a backpack sprayer (prior to tree planting), at a rate of 3.4 kg acid equivalent ha^{-1}. Additional vegetation was removed manually throughout the season. In the high-competition (HC) treatments, the natural vegetation community was allowed to establish. In addition to this, we planted blue wildrye (*Elymus glaucus* Buckley (Clearwater Seed, Spokane, WA, USA)) seeds, a grass species native to the region, at an approximate density of 75 plants m^{-2} in May 2012. Our split-plot factor was the four container volumes. Within the GS and HC treatments, the four container volumes (80, 130, 200, 250 cm^3) were randomly assigned to four rows, each row contained 20 seedlings from one of the four container sizes. Spacing was 1 and 2 m between seedlings and rows, respectively. To minimize browse damage, seedlings were surrounded with 1 m tall yellow mesh protection tubes (Forestry Suppliers, Inc., Jackson, MS, USA); additionally, a 2 m buffer of animal repellant (Plantskydd®, concentration of 0.125 kg L^{-1}, Tree World, St. Joseph, MO, USA) was sprayed around each vegetation treatment using a backpack sprayer.

Existing vegetation cover was quantified using sampling approach described by Daubenmire [31]. A diagonal transect was established running from the northwestern-most corner of the plot to the southeastern-most corner. A 20 × 50 cm frame was placed at 4-, 8-, 12-, 16-, and 20-m intervals along the transect. At each sampling point, total cover of live vegetation in the frame, using the projected cover of foliage onto the ground below, was recorded as a percentage.

2.4. Edaphic and Atmospheric Monitoring

A weather station (model 2900ET, Spectrum Technologies, Inc., Plainfield, IL, USA), installed on site, was used to monitor hourly air temperature (°C), precipitation (mm), and relative humidity (%) during the first growing season. ECH20-TE soil moisture probes (Decagon Devices, Inc., Pullman, WA, USA) were installed in HC and GS plots that most closely represented the diversity of slope, soil moisture, and initial vegetative cover on the site. Volumetric soil moisture (θ, m^3 m^{-3}) measurements were collected hourly at three different soil depths (5, 15, and 30 cm) using an Em50 data logger (Decagon Devices, Inc., Pullman, WA, USA) from June to October 2012. In situ soil calibrations were performed to increase the accuracy of the volumetric soil moisture measurements.

2.5. Survival and Morphology Measurements

Height (cm) and RCD (mm) were measured immediately after planting (early June 2012), at the end of the first growing season (October 2012), and at the end of the second growing season (October 2013). Seedling survival was assessed at the end of each growing season. Height and RCD increment for all surviving trees was calculated by subtracting the initial from the final measurement for each surviving, un-sampled seedling.

2.6. Seedling Gas Exchange and Water Potential Measurements

Both seedling gas exchange and pre-dawn water potential (Ψ_{pd}) were measured, for each species, on one randomly-selected seedling from the two vegetative competition treatments × four container sizes × five replication blocks ($n = 40$, per species), three times throughout the season. These measurements corresponded with pre-drought (7 and 11 July), early drought (7 and 8 August), and late drought (25 and 26 September) periods. A fourth, post-drought (21 October), set of measurements was taken on Douglas fir seedlings only, since western larch seedlings had experienced high levels of mortality, and those surviving the first season had begun to senesce. For each measurement period, a new seedling was chosen and was no longer included in future measurements.

Pre-dawn water potential was measured using a pressure chamber (model 1505D-EXP, PMS Instrument Company, Corvallis, OR, USA) between 0000 and 0400 h. A small woody lateral branch from each seedling was excised and used for measurement (*nb*: seedlings with excised biomass were no longer used in subsequent measurements). Gas exchange measurements were conducted on the same seedlings using a portable photosynthesis instrument (model LI-6400XT, LI-COR Environmental, Lincoln, NE, USA) equipped with a lighted conifer chamber (model 6400–22L), a RGB light source, and a CO_2 injector. Measurements began in the morning once photosynthetically active radiation (PAR) reached $\geq$800 µmol m^{-2} s^{-1} and were completed between 0800 and 1230 h. The upper 7 cm portion of the terminal leader was placed into the conifer chamber while still attached to the seedling for measurements because at the start of the season this was the only portion of the seedling tall enough to reach into the LI-6400 chamber. The chamber environment was set to 1400 µmol m^{-2} s^{-1} PAR, 400 µmol mol^{-1} CO_2 with a flow rate of 400 µmol s^{-1}, as described by Pinto et al. [15]. The temperature was initially set to 25 °C, but was raised, as the outside temperature rose (maximum 28 °C), to extend battery life. The portion of the branch placed inside the conifer chamber during measurements was then severed from the seedling. Leaf tissue was scanned on a flatbed scanner, and quantified using Image J software (Version 10.2, National Institutes of Health, Bethesda, MD, USA).

2.7. Statistical Analysis

All analyses, except where noted, were performed separately for each species. As well, for all analyses of variance (ANOVA), block effects were removed from models once found not to be significant. ANOVA using SAS (Version 9.3, SAS Institute, Inc., Cary, NC, USA) PROC MIXED for a RCBD (four container volumes × five blocks) was used to determine if differences in seedling morphology existed for stocktype treatments ($p < 0.05$) after nursery culture. Residual plots were used to assure data met model assumptions. Post-hoc means separations were performed using the Tukey HSD ($\alpha = 0.05$).

Treatment differences for response variables from the outplanting experiment (seedling height, RCD, height and RCD increments, photosynthesis (A), and Ψ_{pd}) were analyzed using an ANOVA model including four container volumes × two vegetative competition treatments × five blocks within a RCBD split-plot design. Competition level served as the whole-plot factor while container volume was the split-plot factor. The design initially contained 20 seedlings (for height and RCD measurements) for all container × competition × replication combinations ($n = 800$, for each species). Height and RCD measurements one and two years following outplanting were collected from all surviving, un-sampled seedlings in each combination. For the physiological measurements, one tree per container volume × competition × block was used ($n = 40$, per species).

Differences between the effects of vegetative competition, on soil moisture throughout the season were analyzed via repeated measures ANOVA using PROC MIXED. Seedling survival was analyzed using logistic regression and a binomial distribution in PROC GLIMMIX. The model included the effects of container, competition, and their interaction. To avoid complete separation in this model, Douglas fir seedlings in the GS plot × 130 cm^3 container volume treatment were excluded due to 0% mortality. Post-hoc, pair-wise comparisons were made using Tukey HSD ($\alpha = 0.05$).

3. Results

3.1. Nursery Culture

3.1.1. Douglas Fir

Container volume significantly ($p < 0.0001$) influenced seedling morphology (height, RCD, RV, R:S) (Table 2). Seedlings cultivated in the smallest (80 cm^3) containers exhibited the lower height, RCD, and RV than the other three stocktypes. The largest container volume produced the tallest seedlings compared to all others. The two larger stocktypes (200 and 250 cm^3) exhibited higher RCD and RV values compared to the other two sizes and significantly higher R:S values compared to those reported for the smallest containers. The R:S values of the 130 cm^3 stocktype were not significantly different from the other three.

Table 2. Mean (±SE) height, root-collar diameter (RCD), root volume (RV), root dry mass:shoot dry mass (R:S) of Douglas fir and western larch seedlings at the end of one-year nursery culture across container types. Different letters within a species column indicate significant differences at $\alpha = 0.05$.

Container Volume (cm^3)	Height (cm)	RCD (mm)	RV (cm^3)	R:S
		Douglas-fir		
80	18.9 (0.3) a	2.77 (0.1) a	3.78 (0.3) a	0.65 (0.0) a
130	23.8 (0.3) b	3.19 (0.1) b	6.33 (0.3) b	0.69 (0.0) ab
200	24.2 (0.3) b	3.61 (0.1) c	8.67 (0.3) c	0.74 (0.0) b
250	25.4 (0.3) c	3.63 (0.1) c	8.89 (0.3) c	0.80 (0.0) b
	$p < 0.0001$	$p < 0.0001$	$p < 0.0001$	$p < 0.0001$
		western larch		
80	22.2 (0.8) a	3.79 (0.1) a	5.04 (0.3) a	0.77 (0.04)
130	27.3 (0.8) bc	4.21 (0.1) b	6.78 (0.3) b	0.82 (0.04)
200	30.3 (0.8) c	4.96 (0.1) c	9.88 (0.3) d	0.78 (0.04)
250	26.6 (0.8) b	4.47 (0.1) b	8.27 (0.3) c	0.87 (0.04)
	$p < 0.0001$	$p < 0.0001$	$p < 0.0001$	$p = 0.2509$

3.1.2. Western Larch

Container volume had a significant effect on western larch seedling height, RCD, and RV values ($p < 0.0001$; Table 2). Generally, seedlings grown in 80 cm^3 containers were significantly smaller in height, RCD, and RV values compared to all other stocktypes. There was no clear relationship between the increase in container volume and the assessed morphological responses. Seedlings grown in the 200 cm^3 containers exhibited greater height, RCD, and RV values. With regard to height, the differences between the 130 cm^3 and the two largest stocktypes were not statistically significant. Seedling R:S did not differ by container volume ($p = 0.2509$).

3.2. Site Conditions

At the time of planting, air temperature and vapor pressure deficit (VPD) were 13.1 °C and 0.3 kPa, respectively. Maximum air temperature (37.0 °C) and VPD (5.8 kPa) were reached on 8 July 2011. During the study period (25 May to 31 October), air temperature and VPD averaged 15.6 °C and 1.2 kPa, respectively. Mean maximum daily air temperature for the season was 24.2 °C, with mean maximum VPD of 2.6 kPa (Figure 1A,B).

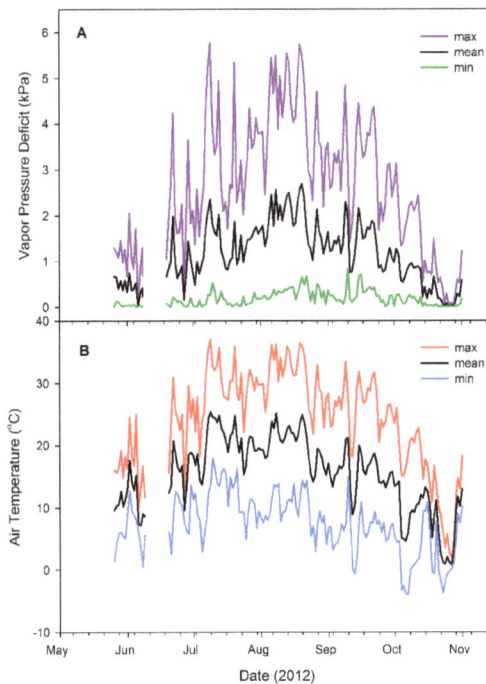

Figure 1. Daily mean, minimum, and maximum vapor pressure deficit (**A**) and air temperature (**B**) conditions during the 2012 growing season at the outplanting site on the University of Idaho Experimental Forest, Moscow, ID, USA. Data from 9–18 June are missing due to ungulate-induced weather station damage.

From 25 May and 31 October, precipitation totaled 104 mm. No precipitation fell from 21 July through 12 October, with 63 mm falling before this drought period and 41 mm after (Figure 2A). Due to animal damage, no data were recorded from the weather station 9–18 June. Volumetric soil moisture content (θ) in the plots were compared at five dates aimed at capturing moisture content differences during the critical establishment period (i.e., seedling planting, pre-, early, late, and post-drought).

For all three soil depths, the interaction of date × vegetative competition treatment was significant ($p < 0.0001$), as were the main effects of date ($p < 0.0001$) and vegetative competition ($p < 0.04$). In June and July 2012, θ did not differ by competition treatment at all depths. However, during the most pronounced drought period in August and September, θ was significantly higher in GS than in HC plots at all three soil depths (Figure 2). After precipitation resumed in October 2012, the two plot types again showed equal measures of θ for all three depths.

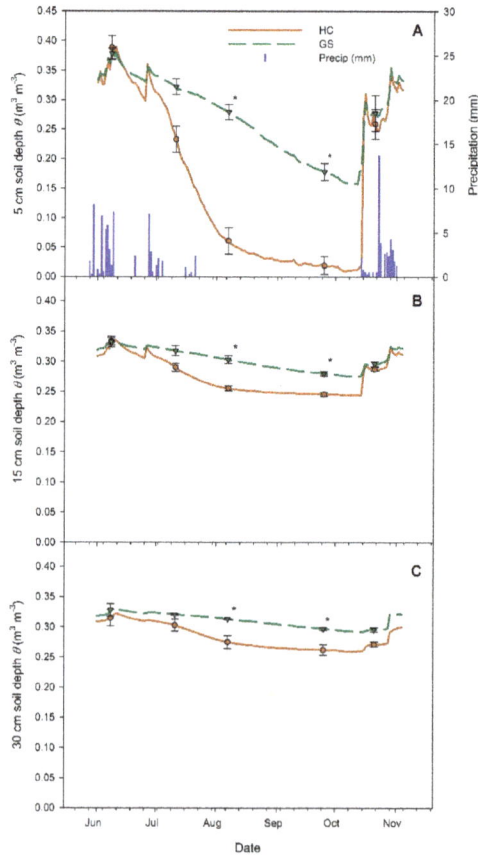

Figure 2. Volumetric soil moisture content (θ) of both vegetative competition treatments (high competition (HC) = solid brown lines, circles; glyphosate-sprayed (GS) = dashed green lines, triangles) at 5 (**A**), 15 (**B**) and 30 cm (**C**) depths, and daily precipitation totals (A, vertical bars on the x-axis) during the 2012 growing season. Points with vertical error bars indicate dates where seedling physiological measurements were performed, as well as seedling planting in June. Points with asterisks (*) indicate a significant difference between HC and GS treatments for that date ($\alpha = 0.05$, $n = 3$). Precipitation data from 9–18 June are missing due to weather station damage.

Differences in vegetation cover, assessed in August, existed between competition treatments. For Douglas-fir, GS plots had significantly less cover (6.6%) than HC plots (45.3%) ($p = 0.0002$). Similarly for western larch, GS plots had significantly less cover (10.1%) than HC plots (59.4%) ($p = 0.0006$).

3.3. Field Survival and Growth

3.3.1. Douglas Fir

By the end of the first growing season in the field, the interaction between container volume × vegetative competition did not significantly affect any of the morphological responses (Table 3). Container volume alone had a significant impact on height, height increment, and RCD, while vegetative competition influenced RCD, RCD increment, and mortality. Height, height increment, and RCD of seedlings cultured in 80 cm^3 cavities were significantly smaller than the other three sizes (Table 3). There were no differences in RCD growth (increment) or mortality among stocktypes. Height and height growth (increment) did not differ significantly due to vegetative competition, but seedlings with larger RCD and RCD increment values were observed in GS plots compared to HC plots.

Table 3. Mean (±SE) height, height increment, root-collar diameter (RCD), and RCD increment of surviving Douglas fir seedlings at end of the first and second field season (October 2012 and 2013) across container types and vegetative competition treatments (GS = glyphosate-sprayed; HC = high competition). The proportion of mortality at the end of the first and second field seasons are also shown. Different letters within a field season column indicate significant differences at α = 0.05.

	Height (cm)	Height Inc. (cm)	RCD (mm)	RCD Inc. (mm)	Proportion Mortality
	First field season (October 2012)–Container Volume (cm^3)				
80	24.7 (0.8) a	5.9 (0.7) a	4.4 (0.2) a	1.3 (0.2)	0.11 (0.1)
130	31.1 (0.8) b	8.7 (0.7) b	5.3 (0.2) b	1.5 (0.2)	0.10 (0.1)
200	30.4 (0.8) b	8.1 (0.7) b	5.4 (0.2) b	1.5 (0.2)	0.07 (0.0)
250	31.5 (0.8) b	7.7 (0.8) b	5.5 (0.2) b	1.4 (0.2)	0.10 (0.0)
	First field season (October 2012)–Vegetative Competition Treatment				
GS	29.8 (0.9)	8.1 (0.9)	5.9 (0.3) a	2.4 (0.3) a	0.03 (0.0) a
HC	29.0 (0.9)	7.1 (0.9)	4.5 (0.3) b	1.0 (0.3) b	0.28 (0.1) b
Container effect	$p < 0.0001$	$p < 0.0001$	$p < 0.0001$	$p = 0.2366$	$p = 0.5233$
Competition effect	$p = 0.5452$	$p = 0.4645$	$p = 0.0196$	$p = 0.0165$	$p = 0.0068$
Interaction	$p = 0.4461$	$p = 0.7935$	$p = 0.6395$	$p = 0.4188$	$p = 0.8799$
	Second field season (October 2013)–Container Volume (cm^3)				
80	32.7 (2.4) a	8.2 (1.7) ab	7.7 (0.6) a	3.2 (0.4) ab	0.36 (0.1)
130	42.5 (2.4) b	10.9 (1.7) b	9.4 (0.6) c	3.9 (0.4) b	0.38 (0.1)
200	38.0 (2.4) b	7.7 (1.7) a	8.3 (0.6) ab	2.9 (0.4) a	0.28 (0.1)
250	42.0 (2.4) b	10.1 (1.7) ab	9.1 (0.6) bc	3.5 (0.4) ab	0.35 (0.1)
	Second field season (October 2013)–Vegetative Competition Treatment				
GS	41.5 (2.8)	11.8 (2.0)	10.1 (0.7)	4.2 (0.4)	0.15 (0.1) a
HC	36.1 (3.3)	6.7 (2.3)	7.1 (0.8)	2.5 (0.5)	0.61 (0.2) b
Container effect	$p < 0.0001$	$p = 0.0374$	$p = 0.0005$	$p = 0.0012$	$p = 0.4722$
Competition effect	$p = 0.2959$	$p = 0.1887$	$p = 0.0585$	$p = 0.0725$	$p = 0.0286$
Interaction	$p = 0.9678$	$p = 0.9388$	$p = 0.2755$	$p = 0.0856$	$p = 0.1414$

Two years following outplanting, the combined effect of container volume × vegetative competition, and competition alone did not influence seedling morphology (Table 3). However, seedling height, height increment, RCD, and RCD increment were significantly influenced by container volume. Seedlings produced in the smallest containers remained significantly shorter compared to the other stocktypes. The smallest containers also yielded the lowest mean RCD, but did not differ from the 200 cm^3 stocktype; while the 130 cm^3 stocktype had the largest RCD and was similar to the 250 cm^3 stocktypes. Height and RCD growth (increment) was significantly higher for the 130 cm^3 compared to 200 cm^3 stocktypes, while the smallest and largest container volumes were statistically indistinguishable from the others.

Mortality following the first growing season was significantly influenced by vegetative competition, evidenced by lower mortality among seedlings planted in the GS plots compared to those in HC plots (3% and 28%, respectively). At the end of the second field growing season, mortality increased two- and five-fold among the HC and GS treatments (61% and 15%, respectively).

3.3.2. Western Larch

At the end of the first field-growing season, mortality among larch vegetation treatments was high; therefore, our analyses were adjusted accordingly. In the full model analysis (mortality = container, vegetative competition, container × vegetative completion), there was no interaction or container volume effect on seedling mortality ($p > 0.5172$). Seedling mortality was affected by vegetative competition, with seedlings grown in HC plots exhibiting significantly higher mortality than those in GS plots ($p < 0.0001$) (98% and 2%, respectively). As a result of the near complete mortality in the HC plots, a priori analyses on the full statistical model for height, height increment, RCD, and RCD increment was not possible; thus, post hoc analyses focused only the effect of container volume within GS plots. Based on the surviving seedlings in these plots, container volume had a significant effect on height, RCD, and their growth increments ($p < 0.0107$; Table 4). Height and RCD were significantly lower among the smallest stocktype compared to all other sizes. The 130 and 200 cm^3 containers produced the tallest seedlings, while the RCD values for the 130, 200, 250 cm^3 stocktypes were statistically indistinguishable from each other. For height and RCD growth increments, the largest stocktype was significantly smaller than the 130 cm^3 stocktype, whereas all other container volume treatments did not differ significantly.

Table 4. Mean ± (SE) height, height increment, root-collar diameter (RCD), and RCD increment of surviving western larch seedlings at end of the first and second field seasons (October 2012 and 2013). Data is from glyphosate-sprayed (GS) plots only due to near complete mortality within the high competition (HC) plots. Different letters within a field season column indicate significant differences at $\alpha = 0.05$.

	Height (cm)	Height Inc. (cm)	RCD (mm)	RCD Inc. (mm)
First field season (October 2012)–Container Volume (cm^3)				
80	39.4 (1.8) a	18.7 (1.6) ab	6.64 (0.3) a	3.16 (0.3) ab
130	49.7 (1.8) c	22.5 (1.6) b	7.83 (0.3) b	3.81 (0.3) b
200	49.5 (1.8) c	18.5 (1.6) ab	8.01 (0.3) b	3.37 (0.3) ab
250	45.2 (1.8) b	17.6 (1.8) a	7.37 (0.3) b	3.05 (0.3) a
Container effect	$p < 0.0001$	$p = 0.0107$	$p < 0.0001$	$p = 0.0163$
Second field season (October 2013)–Container Volume (cm^3)				
80	60.8 (2.5) a	21.6 (2.4)	10.9 (0.4) a	4.2 (0.2) a
130	73.7 (2.4) b	23.0 (2.4)	13.3 (0.3) b	5.3 (0.2) b
200	74.5 (2.4) b	24.8 (2.3)	12.9 (0.3) b	4.9 (0.2) b
250	68.8 (2.5) b	23.1 (2.4)	12.4 (0.4) b	5.0 (0.2) b
Container effect	$p < 0.0001$	$p = 0.5697$	$p < 0.0001$	$p = 0.0071$

As with the first year outplanting data, the full model analysis on mortality indicated no container volume × vegetative competition interaction ($p = 0.5819$), or container volume effects ($p = 0.9227$) two years following outplanting. Significant differences as a result of vegetative competition persisted at the end of the second field season. Mortality increased seven-fold between the first and second year within the GS plots, but still only amounted to 14% of the total seedlings planted; HC mortality totaled 99%. Again, because of the high mortality, post hoc analyses focused on GS container effects only. Two years following outplanting, container volume significantly influenced seedling morphology and growth (Table 4). Seedlings grown in the 80 cm^3 containers exhibited the smallest height, RCD, and RCD increment; however, differences in height increment no longer varied among containers.

3.4. Physiology

3.4.1. Douglas Fir

For both field-measured physiological variables (i.e., A and Ψ_{pd}), there were no significant interactions between container volume $\times$ vegetative competition ($p > 0.0748$). The main effect of container volume was also not significant ($p > 0.1123$). Effects of vegetative competition on A were not significant in July ($p = 0.2256$), but were significant for August, September, and October 2012 measurements ($p < 0.0056$) (Figure 3). At these measurement points, seedlings in the GS plots exhibited significantly higher A rates than those in the HC plots (Figure 3A). Overall, GS plots exhibited a 128% increase in A from July to September, when soils were rapidly drying in the upper soil profile (Figure 2A). Conversely, HC plots exhibited a 39% decrease in A during the same period.

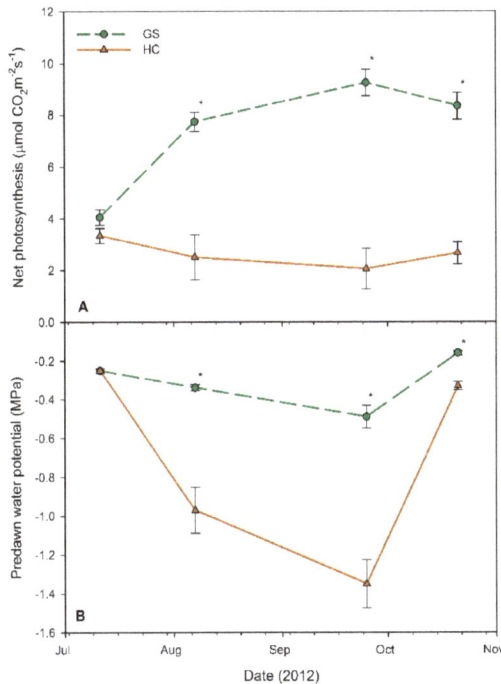

Figure 3. (**A**) Net photosynthesis and (**B**) predawn water potential of Douglas fir seedlings at four measurement dates in 2012. Each point represents the mean ($n = 19$–20) across all container sizes, within each vegetative treatment, high competition (HC: solid brown lines, triangles), and glyphosate sprayed (GS: dashed green lines, circles). Error bars represent the standard error of the mean. Points with asterisks (*) indicate a significant difference between HC and GS treatments for that date ($\alpha = 0.05$).

No differences in Ψ_{pd} between GS and HC were detected at the first measurement date in July ($p = 1.000$), but HC plots exhibited significantly lower values than GS for the remaining three measurement points (August, September, and October 2012; $p < 0.0301$), during and after the drought period. GS decreased 0.24 MPa (96%) from July to September, while HC decreased 1.10 MPa (440%). The September Ψ_{pd} values were most negative for both vegetative competition levels, respectively, but they both recovered to near July values by 21 October (Figure 3B), after 4.6 mm of precipitation fell between 13–21 October (Figure 2).

3.4.2. Western Larch

Physiological data for western larch was only collected for the first three measurement points (i.e., pre-, early, and late drought) due to high seedling mortality and the onset of fall senescence at the last measurement date (October 2012). For all three measurement points, both net photosynthesis and Ψ_{pd} showed no significant interactions between container volume and vegetative competition ($p > 0.1580$); container volume alone had no impact ($p > 0.0744$). Conversely, vegetative competition influenced both photosynthesis and Ψ_{pd} measurements in August ($p = 0.0073$ and $p = 0.0068$) and September ($p = 0.0125$ and $p = 0.0018$) (Figure 4). Seedlings in the HC plots exhibited a 78% decrease in net photosynthesis from July to September, while those in GS plots experienced a 68% increase in photosynthesis during the same period (Figure 4A). The Ψ_{pd} values stayed relatively similar through the three measurement periods for GS seedlings, changing a maximum of 0.06 MPa (20%), while those in the HC plots became significantly more negative during the August and September measurement points, changing a maximum of 1.98 MPa (615%) (Figure 4B). Seedlings in both vegetative competition treatments had their most negative Ψ_{pd} measurements in September.

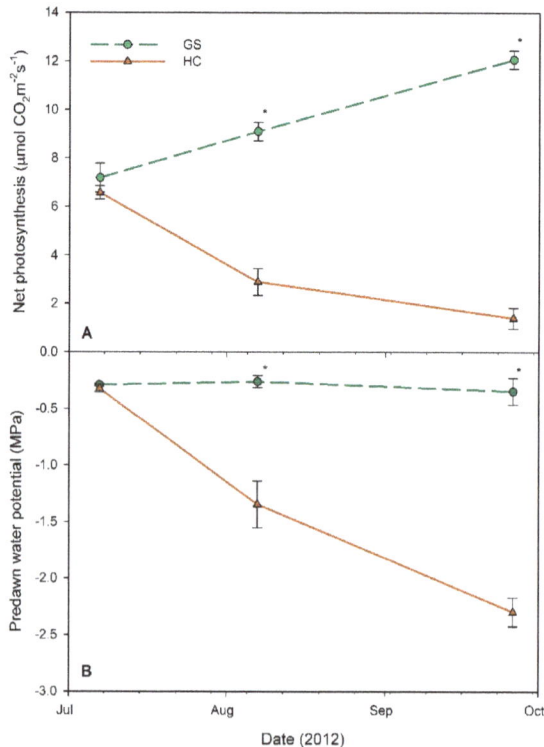

Figure 4. (**A**) Net photosynthesis and (**B**) predawn water potential of western larch seedlings at three measurement dates in 2012. Each point represents the mean (n = 13–20) across all container sizes, within each vegetative treatment, high competition (HC: solid brown lines, triangles), and glyphosate sprayed (GS: dashed green lines, circles). Error bars represent the standard error of the mean. Points with asterisks (*) indicate a significant difference between HC and GS treatments for that date ($\alpha = 0.05$).

4. Discussion

Following nursery culture, the positive correlation between seedling size and increasing container volume is well-documented [7,10,17,32]. In this study, we detected the same trend in Douglas fir and western larch with regard to the morphological traits measured. However, the largest containers were not often statistically set apart from the next smallest size. In fact, western larch seedlings followed a slightly different pattern of growth with container size. In general, morphological traits increased from the 80 to 200 cm^3 container sizes, but then decreased slightly for the largest size (250 cm^3). Despite our best efforts to equilibrate seedling quality across container types, according to the recommendations by Pinto et al. [27], we suspect the largest stocktype may not have reached its fullest size potential during nursery culture (for either species). It is difficult to know whether growth times or nutrient delivery contributed to this scenario. In some cases, situations like this can make outplanting performance interpretation tenuous. Lamhamedi et al. [33] experienced similar issues with large experimental stocktypes. The subsequent performance attributes show that there may have been a lasting effect among the largest stocktype in our study, but for only the first year of growth.

Despite the initial height and RCD differences among Douglas fir stocktypes at planting, the stratification of these traits began to dissipate after one field season. After two seasons, height, RCD, and their growth increments were not statistically different amongst stocktypes, with the exception of the smallest. Other studies with Douglas fir show similar trends where growth rates, as well as absolute morphological characteristics of different stocktypes, tend to equilibrate over time [11,34]. Despite the growth similarities among the Douglas fir stocktypes, we observed noticeable height and diameter growth among seedlings produced in the 130 cm^3 containers following the second growing season. Although not statistically different, this may suggest that relative growth rates of smaller seedlings can be as good as, or greater than, those of larger seedlings, reducing the seedling size gap over time [19,35,36].

While seedling height and shoot growth appear to be more influenced by container volume, survival and below-ground production are likely influenced by competition. After one growing season, the positive effect of reducing vegetative competition on Douglas fir size and growth was seen only in terms of RCD and seedling survival, but not seedling height. This supports previous findings that seedling height is not a strong predictor of performance in all cases [25,26,37,38]. Conversely, RCD has been used as a better predictor of early seedling health and survival [24,39]. Seedlings with greater RCD, which is correlated to higher root mass, have been shown to exhibit increased hydraulic conductivity, reduced transplanting shock, and improved survival under moisture stress [40–44].

For western larch seedlings, the role of container volume and its interaction with vegetative competition could not be fully assessed due to near 100% mortality under high competition in the first year. Our investigation of the larch stocktypes grown with reduced competition (GS), yielded similar results to the Douglas fir, where the initial size differences between stocktypes largely dissipated, and growth mostly equilibrated over two years. However, despite reduced soil moisture stress as a result of lower competition, the smallest stocktype consistently underperformed and remained small.

Generally, western larch has been found to be less drought tolerant than Douglas fir [45]. This may partly explain why drought has been attributed to high levels of mortality in western larch during the first season following outplanting [46,47]. In our study, the period between July and September was likely accountable for the majority of seedling mortality in the first year. During this period, the Ψ_{pd} values fell below -2.0 MPa, which may have led to stomatal closure and the subsequent reduction in net photosynthesis to below 4 μmol CO_2 m^{-2} s^{-1}, indicating that seedlings were unable to effectively manage moisture stress. Pinto et al. [48] modeled similar conditions for ponderosa pine seedlings and predicted that dry soils in the upper soil profile (0–15 cm) would reduce seedling photosynthesis and transpirational relationships.

While our photosynthesis rates measured were comparable with those reported in the literature [49,50], we did not observe a combined effect of vegetative competition × container volume, or container volume alone on the net photosynthesis measurements of either species. Although seedling stocktypes differed in initial size, their carbon allocation rates did not. This is contrary to several comparable studies that observed a significant influence of stocktype × competition interaction on photosynthesis rates [14,15,25,26]. However, despite the significance of the interaction there is uniform correlation that is shared by these studies. For example, Cuesta et al. [14] observed higher photosynthesis rates for larger versus smaller stocktypes under high competition, with no differences in performance in the absence of vegetative competition—an outcome expected in this study. Mohammed et al. [26] saw increased photosynthesis rates in a smaller (57 cm^3) stocktype under reduced competition, while a larger (98 cm^3) stocktype exhibited no change between competition treatments. Lamhamedi et al. [25] saw no difference between black spruce produced in 110 or 300 cm^3 containers either with or without competition, but noted that in the presence of vegetative competition the largest (700 cm^3) stocktype performed worse compared to the smaller sizes. Finally, Pinto et al. [15] reported that high levels of vegetative competition resulted in complete mortality among smaller stocktypes (60 and 90 cm^3) of ponderosa pine, whereas the largest stocktype (120 cm^3) exhibited increased survival.

Soil moisture in the upper profile was not markedly different between competition treatments early in the growing season. As the season progressed, however, differences between the treatments became pronounced in the upper soil profile, and were larger than those at greater depths (>15 cm; Figure 2). Due to the drastic changes observed in seedling physiology over time (seedling A and Ψ_{pd}), and consequent survival (for western larch), there is evidence to suggest that a large portion of the seedlings' functional root system was in the upper profile. For example, our reported Ψ_{pd} measurements for Douglas fir are similar to those observed in seedlings under high moisture stress [13,51]. This corroborates findings of Pinto et al. [48], which show a strong correlation between rooting depth, soil moisture, and physiological functioning. This also shows the importance of considering the interaction of edaphic characteristics of a site with stocktype selection and site preparation.

Successful seedling establishment depends on the ability of seedlings to quickly become coupled to the site hydrological cycle and respond to environmental and silvicultural conditions [52]. The low levels of photosynthesis and Ψ_{pd}, along with increased mortality (especially in western larch) under a high level of competition indicate that the seedlings did not have ample resources and time to establish prior to the onset of summer drought. Similar findings have been observed among ponderosa pine, Aleppo pine, and Holm oak [14,15]. Along these lines, root growth is sensitive to both plant moisture stress [22,53] and soil temperature [52,54]. Our data suggests that seedlings outplanted into reduced competition conditions were able to access soil water in the upper profile, which had significantly greater θ than those observed in the high competition plots, which promoted greater root egress into the soil. Pinto et al. [15] reported that out of the three tested stocktypes that varied in volume and length, only the largest, longest stocktype survived in high competition. That same stocktype would have had roots that extended well into the zone of higher soil moisture measured within the scope of our study, even under high competition. Thus, future work should focus on examining the performance of taller containers, especially in conditions where vegetation control methods are not available.

Interestingly, Douglas fir seedlings in HC plots experienced a recovery to pre-drought levels of Ψ_{pd} in October, after rainfall resumed, but this was not accompanied by an increase in photosynthesis rates. This contrasts findings by Pinto et al. [15], who reported an increase in post-drought photosynthesis rates for the largest stocktype of ponderosa pine seedlings in high competition plots in October. The difference in these reports could be related to the number of precipitation events. Pinto et al. [15] stated that seedlings received several small rain events throughout the drought season, and prior to the final gas exchange measurement. Drought conditions within our study were more pronounced, with a complete absence of precipitation lasting 87 days during the first growing season. In addition to species-specific drought response strategies [45], other studies have shown that photosynthetic recovery is strongly influenced by the length and severity of drought [55]. The mechanism behind this damage is attributed to restricted CO_2 diffusion into the plant, a depletion of carbon reserves, and general metabolic impairment over time [56,57]. However, less severe intermittent drought, followed by increased water availability, may allow for seedling acclimatization to drought and eliminate the reduction in photosynthesis rates [58].

5. Conclusions

On sites with high vegetative competition resulting in depleted upper-profile soil moisture, seedling mortality can be excessively high leading to establishment failure. This study used vegetative competition to further increase summer drought conditions in order to isolate stocktype- and species-driven differences in performance following outplanting while using uniformly-cultured nursery seedlings, in an effort to reduce confounding. Our results indicate that the largest operational stocktype evaluated in this study did not offer an advantage to overcoming drought conditions observed in the first two growing seasons. When competing vegetation was controlled, ample soil moisture remained in the upper profile, which refuted the initial differences in seedling size for all but the smallest containers. Thus, western larch and, to a lesser extent, Douglas fir seedlings planted on sites with moisture limitations require control of vegetative competition prior to spring planting. Furthermore, in the Inland Northwest, Douglas fir and western larch seedlings selected for spring planting may not require container volumes beyond 130 cm^3, especially in the absence of vegetative competition.

Author Contributions: B.A.M., A.S.D., and J.R.P. conceived and designed the experiment; B.A.M. and J.R.P. implemented the experiment; B.A.M, J.R.P., and O.A.K. analyzed data; B.A.M. wrote the original paper as part of her thesis; A.S.D. coordinated the work; and all authors contributed to the final version of the manuscript.

Acknowledgments: We thank the Potlatch Corporation for funding and providing seed for the project, and Abbie Acuff, in particular, for the support of this work; University of Idaho Center for Forest Nursery and Seedling Research for nursery culturing space and logistical support; and University of Idaho Experimental Forest for the planting site. We are grateful for the data collection assistance in the nursery and field provided by Don Regan, Jake Kleinknecht, and Shannin Murphy. Thanks to Amy Ross-Davis, who contributed to an extensive revision of the manuscript in an earlier form; Douglass Jacobs and Randy Brooks provided support and insight during the study development. Further support was provided by the USDA Forest Service, Rocky Mountain Research Station (RMRS) and the National Center for Reforestation, Nurseries, and Genetic Resources. The views expressed are strictly those of the authors and do not necessarily represent the positions or policy of their respective institutions.

Conflicts of Interest: The authors declare no conflict of interest.

References

1. Landis, T.D.; Tinus, R.W.; McDonald, S.E.; Barnett, J.P. Containers and growing media. In *The Container Tree Nursery Manual*; USDA Forest Service Agricultural Handbook 674: Washington, DC, USA, 1990; Volume 2, p. 88.

2. Rose, R.; Carlson, W.C.; Morgan, P. The target seedling concept. In *Target Seedling Symposium: Proceedings, Combined Meeting Western Forest Nursery Associations, Roseburg, OR, USA, 13–17 August 1990*; Rose, R., Campbell, S.J., Landis, T.D., Eds.; USDA Forest Service General Technical Report RM-200; U.S. Department of Agriculture, Forest Service, Rocky Mountain Forest and Range Experiment Station: Fort Collins, CO, USA, 1990.

3. Thiffault, N. Stock type in intensive silivculture: A (short) discussion about roots and size. *For. Chron.* **2004**, *8*, 463–468. [CrossRef]

4. Endean, F.; Carlson, L.W. The effect of rooting volume on the early growth of lodgepole pine seedlings. *Can. J. For. Res.* **1975**, *5*, 55–60. [CrossRef]

5. Jinks, R.; Mason, B. Effects of seedling density on the growth of Corsican pine (*Pinus nigra* var. *maritima* Melv.), Scots pine (*Pinus sylvestris* L.) and Douglas-fir (*Pseudotsuga menziesii* Franco) in containers. *Ann. For. Sci.* **1998**, *55*, 407–423. [CrossRef]

6. Chirino, E.; Vilagrosa, A.; Hernández, E.I.; Matos, A.; Vallejo, V.R. Effects of a deep container on morpho-functional characteristics and root colonization in *Quercus suber* L. seedlings for reforestation in Mediterranean climate. *For. Ecol. Manag.* **2008**, *256*, 779–785. [CrossRef]

7. Carlson, L.W.; Endean, F. The effect of rooting volume and container configuration on the early growth of white spruce seedlings. *Can. J. For. Res.* **1976**, *6*, 221–224. [CrossRef]

8. Simpson, D. Nursery growing density and container volume affect nursery and field growth of Douglas-fir and lodgepole Pine seedlings. In *National Proceedings: Forest and Conservation Nursery Associations, Williamsburg, VA, USA, 11–14 July 1994*; Landis, T.D., Dumroese, R.K., Tech. Coords., Eds.; USDA Forest Service General Technical Report RM-GTR-257; U.S. Department of Agriculture, Forest Service, Rocky Mountain Forest and Range Experiment Station: Fort Collins, CO, USA, 1994.

9. Aphalo, P.; Rikala, R. Field performance of silver-birch planting-stock grown at different spacing and in containers of different volume. *New For.* **2003**, *25*, 93–108. [CrossRef]

10. Dominguez-Lerena, S.; Herrerosierra, N.; Carrascomanzano, I.; Ocanabueno, L.; Penuelasrubira, J.; Mexal, J. Container characteristics influence seedling development in the nursery and field. *For. Ecol. Manag.* **2006**, *221*, 63–71. [CrossRef]

11. Haase, D.L.; Rose, R.; Trobaugh, J. Field Performance of Three Stock Sizes of Douglas-fir Container Seedlings Grown with Slow-release Fertilizer in the Nursery Growing Medium. *New For.* **2006**, *31*, 1–24. [CrossRef]

12. Puértolas, J.; Jacobs, D.F.; Benito, L.F.; Peñuelas, J.L. Cost–benefit analysis of different container capacities and fertilization regimes in *Pinus* stock-type production for forest restoration in dry Mediterranean areas. *Ecol. Eng.* **2012**, *44*, 210–215. [CrossRef]

13. Carlson, W.C.; Miller, D.E. Target seedling root system size, hydraulic conductivity, and water use during seedling establishment. In *Target Seedling Symposium: Proceedings, Combined Meeting Western Forest Nursery Associations, Roseburg, OR, USA, 13–17 August 1990*; Rose, R., Campbell, S.J., Landis, T.D., Eds.; USDA Forest Service General Technical Report RM-200; U.S. Department of Agriculture, Forest Service, Rocky Mountain Forest and Range Experiment Station: Fort Collins, CO, USA, 1990.

14. Cuesta, B.; Villar-Salvador, P.; Puértolas, J.; Jacobs, D.F.; Rey Benayas, J.M. Why do large, nitrogen rich seedlings better resist stressful transplanting conditions? A physiological analysis in two functionally contrasting Mediterranean forest species. *For. Ecol. Manag.* **2010**, *260*, 71–78. [CrossRef]

15. Pinto, J.R.; Marshall, J.D.; Dumroese, R.K.; Davis, A.S.; Cobos, D.R. Photosynthetic response, carbon isotopic composition, survival, and growth of three stock types under water stress enhanced by vegetative competition. *Can. J. For. Res.* **2012**, *42*, 333–344. [CrossRef]

16. Villar-Salvador, P.; Puértolas, J.; Cuesta, B.; Peñuelas, J.L.; Uscola, M.; Heredia-Guerrero, N.; Rey Benayas, J.M. Increase in size and nitrogen concentration enhances seedling survival in Mediterranean plantations. Insights from an ecophysiological conceptual model of plant survival. *New For.* **2012**, *43*, 755–770. [CrossRef]

17. Sutherland, C.; Day, R.J. Container volume affects survival and growth of white spruce, black spruce, and jack pine seedlings: A literature review. *North. J. Appl. For.* **1988**, *5*, 185–189. [CrossRef]

18. Jobidon, R.; Roy, V.; Cyr, G. Net effect of competing vegetation on selected environmental conditions and performance of four spruce seedling stock sizes after eight years in Québec (Canada). *Ann. For. Sci.* **2003**, *60*, 691–699. [CrossRef]

19. Close, D.C.; Paterson, S.; Corkrey, R.; McArthur, C. Influences of seedling size, container type and mammal browsing on the establishment of *Eucalyptus globulus* in plantation forestry. *New For.* **2010**, *39*, 105–115. [CrossRef]

20. Van den Driessche, R. Relationship between spacing and nitrogen fertilization of seedlings in the nursery, seedling mineral nutrition, and outplanting performance. *Can. J. For. Res.* **1984**, *14*, 431–436. [CrossRef]

21. Pinto, J.R.; Marshall, J.D.; Dumroese, R.K.; Davis, A.S.; Cobos, D.R. Establishment and growth of container seedlings for reforestation: A function of stocktype and edaphic conditions. *For. Ecol. Manag.* **2011**, *261*, 1876–1884. [CrossRef]

22. Burdett, A.N. Physiological processes in plantation establishment and the development of specifications for forest planting stock. *Can. J. For. Res.* **1990**, *20*, 415–427. [CrossRef]

23. Ferguson, S.A. *Climatology of the Interior Columbia River Basin*; USDA Forest Service General Technical Report PNW-GTR-445; USDA Forest Service: Portland, OR, USA, 1999.

24. Grossnickle, S.C. Why seedlings survive: Influence of plant attributes. *New For.* **2012**, *43*, 711–738. [CrossRef]

25. Lamhamedi, M.S.; Bernier, P.Y.; Hébert, C.; Jobidon, R. Physiological and growth responses of three sizes of containerized Picea mariana seedlings outplanted with and without vegetation control. *For. Ecol. Manag.* **1998**, *110*, 13–23. [CrossRef]

26. Mohammed, G.H.; Noland, T.L.; Wagner, R.G. Physiological perturbation in jack pine (*Pinus banksiana* Lamb.) in the presence of competing herbaceous vegetation. *For. Ecol. Manag.* **1998**, *103*, 77–85. [CrossRef]

27. Pinto, J.R.; Dumroese, R.K.; Davis, A.S.; Landis, T.D. Conducting seedling stocktype trials: A new approach to an old question. *J. For.* **2011**, *109*, 293–299.

28. Dumroese, R.K.; Montville, M.E.; Pinto, J.R. Using container weights to determine irrigation needs: A simple method. *Native Plants J.* **2015**, *16*, 67–71. [CrossRef]

29. Burdett, A.N. A nondestructive method for measuring the volume of intact plant parts. *Can. J. For. Res.* **1979**, *9*, 120–122. [CrossRef]

30. Web Soil Survey: Natural Resources Conservation Service (NRCS) Soil Survey Staff. United States Department of Agriculture. Available online: http://websoilsurvey.nrcs.usda.gov/ (accessed on 8 February 2012).

31. Daubenmire, R. A canopy-coverage method of vegetational analysis. *Northwest Sci.* **1959**, *33*, 43–64.

32. Paterson, J. Growing environment and container type influence field performance of black spruce container stock. *New For.* **1997**, *13*, 329–339. [CrossRef]

33. Lamhamedi, M.S.; Bernier, P.Y.; Herbert, C. Effect of shoot size on the gas exchange and growth of containerized *Picea mariana* seedlings under different watering regimes. *New For.* **1997**, *13*, 207–221. [CrossRef]

34. Rose, R.; Haase, D.L.; Kroiher, F.; Sabin, T. Root volume and growth of Ponderosa pine and Douglas-fir seedlings: A summary of eight growing seasons. *West. J. Appl. For.* **1997**, *12*, 69–73. [CrossRef]

35. Van den Driessche, R. Absolute and relative growth of Douglas-fir seedlings of different sizes. *Tree Physiol.* **1992**, *10*, 141–152. [CrossRef] [PubMed]

36. Faure-Lacroix, J.; Tremblay, J.P.; Thiffault, N.; Roy, V. Stock type performance in addressing top-down and bottom-up factors for the restoration of indigenous trees. *For. Ecol. Manag.* **2013**, *307*, 333–340. [CrossRef]

37. Overton, W.S.; Ching, K.K. Analysis of differences in height growth among populations in a nursery selection study of Douglas-fir. *For. Sci.* **1978**, *24*, 497–509. [CrossRef]

38. Rose, R.; Gleason, J.F.; Atkinson, M. Morphological and water-stress characteristics of three Douglas-fir stocktypes in relation to seedling performance under different soil moisture conditions. *New For.* **1993**, *7*, 1–17. [CrossRef]

39. Chavasse, C.G.R. The significance of planting height as an indicator of subsequent seedling growth. *N. Z. J For. Sci.* **1977**, *22*, 283–296.

40. Carlson, W.C. Root system considerations in the quality of loblolly pine seedlings. *South. J. Appl. For.* **1986**, *10*, 87–92. [CrossRef]

41. Mexal, J.G.; Landis, T.D. Target seedling concepts: Height and diameter. In *Target Seedling Symposium: Proceedings, Combined Meeting Western Forest Nursery Associations, Roseburg, OR, USA, 13–17 August 1990*; Rose, R., Campbell, S.J., Landis, T.D., Eds.; USDA Forest Service General Technical Report RM-200; U.S. Department of Agriculture, Forest Service, Rocky Mountain Forest and Range Experiment Station: Fort Collins, CO, USA, 1990; pp. 17–35.

42. Haase, D.L.; Rose, R. Soil moisture stress induces transplant shock in stored and unstored 2 + 0 Douglas-fir seedlings of varying root volumes. *For. Sci.* **1993**, *39*, 275–294.

43. South, D.; Harris, S.; Barnett, J.; Hainds, M.; Gjerstad, D. Effect of container type and seedling size on survival and early height growth of seedlings in Alabama, U.S.A. *For. Ecol. Manag.* **2005**, *204*, 385–398. [CrossRef]

44. McDowell, N.; Pockman, W.T.; Allen, C.D.; Breshears, D.D.; Cobb, N.; Kolb, T.; Plaut, J.; Sperry, J.; West, A.; Williams, D.G.; et al. Mechanisms of plant survival and mortality during drought: Why do some plants survive while others succumb to drought? *New Phytol.* **2008**, *178*, 719–739. [CrossRef] [PubMed]

45. Piñol, J.; Sala, A. Ecological implications of xylem cavitation for several Pinaceae in the Pacific Northern USA. *Funct. Ecol.* **2000**, *14*, 538–545. [CrossRef]

46. Schmidt, W.C.; Shearer, R.C.; Roe, A.L. *Ecology and Silviculture of Western Larch Forests*; Technical Bulletin. 1520; U.S. Department of Agriculture, Forest Service: Washington, DC, USA, 1976; 96p.

47. Schmidt, W.C. *Larix occidentalis* . In *Silvics of North America: Volume 1. Conifers*; Burns, R.M., Honkala, B.H., Eds.; Agricultural Handbook 654; USDA Forest Service : Washington, DC, USA, 1990; p. 877.

48. Pinto, J.R.; Marshall, J.D.; Dumroese, R.K.; Davis, A.S.; Cobos, D.R. Seedling establishment and physiological responses to temporal and spatial soil moisture changes. *New For.* **2016**, *47*, 223–241. [CrossRef]

49. Rosenthal, S.I.; Camm, E.L. Photosynthetic decline and pigment loss during autumn foliar senescence in western larch (*Larix occidentalis*). *Tree Physiol.* **1997**, *17*, 767–775. [CrossRef] [PubMed]

50. Robertson, N.D.; Davis, A.S. Sulfometuron methyl influences seedling growth and leaf function of three conifer species. *New For.* **2011**, *43*, 185–195. [CrossRef]

51. Dumroese, R.K.; Haase, D.L.; Landis, T.D. Seedling processing, storage, and outplanting. In *The Container Tree Nursery Manual*; USDA Forest Service Agricultural Handbook: Washington, DC, USA, 2010; Volume 7, p. 674.

52. Rietveld, W.J. Transplanting stress in bareroot conifer seedlings: Its development and progression to establishment. *North. J. Appl. For.* **1989**, *6*, 99–107. [CrossRef]

53. Grossnickle, S.C. Importance of root growth in overcoming planting stress. *New For.* **2005**, *30*, 273–294. [CrossRef]

54. Sayer, M.S.; Brissette, J.C.; Barnett, J.P. Root growth and hydraulic conductivity of southern pine seedlings in response to soil temperature and water availability after planting. *New For.* **2005**, *30*, 253–272. [CrossRef]

55. Rouhi, V.; Samson, R.; Lemeur, R.; Van Damme, P. Photosynthetic gas exchange characteristics in three different almond species during drought stress and subsequent recovery. *Environ. Exp. Bot.* **2007**, *59*, 117–129. [CrossRef]

56. Flexas, J.; Bota, J.; Galmes, J.; Medrano, H.; Ribas-Carbo, M. Keeping a positive carbon balance under adverse conditions: Responses of photosynthesis and respiration to water stress. *Physiol. Plant.* **2006**, *127*, 343–352. [CrossRef]

57. Niinemets, U. Responses of forest trees to single and multiple environmental stresses from seedlings to mature plants: Past stress history, stress interactions, tolerance and acclimation. *For. Ecol. Manag.* **2010**, *260*, 1623–1639. [CrossRef]

58. Stewart, J.D.; Zine el Abidine, A.; Bernier, P.Y. Stomatal and mesophyll limitations of photosynthesis in black spruce seedlings during multiple cycles of drought. *Tree Physiol.* **1995**, *15*, 57–64. [CrossRef] [PubMed]

![forests logo] *forests*

MDPI

Article

Biochar Can Be a Suitable Replacement for Sphagnum Peat in Nursery Production of *Pinus ponderosa* Seedlings

R. Kasten Dumroese [1,*], Jeremiah R. Pinto [1], Juha Heiskanen [2], Arja Tervahauta [3], Katherine G. McBurney [1], Deborah S. Page-Dumroese [1] and Karl Englund [4]

[1] U.S. Department of Agriculture Forest Service, Rocky Mountain Research Station, 1221 South Main Street, Moscow, ID 83843, USA; jpinto@fs.fed.us (J.R.P.); kgmcburney@gmail.com (K.G.M.); ddumroese@fs.fed.us (D.S.P-D.)

[2] Natural Resources Institute Finland, Soil Ecosystems, Neulaniementie 5, FI-70210 Kuopio, Finland; juha.heiskanen@luke.fi

[3] Natural Resources Institute Finland, Soil Ecosystems, Latokartanonkaari 9, FI-00790 Helsinki, Finland; arja.tervahauta@luke.fi

[4] Composite Materials & Engineering Center, Washington State University, Pullman, WA 99164-2262, USA; englund@wsu.edu

* Correspondence: kdumroese@fs.fed.us; Tel.: +1-208-883-2324

Received: 27 March 2018; Accepted: 24 April 2018; Published: 27 April 2018

Abstract: We replaced a control peat medium with up to 75% biochar on a volumetric basis in three different forms (powder, BC; pyrolyzed softwood pellets, PP; composite wood-biochar pellets, WP), and under two supplies of nitrogen fertilizer (20 or 80 mg N) subsequently grew seedlings with a comparable morphology to the control. Using gravimetric methods to determine irrigation frequency and exponential fertilization to ensure all treatments received the same amount of N at a given point in the growing cycle, we successfully replaced peat with 25% BC and up to 50% PP. Increasing the proportion of biochar in the media significantly increased pH and bulk density and reduced effective cation exchange capacity and air-filled porosity, although none of these variables was consistent with resultant seedling growth. Adherence to gravimetric values for irrigation at an 80% water mass threshold in the container revealed that the addition of BC and WP, but not PP, required adjustments to the irrigation schedule. For future studies, we encourage researchers to provide more details about bulk density, porosity, and irrigation regime to improve the potential inference provided by this line of biochar and growing media work.

Keywords: bulk density; nursery production; growing media; nutrients; porosity; reforestation

---

1. Introduction

Deforestation is a global crisis [1–3]. As Haase and Davis [4] note, mitigating deforestation and other forms of forest degradation often requires active afforestation and reforestation, especially the outplanting of seedlings grown in nurseries. In addition, the practice of reforestation is recognized as having, among management options relying on natural pathways, the greatest potential to mitigate changes in climate [5]. Growing seedlings for reforestation in nurseries using containers is a common practice worldwide, and a prominent method in, for example, Canada, Finland, Chile, and other countries with intensive forest management activities.

While producing reforestation seedlings efficiently and economically has long been the prevailing practice, a conundrum for nursery managers is how to do so while reducing impacts to the environment. Recently, several techniques have emerged to diminish the environmental impacts of seedling production. For example, reducing irrigation needs through sub-irrigation [6,7] and efficiently applying nutrients through controlled-release fertilizer [8] or exponential fertilization [9] can reduce runoff and potential negative impacts on ground and surface water [10–12]. Using light-emitting diodes rather than more traditional energy-consuming light sources works well [13–15]. In addition, employing more sustainable organic materials to grow reforestation seedlings, such as coir [16], sawdust [17], compost [18], or composted wood bark [19] are gaining interest as growing media because they are perceived as a way to avoid issues (e.g., reduced biodiversity, increased carbon emissions) associated with traditional Sphagnum peat moss harvesting [20,21]. Moreover, local alternatives for some inorganic components of growing media, such as vermiculite or perlite that are mined and often shipped great distances, are also being sought, especially given that the costs of some commonly used amendments, such as vermiculite, continue to climb [22].

One alternative to inorganic and organic constituents in growing media for container plants is biochar. Biochar is a carbon-rich byproduct consisting of the fine-granular material remaining after pyrolysis, the process of combusting a biomass feedstock rapidly in the absence of oxygen [23]. In general, biochar properties appear conducive to plant growth in container nursery systems [24], and have shown promising potential as a replacement for peat [21,25–27] and inorganic components of media [24,28,29] in the production of container crops, including forest trees. In addition to its role as a suitable component of growing media, biochar can also provide the extra benefit of sequestering carbon (C) belowground; in addition to C storage, buried C provides enumerable ecosystem benefits through the enhancement of many biogeochemical processes [30]. As noted by Dumroese et al. [24], incorporating biochar into the growing medium becomes part of the seedling root plug, and therefore most of the expense of the transportation and burial of the carbon, a significant hindrance in many agricultural and forest situations [31,32], is already included in the overall cost of outplanting seedlings.

We previously described the potential of using pelleted biochar to grow seedlings in containers, suggesting that pelletizing biochar may be a means to avoid both the nuisance dust associated with it and its non-uniform distribution in small-volume containers typical of reforestation seedlings [24]. Our primary study objective was to evaluate different modes of biochar delivery to amend and replace Sphagnum peat moss in the production of nursery plants in containers. Therefore, we report on the growth of ponderosa pine (*Pinus ponderosa*) seedlings grown with three types of biochar (fine biochar powder, pelletized fine biochar powder as described in Dumroese et al. [24], and pyrolyzed softwood pellets) under two different supplies of nitrogen.

2. Materials and Methods

To satisfy the objectives, we grew *Pinus ponderosa* seedlings (Lolo National Forest, MT, USA, 730 m elevation) at the U.S. Department of Agriculture Forest Service, Rocky Mountain Research Station in Moscow, ID, USA (lat 46.723179, long -117.002753) in various mixtures of Sphagnum peat (peat) amended with either fine biochar powder, composite wood-biochar pellets, or pyrolyzed softwood pellets.

2.1. Media Components and Analysis of Individual Medium

The peat was a fine-textured, non-fertilized horticultural grade without a wetting agent (Sunshine grower grade green, Sun Gro Horticulture Ltd., Vancouver, BC, Canada). Biochar powder (BC) was created as a byproduct of fast pyrolysis that was produced from 1 to 2 mm particles of cellulosic biomass from mixed hardwood residues with <10% moisture, pyrolyzed at 450 to 500 °C (C-Quest biochar, Dynamotive Energy Systems Corp., Richmond, BC, Canada), and with 69% C content, 9% ash, and 2.8 m^2 g^{-1} surface area [33]. Composite wood–biochar pellets (WP) were produced at the Composite Materials and Engineering Center (Washington State University, Pullman, WA, USA) by dry blending 43% BC, 43% finely-ground *Pinus strobus* wood flour, 7% polylactic acid, and 7% wheat starch in a ribbon mixer and feeding that into a 75 kW (100 hp) commercial pellet mill fitted with a parabolic entry die with an overall length of 63.5 mm. The mill extruded random length (4 to 25 mm) pellets with an output diameter of 5.4 mm (see [24] for additional detail on material specifications and pellet output). Pyrolyzed pellets (PP) were the result of wood pellets (6 mm diameter; 5 to 15 mm length) comprised primarily of *Pseudotsuga menziesii* and *Tsuga heterophylla* that were pyrolyzed at 500 °C for 10 min (Sonofresco, Burlington, WA, USA). By hand and on a volume basis (0, 25, 50, 75, and 100%), we combined peat with BC, WP, or PP to form 13 distinct growing media (Table 1). All chemical and physical assessments were conducted at the Natural Resources Institute Finland (LUKE) facilities in Vantaa and Suonenjoki, respectively.

Table 1. Initial, mean ($n = 5$) pH, bulk density (Db), and effective cation exchange capacity (ECEC) for peat amended with biochar (BC), pyrolyzed softwood pellets (PP), and composite wood-biochar pellets (WP) at rates of 0, 25, 50, 75, and 100% ($v\,v^{-1}$). Different letters within a column indicate significant differences at $\alpha = 0.05$.

Growing Media Designation	($v\,v^{-1}$)		($w\,w^{-1}$) [a]	pH		Db ($g \cdot cm^{-3}$)		ECEC ($cmol \cdot kg^{-1}$)	
	Peat (%)	Biochar Amendment (%)	Biochar Amendment (%)						
Peat									
Peat (control)	100	0	-	3.9	g	0.099	j	49.6	a
Peat + biochar (BC)									
BC25	75	25	10	5.0	e	0.173	i	31.0	b
BC50	50	50	70	5.9	c	0.251	g	23.8	c
BC75	25	75	90	6.7	b	0.294	f	15.4	de
BC100	0	100	100	-		0.331	d	7.2	gh
Peat + pyrolized softwood pellets (PP)									
PP25	75	25	7	4.5	f	0.179	i	31.8	b
PP50	50	50	69	5.4	d	0.264	g	17.8	d
PP75	25	75	90	7.0	a	0.313	e	11.1	f
PP100	0	100	100	-		0.318	de	5.2	h
Peat + wood-biochar pellets (WP)									
WP25	75	25	44	4.4	f	0.223	h	22.7	c
WP50	50	50	81	4.7	ef	0.387	c	16.8	de
WP75	25	75	94	5.2	de	0.469	b	13.2	ef
WP100	0	100	100	-		0.527	a	10.4	fg
P values				<0.0001		<0.0001		<0.0001	

[a] Estimated from bulk density measurements.

2.1.1. Physical Properties

The particle size distribution for individual media components (peat, BC, WP, and PP) was measured using a series of sieves (0.5 to 5 mm; $n = 3$). We determined bulk density as the ratio of dry mass (dried at 105 °C) to saturated volume ($n = 5$) [34]. Particle density was estimated using an average density of 2.65 g cm^{-3} for mineral and 1.5 g cm^{-3} for organic components [34,35]. Loss-on-ignition at 550 °C for 2 h provided an approximate estimate of the organic matter for each growing medium ($n = 5$) [36]. We measured the water uptake and volume change of the growing media directly from the bag using metal cylinders (height 60 mm, diameter 58 mm) filled with each media; cylinders were placed into water kept 5 to 10 mm deep ($n = 3$) [24]. Volumetric water content (VWC) at decreasing matric potentials (i.e., desorption water retention characteristics) was measured using a pressure plate apparatus (Soilmoisture Equipment Corp., Santa Barbara, CA, USA) and standard methods [37,38]—similar metal cylinders were filled with each growing media, saturated, allowed to drain freely (to about −0.3 kPa), and then exposed to successive matric potentials of −1, −5, and −10 kPa ($n = 5$). Water content was reassessed gravimetrically at each matric potential. Our initial suction was 1 kPa because this value reflects the "container capacity", the upper limit of plant available water retained in the container following saturation and subsequent free draining of the medium [39,40].

Total porosity (TP) was estimated using:

$$TP = (Dp - Db)/Dp$$

where Dp is the particle density of the material and Db is the bulk density.

Air-filled porosity (AFP) was estimated using:

$$AFP = TP - VWC$$

where VWC is the volumetric water content at -1 kPa matric potential, assumed to be container capacity.

Unsaturated hydraulic conductivity was measured using an automated evaporation ku-pF apparatus (UGT GmbH, Müncheberg, Germany), where sample cylinders ($n = 2$) were sealed on the bottom and the top of the core was allowed to evaporate at room temperature [41,42]. Cylinders were measured every 10 min with moisture tensiometers.

2.1.2. Chemical Properties

Our measurements of total, soluble, and press water nutrient concentrations, as well as effective cation exchange capacity, were replicated 5 times. We measured total C and nitrogen (N) from sieved and air-dried samples on a CHN analyzer (LECO-1000, LECO Corp., St. Joseph, MI, USA). Samples for other elements were digested by the closed wet HNO_3-HCl digestion method in a microwave (CEM MDS-2000; CEM Corp., Matthews, NC, USA) and the extract was analyzed on an iCAP 6500 Duo ICP-emission spectrometer (Thermo Scientific Ltd., Cambridge, UK).

To assess soluble nutrients, we wetted samples of each medium and allowed them to incubate for 1, 15, or 29 days at room temperature to see how amounts of soluble nutrients change over time, especially N forms (see [24]). To mimic the wetting and drying cycles found under normal nursery cultural practices, we remoistened the samples about twice each week. For each sample date, acid ammonium acetate (pH 4.65) was used to gather soluble cations and easily soluble phosphorus (P). We quantified the cations in the filtrate using the previously described ICP-emission spectrometer. Soil ammonium (NH_4-N), nitrate (NO_3-N), and total N were determined from a KCl-extract on a FIA-analyzer (Lachat QuickChem 8000, Lachat Instruments, Milwaukee, WI, USA). Using a microwave (CEM MDS-2000 described above), we used the hot water refluxing method to extract easily soluble boron [43], quantified using the previously described ICP-emission spectrometer.

For cation exchange capacity, substrates were prepared as described for soluble nutrients. We used a 0.1 M $BaCl_2$ solution to extract exchangeable cations, and their total concentrations in the filtrate were determined using the previously described ICP-emission spectrometer. To determine exchangeable acidity, the 0.1 M $BaCl_2$ extract was titrated with a 0.05 M NaOH solution up to pH 7.8. Effective cation exchange capacity [ECEC(cmol·kg^{-1})] was then calculated using:

$$ECEC(cmol \cdot kg^{-1}) = Na(cmol \cdot kg^{-1}) + K(cmol \cdot kg^{-1}) + Ca(cmol \cdot kg^{-1}) + Mg(cmol \cdot kg^{-1}) + ACI\_E(cmol \cdot kg^{-1})$$

where ACI_E is exchangeable acidity from $BaCl_2$ extract. Percentage base saturation was calculated as the sum of the bases (Na, K, Ca, Mg) divided by ECEC.

To determine the nutrients in a press water extract after the incubation periods described above, we pressed each growing media sample in a custom apparatus consisting of a cylindrical chamber and a vertical piston that, when deployed, delivered a constant 300 kPa pressure. The resulting extracts were measured for pH and electrical conductivity, filtered, and analyzed for dissolved micro and macro elements on the previously described spectrometer. Concentrations of dissolved NH_4-N, NO_3-N, and dissolved total N were determined on the FIA-analyzer described above. Because our analysis of NO_3-N included NO_2-N, we estimated organic N (ON) using:

$$ON = N_{total} - NH_4\text{-}N - NO_3\text{-}N.$$

2.2. Seedling Culture

Our original study plan only included peat, BC, and WP; these were tested the first year. As we had the opportunity to obtain PP, we repeated the experiment the second year but limited the treatments to peat and PP because of limited resources. In neither year were seedlings grown in media comprised of 100% BC, PP, or WP.

2.2.1. Year One

In early April (Julian dates 98 and 99, hereafter Julian), each medium was hand loaded into 3 trays that each held 98 Ray Leach SC-10 Super "Cone-tainers"™ (hereafter, cell; each 3.8 cm diameter, 21 cm depth, 164 ml, 528 seedlings m^{-2}) and irrigated to container capacity. On Julian 111, three seeds were sown per cell. After germination (Julian 127), germinants were thinned to one per cell and 240 individual cells from each medium were evenly dispersed across eight trays to faciliate irrigation and fertigation (irrigation with soluble fertilizer added). Subsequently, four trays (120 seedlings) were randomly assigned to each of two soluble N treatments: 20 (low N) or 80 (based on a typical rate [17]) mg N seedling^{-1} for the growing season. Daytime greenhouse temperatures ranged from 21 to 29 °C and nighttime low temperatures were kept above 16 °C.

To avoid confounding N application and irrigation, we used exponential fertilization [17] and determined the irrigation frequency and amount gravimetrically [44]. The basic exponential fertilization equation was:

$$N_T = N_S \times (e^{rt} - 1)$$

where r is the relative addition rate required to increase N_S (initial level of N in plant) and N_T is the desired amount to be added during t, the number of fertilizer applications [45]. For both N rates, $t = 150$ (the number of days between the first and last fertigation during the growing season) and N_S was assumed to be 0.5 mg N. For the $N_T = 80$ mg N treatment, $r = 0.03388$ whereas for $N_T = 20$, $r = 0.02476$. The amount to apply on a specific day was calculated using:

$$N_T = N_S \times (e^{rt} - 1) - N_{t-1}$$

where N_T is the amount of N to apply daily, N_{t-1} is the cumulative amount of N applied, and t goes from 1 to 150. For each application, we custom-blended fertilizers, including micronutrients (Peters Professional® S.T.E.M.™. The Scotts Company, Marysville, OH, USA) and chelated Fe (Sprint 330; 10% Fe; Becker Underwood, Inc., Ames, IA, USA) to achieve these nutrient ratios: 100N ($54NO_3^-$: $46NH_4^+$): 90P: 109K: 68S: 33Mg: 3Fe: 0.3Cu: 0.3Mn: 0.7Zn: 0.2B: 0.006Mo.

For gravimetric water content, we determined the average mass of an empty tray, 30 empty cells, and their oven-dry growing medium (60 °C for 72 h). On Julian 102, each tray was weighed approximately 60 min after watering to container capacity; the mass of the container at container capacity minus the container and media mass equaled the mass of the water. Between Julian 103 and 131, cells were weighed daily at 0800 and irrigated when the water mass reached a threshold of 80% (±5 percentage points) of the water mass at container capacity [44]. Container capacity mass was recalculated monthly to adjust for media shrinkage and plant biomass. Beginning on Julian 131, seedlings were fertilized during each irrigation (fertigation). The necessary amount of fertilizer (cumulative daily amounts since the prior irrigation) was diluted in the calculated amount of water required to recharge the medium to container capacity. Fertigation solutions were carefully applied by hand to individual seedlings to ensure an even distribution of nutrients and minimize leaching. From the end of the fertigation period (early October; Julian 281) until harvest, seedlings were irrigated when the water mass reached 75% (±5 percentage points). Fourteen days after the last fertigation, greenhouse temperatures were allowed to go ambient but above freezing (4 to 10 (day)/2 to 4 °C (night)).

Eight randomly-selected seedlings (two from each tray) from each medium × fertilizer combination were sampled on Julian 328. We measured height and stem diameter at the root collar (RCD). Shoots were separated from roots, roots were gently washed free of media, and roots and shoots were dried 72 h at 60 °C to determine biomass. Tissue samples were analyzed for macro-and micro-nutrient concentrations by JR Peters Laboratory (Allentown, PA, USA).

2.2.2. Year Two

We used the same seed and peat sources and followed the methods described above, except that BC was not repeated and PP replaced WP. Due to logistical constraints, seeds were sown on Julian 165 and fertigation commenced on Julian 182. Therefore, the exponential fertilization period was shortened to t = 93; thus r = 0.0546 for N_T = 80, and r = 0.0399 for N_T = 20. On Julian 311 seedlings were sampled and analyzed as described above.

2.3. Statistical Analyses and Visualizations

We used generalized linear mixed models (GLIMMIX) within SAS (version 9.4 Software; SAS, Inc., Cary, NC, USA) to compare treatment means using the Gaussian response distribution and the default covariance matrix format. Type III tests were utilized. We used Tukey–Kramer adjustments for post-hoc multi-comparison tests of the differences between model means.

GLIMMIX tested for differences among the biochar types (BC, PP, WP) and peat for media physical and chemical properties. For seedlings, we previously speculated [24] that peat amended with ≥50% WP would likely experience too much expansion when wetted to be a valid treatment in a nursery. Indeed, when wetted in the current experiment, WP ≥50% expanded and split the cells. Subsequently, we were unable to control water loss (evaporation as well as fertigation) through the ruptures, and although we continued to culture the seedlings, the result was extremely poor growth. Thus, seedling growth in WP50 and WP75 was excluded from analysis.

Seedling biomass and soil chemistry data was relativized using response ratios in order to reduce variation between the two years [46]. The response ratio is the difference between the natural logarithm for each biomass variable (shoot height, stem diameter at the root collar, shoot and root dry biomass) and soil chemistry variable (media C, N, pH and electrical conductivity (EC)) and the natural logarithm for each biomass, soil chemistry, or VWC control (100% peat treatments). Seedling biomass response ratios were analyzed using GLIMMIX, accounting for the split-plot design by including the nitrogen treatment as the whole plot followed by media treatment as the split-plot (*n* = 9) before comparing variable means.

Visualizations, including vector diagrams that allow for the robust presentation and comparison of relative values [47], were created using SigmaPlot (version 13.0; Systat Software, San Jose, CA, USA).

3. Results

3.1. Media Characteristics

3.1.1. Physical Properties

The mean particle sizes of peat were the most evenly distributed, with all size classes well represented except for >5 mm (Table 2). In contrast, most (99%) of the BC had a particle size ≤1 mm, whereas for pellets (PP and WP) most (85%+) of the particles were >2 mm, and for PP nearly half were >5 mm. Peat had the lowest Db (0.099 g cm^{-3}) and BC and PP had a similar Db at each added proportion, ranging from about 0.176 g cm^{-1} at the 25% level to about 0.323 g cm^{-3} at 100%; and WP had the highest Db at each added proportion, ranging from 0.223 to 0.527 g cm^{-3} as the proportion of WP increased in the media from 25 to 100%, respectively (Table 1). Organic matter (%) significantly decreased as the amount of peat replaced by individual biochar-based components increased (Figure 1). Across the components, peat had the greatest level of organic matter, followed by WP, and finally BC and PP.

Table 2. Mean particle size distribution (%) of the peat, biochar powder (BC), pyrolyzed softwood pellets (PP), and composite wood−biochar pellets (WP) (*n* = 3).

	Mean Particle Size Distribution (%)				
	(mm)				
	<0.5	0.5−1	1−2	2−5	>5
Peat	30.8	22.7	27.6	13.3	5.6
BC	92.5	6.6	0.7	0.2	0.0
PP	4.7	2.5	2.4	44.9	45.5
WP	8.0	2.7	4.3	65.9	19.2

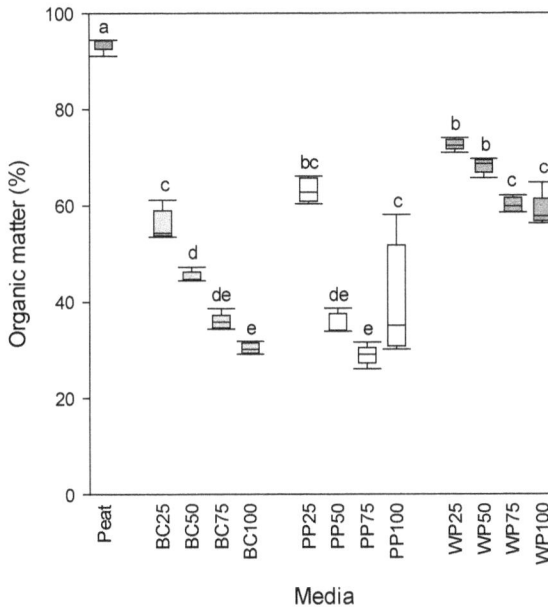

Figure 1. Organic matter (*n* = 5) for peat and peat amended with biochar powder (BC), pyrolyzed softwood pellets (PP), and composite wood−biochar pellets (WP) at rates of 25, 50, 75, and 100% ($v\,v^{-1}$). Vertical boxes represent approximately 50% of the observations and lines extending from each box are the upper and lower 25% of the distribution. The solid horizontal line in the center of each box is the median value. Different letters indicate significant differences at α = 0.05.

When initially exposed to water, all growing media absorbed water with the exception of BC100 (data not shown). During the first 5 min, BC25 and BC50 absorbed only about one-fourth and one-fifth that of peat, respectively. Conversely, absorption doubled or tripled for PP ≤75 compared to peat and absorption values for WP25 and WP50 were similar to peat. Upon initial wetting of the media to container capacity, only WP50, WP75, and WP100 showed an increase in volume (≈12 to 27%) (Figure 2). Conversely, the shrinkage in peat was about 9%. The addition of BC ≤75% and any addition of PP (except PP50) decreased the shrinkage relative to 100% peat.

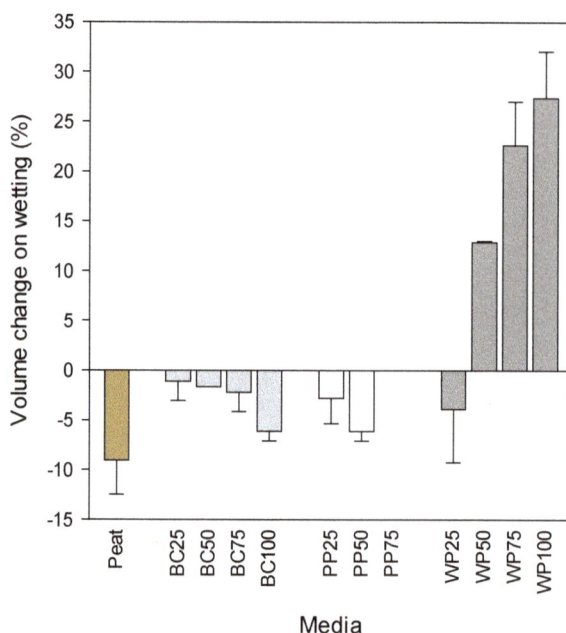

Figure 2. Change (percentage points) of bale-dry sample volumes during wetting in cylinders from below (n = 3; mean ± standard deviation). Peat was amended with biochar powder (BC), pyrolyzed softwood pellets (PP), and composite wood−biochar pellets (WP) at rates of 0, 25, 50, 75, and 100% ($v\,v^{-1}$). PP75 had no change (all values were zero) and PP100 was not measured.

For peat, the water conductivity occurred at the highest matric potential (−0.3 kPa) but the rate was variable (1 to 10 cm day^{-1}), declining steadily once the matric potential dropped to −10 kPa (Figure 3). BC50 and WP25 also showed consistent conductivity of about 1 cm day^{-1} at the highest potential. While BC50 followed a similar trend to peat, conductivity in WP25 began a steady decline at about −10 kPa. Water moved about 1 cm day^{-1} in PP50 at matric potentials between −1 and −10 kPa. BC25 and PP25 had little conductivity at matric potentials <−7 kPa, whereas WP50 had little conductivity at matric potentials <−5 kPa.

Once brought to container capacity, the subsequent volumes of the media during drying from −1 to −10 kPa varied. The volume of peat at each matric potential decreased (94.2 to 90.7 to 89.1% for −1, −5, and −10 kPa, respectively), and each volume was significantly lower than any biochar-amended media (Figure 4). BC25 and WP25 displayed the next greatest amount of shrinkage, significantly more than the other BC and WP rates, and all PP. In general, when the proportion of any biochar was ≥50%, the changes in volume were small (<5% shrinkage to <4% swelling). At −1 kPa, VWC, in general, decreased as the amount of biochar amendment increased (Figure 5). Amending peat with BC significantly reduced air-filled porosity (AFP) compared to all other treatments (about a 65% reduction compared to peat). AFP in peat, peat amended with up to 50% PP, and all rates of WP were fairly similar (28 to 38%); higher rates of PP (75 and 100%) increased the AFP by about 34 and 75%, respectively, compared to peat.

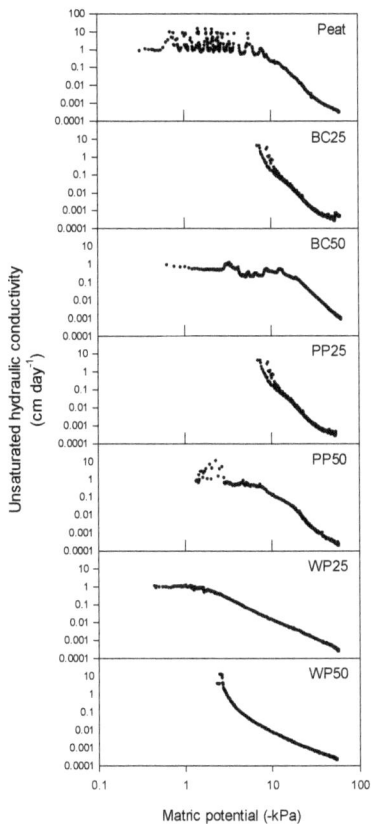

Figure 3. Unsaturated hydraulic conductivity (*n* = 2) for peat amended with biochar powder (BC), pyrolyzed softwood pellets (PP), and composite wood−biochar pellets (WP) at rates of 0, 25, and 50% ($v\,v^{-1}$).

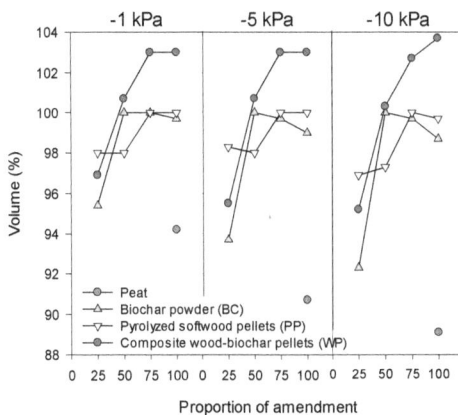

Figure 4. Media sample volumes at three matric potentials in relation to the initial wet volumes (=100%) (*n* = 5) for peat amended with biochar powder (BC), pyrolyzed softwood pellets (PP), and composite wood−biochar pellets (WP) at rates of 0, 25, 50, 75, and 100% ($v\,v^{-1}$).

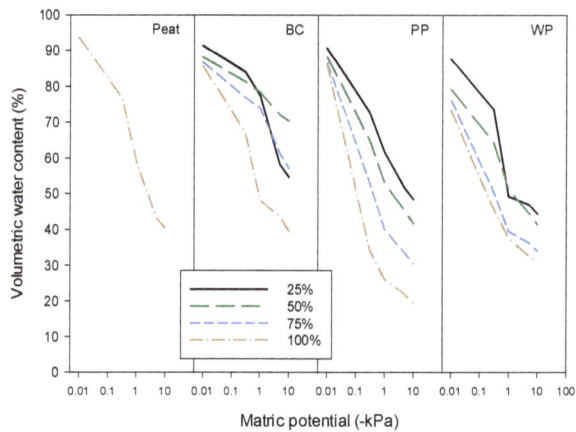

Figure 5. Mean desorption water retention characteristics of the growing media (peat amended with biochar powder (BC), pyrolyzed softwood pellets (PP), and composite wood–biochar pellets (WP) at rates of 0, 25, 50, 75, and 100% ($v\,v^{-1}$) in relation to the initial wet volume means (n = 5). At estimate of total porosity is plotted as water content at −0.01 kPa, with air-filled porosity determined as total porosity less volumetric water content at each matric potential.

3.1.2. Chemical Properties

All four media components (peat, BC, PP, and WP) had significantly ($P < 0.0001$) different amounts of C (53, 74, 91, and 59% for peat, BC, PP, and WP, respectively). For N, peat had the greatest concentration (1.3%), significantly ($P < 0.0001$) more than BC and PP, which had similar values of 0.37 and 0.45%, respectively, which were statistically greater than WP (0.23%).

Peat had an initial pH of 3.9 (Table 1). Additions of biochar in any form increased the pH and the media with the most biochar also had the highest pH. Nitrogen content varied among media (Table 3); total N in the media containing pure biochar (either BC or PP) followed the same trend, with total N decreasing with increasing amounts of amendment. The opposite result was noted for WP. PP had, in general, greater total N and more ammonium at each amendment rate than BC. WP had minor amounts of ammonium regardless of the amendment rate. Conversely, WP had higher amounts of organic N compared to either BC or WP, which had similar amounts. Low levels of nitrate were observed across all media and amendment levels. The levels of soluble elements varied by media. Compared to peat, amending with biochar in any form reduced the amounts of calcium, magnesium, manganese (Mn), and sulfur and increased the levels of boron (B) and potassium (K) (Table 4). Low levels of heavy metals (cadmium, chromium, copper, nickel, and lead) were observed in the press water extract regardless of the amendment level (Table 4).

The effective cation exchange capacity (ECEC) was greatest in pure peat (Table 1). The addition of 25% $v\,v^{-1}$ of any amendment significantly decreased ECEC by 37 to 46%, and each additional 25% $v\,v^{-1}$ increase further decreased ECEC. Pure amendment had, on average, just 15% of the total ECEC of pure peat.

Table 3. Presswater extracts of ammonium (NH_4), total nitrogen, nitrate (NO_3), and organic nitrogen. All measured mg L^{-1}. Ammonium and total N ($n = 15$ for each media) include all sampling days as the incubation day and the interaction with the media type was not significant ($P > 0.05$). Nitrate and organic N did have significant interactions between the media and date ($P < 0.05$), so the differences between each media treatment are shown for the three incubation dates. Different letters within a column indicate significant differences at $\alpha = 0.05$.

Media	Total N		NH$_4$		NO$_3$						Organic N					
---	---	---	---	---	Day 1		Day 15		Day 29		Day 1		Day 15		Day 29	
Peat	17.9	a	13.0	a	1.60	a	0.70	a	0.27	a	5.04	d	3.85	d	3.19	c
BC25	7.5	bc	1.1	bc	1.53	ab	0.01	b	0.01	b	5.60	d	5.91	cd	6.30	b
BC50	6.9	bc	0.3	c	0.93	abc	0.01	b	0.01	b	4.56	d	7.00	cd	7.43	b
BC75	3.5	c	<0.1	c	0.43	c	0.01	b	0.01	b	2.52	e	3.99	d	3.37	c
PP25	19.7	a	15.3	a	0.03	c	0.03	b	0.03	ab	5.73	d	3.81	d	3.63	c
PP50	10.8	b	6.7	b	0.02	c	0.02	b	0.04	ab	4.66	d	3.74	d	3.75	c
PP75	5.5	bc	1.8	bc	0.07	c	0.04	b	0.03	ab	4.24	de	3.79	d	2.99	c
WP25	9.3	bc	<0.1	c	0.49	bc	0.01	b	0.01	b	10.67	c	8.76	c	7.96	b
WP50	17.7	a	0.2	c	1.53	ab	0.05	b	0.06	ab	22.20	b	13.51	b	15.31	a
WP75	19.7	a	0.1	c	1.02	abc	0.26	ab	0.07	ab	27.45	a	17.70	a	14.28	a

Table 4. Mean total element concentrations (mg kg^{-1}) in peat, biochar powder (BC), pyrolyzed softwood pellets (PP), and composite wood−biochar pellets (WP) prior to mixing the growing media ($n = 5$); soluble nutrients (mg kg^{-1}) in each growing media after 29 days of moist incubation ($n = 5$); and elements in the press water extract (mg L^{-1}) of each growing media after 29 days of moist incubation ($n = 5$).

	Al	B	Ca	Cd	Cr	Cu	Fe	K	Mg	Mn	Na	Ni	P	Pb	S	Zn
Total																
Peat	1036	5.5	6615	0.11	1.4	2.3	1619	446	1131	150	82	1.8	523	2.2	2111	22
BC	164	17.0	4694	<0.11	98.4	8.0	1108	4340	509	139	82	10.7	179	3.3	117	16
PP	178	8.2	999	<0.11	0.9	3.1	230	1868	165	100	856	0.5	238	<2.1	45	9.6
WP	93	10.6	2642	<0.11	98.7	5.8	1168	2540	363	98	148	5.2	277	2.6	140	18
Soluble																
Peat	26	2.8	5347	-	-	-	9.5	361	1045	1344	136	-	91	-	338	-
BC25	18	5.8	3666	-	-	-	8.9	2570	435	76	79	-	44	-	156	-
BC50	16	6.6	3062	-	-	-	12.8	3044	274	62	64	-	36	-	104	-
BC75	13	7.0	2440	-	-	-	22.2	3040	177	53	52	-	32	-	77	-
BC100	3	7.1	1522	-	-	-	93.9	2890	120	41	43	-	25	-	51	-
PP25	11	3.9	3614	-	-	-	3.7	1081	780	101	426	-	106	-	254	-
PP50	6	4.5	1928	-	-	-	1.2	1512	333	65	586	-	103	-	152	-
PP75	6	5.1	1074	-	-	-	0.6	1580	168	48	634	-	114	-	102	-
PP100	11	5.4	809	-	-	-	2.3	1524	114	43	621	-	117	-	91	-
WP25	20	5.7	2492	-	-	-	11.8	1408	360	67	140	-	87	-	109	-
WP50	17	6.2	1964	-	-	-	14.6	1680	267	57	140	-	99	-	72	-
WP75	13	6.2	1456	-	-	-	31.9	1690	191	46	131	-	106	-	49	-
WP100	5	6.2	1186	-	-	-	48.3	1590	137	39	128	-	108	-	57	-
Press water extract																
Peat	0.38	0.1	20	<0.001	0.01	0.000	0.5	11	7	1	7	0.00	5	<0.015	34	0.04
BC25	0.53	0.5	7	<0.001	0.05	0.013	1.7	93	1	<1	5	<0.01	8	<0.015	20	0.02
BC50	0.67	0.5	12	<0.001	0.12	0.037	2.9	155	1	<1	6	0.02	10	<0.015	7	0.02
BC75	0.12	0.3	42	<0.001	0.07	0.005	1.7	250	7	1	8	0.01	4	<0.015	3	<0.01
BC100	-	-	-	-	-	-	-	-	-	-	-	-	-	-	-	-
PP25	0.39	0.3	6	<0.001	<0.01	0.006	0.5	60	2	<1	39	<0.01	21	<0.005	44	0.02
PP50	0.40	0.4	4	<0.001	<0.01	0.007	0.5	78	1	<1	51	<0.01	30	<0.005	38	0.02
PP75	0.30	0.3	3	<0.001	<0.01	0.007	0.3	75	1	<1	52	<0.01	19	<0.005	24	0.02
PP100	-	-	-	-	-	-	-	-	-	-	-	-	-	-	-	-
WP25	0.58	0.4	23	<0.001	0.04	0.009	2.7	135	5	1	16	0.01	16	<0.015	27	0.07
WP50	1.75	0.6	53	<0.001	0.12	0.005	6.3	368	14	1	38	0.03	36	<0.015	28	0.18
WP75	1.12	0.6	77	<0.001	0.11	0.008	6.9	430	20	2	4	0.04	30	<0.015	20	0.26
WP100	-	-	-	-	-	-	-	-	-	-	-	-	-	-	-	-

3.2. Seedling Growth

Although the media and N fertilization rate interacted to affect RCD, shoot biomass, and root biomass measured at the end of the experiment (Table 5), N fertilization as an independent variable was not significant. This is likely an artifact of analysis because the morphological values of seedlings from the biochar-amended media were normalized to the control for each year, and the pattern of growth was similar for each level of N (Figure 6). We noted no significant differences in the morphological attributes for the control and seedlings grown with ≤50% biochar (all $P > 0.05$), with the exception of WP, where a 25% addition dramatically reduced all morphological parameters relative to the 100% peat control. For BC, the higher rate of N in combination with a 25% addition yielded similar results (95 to 108%) to the control for all morphological traits, as did the addition of PP at either 25% or 50% (91 to 107%). Moreover, with the higher N rate, BC25, BC50, and PP25 had similar shoot N concentrations (96 to 100% of the control), whereas PP50 had 86% of the control.

Table 5. *P*-values for final seedling morphological characteristics.

Independent Variables	Height	Stem Diameter	Shoot Biomass	Root Biomass
N fertilization (F)	0.2672	0.1341	0.0784	0.6250
Medium (M)	<0.0001	<0.0001	<0.0001	<0.0001
F × M	0.6368	0.0143	<0.0001	0.0335

For the most part, the concentrations of macro-and micro-nutrients in the shoots, regardless of amendment or N rate, were within the standard ranges for conifer seedlings [48,49]. Iron (Fe), B, and Mn were most affected. Seedlings grown with BC25 and BC75 and 20 mg N had 41 to 340% more Fe than the control, which exceeded (by about 40%) the high end of the recommendation range (200 ppm). All seedlings grown with PP and receiving 80 mg N had B values 10 to 40% higher than the peat control (4 to 22% higher than the 100 ppm recommendation). Mn was high across all treatments; only the two amendments with 75% biochar, BC75 and PP75, fell within the recommended range of 100 to 250 ppm. All others ranged from 300 to 640 ppm. For all treatments, molybdenum fell within the recommended range (0.05 to 5 ppm), but the control peat had the lowest concentrations (0.05 ppm), whereas all biochar-amended treatments ranged from 0.1 to 4.2 ppm with an average of 1.2 ppm.

The substrates affected the number of irrigations required using a water mass threshold of 80% of container capacity. We observed less irrigation events for the BC and WP treatments, whereas PP required about the same number as pure peat (Table 6). The BC irrigation frequency was reduced from 12 to 25%, with the greatest reduction noted when 50% of the peat was replaced.

Table 6. Relative number of irrigation events for peat, biochar powder (BC), pyrolyzed softwood pellets (PP), and composite wood−biochar pellets (WP) substrates using a target water mass of 75%.

	Percentage of Peat Replaced ($v\,v^{-1}$)			
	0	25	50	75
Peat	100			
BC		88	75	88
PP		101	102	100
WP		58	—	—

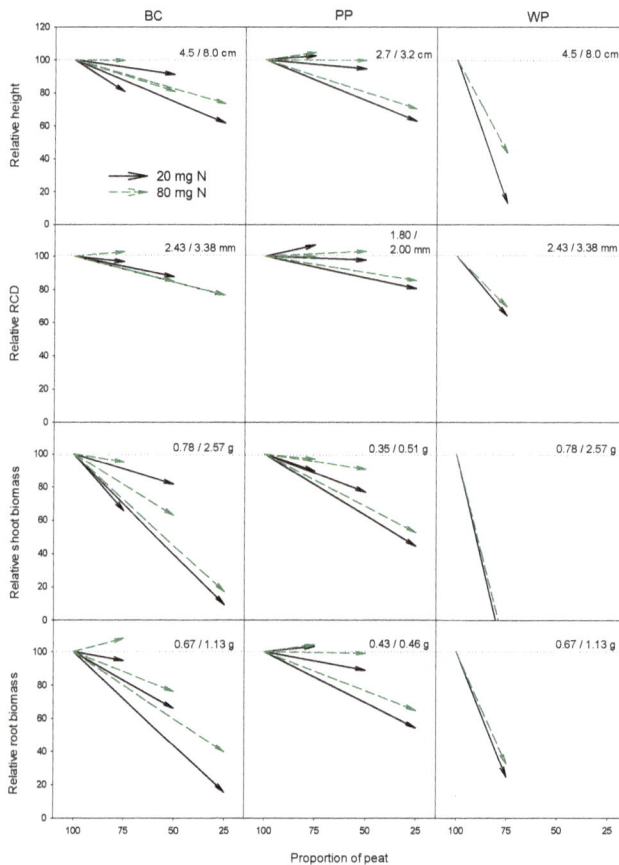

Figure 6. Vectors representing the changes in relative morphological values (height; root-collar diameter, RCD; and root and shoot biomass) for seedlings grown in peat (endpoint) and peat amended with biochar powder (BC), pyrolyzed softwood pellets (PP), and composite wood−biochar pellets (WP) at rates of 25, 50, and 75% ($v\,v^{-1}$; arrows) and supplied with either 20 (solid black line) or 80 (dashed green line) mg N. The control means for each N rate (low/high) are provided.

4. Discussion

For healthy seedling growth, media pH, CEC, inherent fertility, and porosity are some of the most important aspects. Our peat substrate had the lowest pH (3.9), but was typical for Sphagnum [50]. The lowest rates of biochar (25%) and peat had pH values lower than the range of 5.5 to 6.5 recommended for most woody plants for restoration [48], although it was within the recommended range for conifers grown in pure peat (4.5 to 5.5 [51]). Replacement of 50% of the peat with the biochar amendments yielded values within the Landis et al. [48] range, and replacement of 75% peat exceeded that range. Bunt [50] notes, however, that most plants can be grown across a wide range of pHs provided that nutrients are appropriately supplied.

All of the components within our media were organic. Organic matter (OM) increases the cation exchange capacity of native soils by increasing the number of negative exchange sites available to retain nutrients. Therefore, it is interesting to note the decline in OM across all amendment combinations as compared to the peat (Figure 1). Our samples were analyzed by Ball's [36] loss-on-ignition and this method is routinely used for soil samples. This method may have shown lower levels of OM in the biochar treatments because charcoal is resistant to further heating and mass loss. Biochar (or black carbon) is not easy to volatilize [52] and, therefore, other thermal or chemical methods may be a better way to assess the contribution of carbon to the amendments. Despite not being able to categorize OM adequately, biochar is unique in that it has a high cation exchange capacity, which can significantly increase nutrient retention because of the higher surface charge [53]. However, the direct evidence of biochar's influence on nutrient cycling and retention in soils is inconsistent [54]. For example, biochar may accelerate nutrient cycling in the long-term and serve as a short-term source of highly available nutrients [55]. Many of the changes in nutrient cycling are related to specific biochars (e.g., feedstock, pyrolysis temperatures) and how they age within the soil matrix. Very little is known about the nutrient exchange from biochar in a nursery setting.

During nursery production, a high cation exchange capacity is desired because it mitigates the leaching of nutrients during irrigation, which maintains a high level of substrate fertility [48]. Earlier we reported that replacing 25% ($v\ v^{-1}$) peat with WP reduced the effective cation exchange capacity (ECEC) by about 50% [24]; here we found that replacing 25% peat with either BC or PP only reduced ECEC by about a third (Table 1). These changes in ECEC did not, however, result in large differences in observed shoot nutrient concentrations (data not shown); we believe that our strict adherence to irrigation applied at discrete thresholds, hand application, and the use of exponential fertilization to ensure that all treatments received the same level of N, may have reduced any potential negative effects of nutrient leaching during fertigation [17,44].

Compared to peat, we noted high levels of soluble K when any amount and type of biochar was used (Table 4), as well as a decreases in soluble Mg, and this was also apparent in the press water extracts. High values of K have also been noted by others, with suggestions that biochar may serve as the sole source of K in container production systems [28,56–58]. We noted increases in shoot K concentrations of 6 to 31% when BC or WP replaced peat (which yielded an average value of 0.93% K), but the values when PP was added were more modest (zero to +4%). While using biochar as the sole source of P has also been suggested [56] and increased nutrient concentrations have been observed with 10% $v\ v^{-1}$ [56] and $\leq$35% $w\ w^{-1}$ [58], we only noted increases (of about 15%) with PP concentrations $\leq$50%. While high rates of K were associated with Mg deficiency in *Pinus radiata* [59], we noted that our combination of biochar and fertigation programs yielded shoot Mg concentrations 4 to 50% higher than the peat, with the exception of PP50 and PP75, which had 7 to 11% reductions, respectively. Despite these findings, the values were generally similar to peat (0.12% Mg versus 0.11% Mg) and within the suggested range of Landis et al. [48]. Although we did not specifically test whether biochar could provide sufficient P and K for seedling growth, our varied results across biochars and proportions suggest that when appropriate nutrition is provided through fertigation, addition by biochar are probably not sufficient to be excessive, and that reliance on biochar as a fertilizer will be biochar-specific.

In his review, Heiskanen [60] suggests that an air-filled porosity (AFP) at −1 kPa near 40% is an optimum threshold for container reforestation seedlings, and later determined that 50% of the TP is about optimum WC and AFP for any medium [18]. In this study, the peat had an AFP of about 35%, and replacing the peat with PP yielded media with an AFP ranging from 29 to 47% (increasing with the increasing addition of biochar; Figure 5). These treatments also required similar intervals of irrigation (Table 6), suggesting similar water and air availability to seedlings among the range of amendments. In contrast, the replacement of peat with BC generated media with a very low AFP (14, 10, and 13% as the amendment increased from 25 to 50 to 75%). This higher proportion of water-holding capacity at the expense of air-filled porosity is reflected in the decreased frequency of required irrigation (Table 6); notably the lowest AFP treatment (BC50) required the fewest irrigation events. WP25, despite having a near-optimum AFP (39%), required the least number of irrigations. Heiskanen [60] cautions, however, that water-and air-filled porosities "do not actually or commensurably describe the availability of air or water to the roots in all media". Accordingly, we observed good growth of the seedlings in BC25 given the higher rate of N despite the low AFP, and less satisfactory growth of PP75 seedlings and very poor growth of WP25 seedlings despite a near-optimum AFP. Other factors, such as bulk density (Db), likely have an effect, given that BC25 had a relatively low Db and PP75 had a relatively high Db. Certainly a low Db is important. Vaughn et al. [26], working with cultivars of tomatoes (*Solanum lycopersicum*) and marigolds (*Tagetes erecta*), and biochar substrates ($\leq$15% $v\,v^{-1}$) with fairly similar Db (0.13 to 0.17) and AFP (24 to 29%), observed few differences in plant growth with the exception of tomato height. In a second experiment with the same species, Vaughn et al. [21] found that biochar mixtures with the greatest AFP (about 47%) yielded the highest amount of biomass for each species. In addition, Conversa et al. [61] reported very good seedling growth with biochar additions up to 70% ($v\,v^{-1}$); as the biochar additions increased from zero to 70%, Db shifted upward from 0.13 to 0.16 g cm^{-3} and the AFP increased from 13 to 21%.

Our results, similar to those of several others [21,25–27,61,62], suggest that acceptable plant growth can often be achieved when peat-based substrates are replaced with suitable biochar forms $\leq$50% ($v\,v^{-1}$). In addition, it is important to consider that in an operational setting and on an annual basis, prudent nursery managers adjust cultural practices to ensure target seedling growth [63,64], and a similar approach would be sensible when incorporating biochar into the growth medium. In their review of the association between biochar and plant diseases, Frenkel et al. [65] caution, however, that biochar rates exceeding 3% ($w\,w^{-1}$) were more conducive to disease (our 25% $v\,v^{-1}$ rates ranged from about 7 to 44% $w\,w^{-1}$; see Table 1). The authors note that soil-borne pathogens were commonly enhanced in 83% of the studies they reviewed, but foliar pathogens were enhanced in only 33% of the studies. For forest nurseries in western North America, soil-borne pathogens (i.e., *Cylindrocarpon*, *Fusarium*, and *Pythium*) are ubiquitous (e.g., [66]), but the expression of disease is usually only associated with prolonged, excessive moisture in the growing media (e.g., [66–69]) often due to excessive irrigation. In addition, the basal portion of all containers, post irrigation, experience saturated conditions for some duration, which is a function of plant phenology, container height, and medium porosity [60]. Too frequent irrigation, even if applied to "maintain container capacity", can prolong this saturated condition, particularly for media with lower porosity, as can be found when biochar is added, and the resulting anaerobic conditions can be stressful to seedlings [6,69,70]. Several studies reviewed by Frenkel et al. [65] that show enhanced disease expression with higher rates of biochar provide either scant, ambiguous, or solely qualitative estimates on how irrigation was managed during the experiments. This is unfortunate, given that Heiskanen [18] notes that when peat-based media are amended, particularly with organic components, irrigation should be adjusted for each mixture to achieve the correct water, oxygen, and nutrient availability. Indeed, Matt et al. [27] found that after increasing the volumetric proportions of biochar powder (same as the BC used in this study) in a well-drained, peat-based substrate (3:1:1 v:v:v peat, perlite, vermiculite), the irrigation frequency required to achieve similar water mass across treatments during the course of the experiment was reduced. That is, due to the specific water retention characteristics of the biochar treatments, those biochar treatments required less frequent irrigation (about 40% for the highest rate of biochar) compared to the more well-drained peat-based substrate. Our results were less straightforward, but we still noted a 12 to 25% difference in irrigation frequency among our biochar treatments. Given that frequent irrigation to container capacity of the media with higher water retention increases the risk of waterlogging [71], the elevated occurrence of disease associated with higher rates of biochar (with its subsequent higher water retention) may be a function of poor irrigation management.

While irrigation and fertilization methods are often poorly described in studies evaluating biochar and its impacts on disease expression, the same is true for published studies evaluating seedling performance when grown in biochar-amended substrates. As concluded by Pinto et al. [72], applying nursery culture without regard for the intrinsic nature of the differences provided by the treatments, for example, irrigating plants with a range of biochar additions every three days regardless of water availability, only evaluates the irrigation practice, not the true potential of the treatment (in this example, biochar). Thus, more attention to irrigation and fertilization practices that avoid confounding should be practiced. Irrigation can be easily managed by measuring water mass loss [44] and is an effective technique to reduce confounding irrigation and fertilization in greenhouse trials (e.g., [17]).

5. Conclusions

We evaluated replacing peat with three types of biochar (BC, powder; PP, pyrolyzed softwood pellets; WP, composite wood-biochar powder pellets) up to 75% ($v\,v^{-1}$) and under two exponential fertilization regimes that supplied either 20 or 80 mg N during the course of the experiment. Exponential fertilization and gravimetric determination of water loss from the media were used to avoid confounding these variables across biochar types and proportions. Seedling growth patterns were similar for either N supply, suggesting that biochar alone has little effect on the overall substrate fertility. Additions of 25% (BC) and up to 50% (PP) with concurrent application of 80 mg N yielded seedlings with similar growth to the peat control. Worldwide, studies have demonstrated mixed responses in terms of plant growth when biochar was a component of the growing media. A better understanding of the potential for biochar as a nursery substrate may be achieved through proper irrigation and fertilization techniques and the reporting of basic media characteristics, in particular bulk density and air-filled porosity.

Author Contributions: R.K.D. and J.H. conceived the experiment; R.K.D., J.R.P., J.H., and D.S.P-D. designed the experiment; D.S.P-D. provided biochar; K.E. designed the composite wood-biochar pellets; J.H. completed the physical analyses of the growing media and their components; A.T. completed the chemical analyses of the growing media and their components; R.K.D. and J.R.P. cultured the seedlings; K.G.M. analyzed the data; R.K.D. wrote the first draft; all authors reviewed and provided comments to improve subsequent versions of the manuscript.

Acknowledgments: This research developed from conversations during R.K.D.'s sabbatical to the Finnish Forest Research Institute (METLA; now Natural Resources Institute Finland [LUKE]) in Suonenjoki, Finland, formalized with agreement 09-CO-11221633-158 and subsequently supported through agreements 10-IJ-11221633-192 (METLA), 13-CR-11221633-127 (Washington State University), and 14-JV-11221633-042 (University of Idaho). Primary support was provided by the U.S. Department of Agriculture Forest Service (USFS) Rocky Mountain Research Station (RMRS) and the USFS National Center for Reforestation, Nurseries, and Genetic Resources. We thank Jake Kleinknecht and Janelle Meyers for tending the seedlings and processing samples, and L. Scott Baggett, RMRS statistician, for assistance with data analysis. The views expressed are strictly those of the authors and do not necessarily represent the positions or policy of their respective institutions.

Conflicts of Interest: The authors declare no conflict of interest.

References

1. Chazdon, R.L. Beyond deforestation: Restoring forests and ecosystem services on degraded lands. *Science* **2008**, *320*, 1458–1460. [CrossRef] [PubMed]
2. Stanturf, J.A.; Palik, B.J.; Dumroese, R.K. Contemporary forest restoration: A review emphasizing function. *For. Ecol. Manag.* **2014**, *331*, 292–323. [CrossRef]
3. Stanturf, J.A.; Palik, B.J.; Williams, M.I.; Dumroese, R.K.; Madsen, P. Forest restoration paradigms. *J. Sustain. For.* **2014**, *33*, S161–S194. [CrossRef]
4. Haase, D.L.; Davis, A.S. Developing and supporting quality nursery facilities and staff are necessary to meet global forest and landscape restoration needs. *Reforesta* **2017**, *4*, 69–93. [CrossRef]
5. Griscom, B.W.; Adams, J.; Ellis, P.W.; Houghton, R.A.; Lomax, G.; Miteva, D.A.; Schlesinger, W.H.; Shoch, D.; Siikamäki, J.V.; Smith, P.; et al. Natural climate solutions. *Proc. Natl. Acad. Sci. USA* **2017**, *144*, 11645–11650. [CrossRef] [PubMed]
6. Heiskanen, J. Effect of subirrigation on the growth of Norway spruce container seedlings in a greenhouse: A pilot study. *J. Appl. Irrig. Sci.* **2007**, *42*, 19–28.
7. Schmal, J.L.; Dumroese, R.K.; Davis, A.S.; Pinto, J.R.; Jacobs, D.F. Subirrigation for production of native plants in nurseries—Concepts, current knowledge, and implementation. *Nativ. Plants J.* **2011**, *12*, 81–93. [CrossRef]
8. Shaviv, A.; Mikkelsen, R.L. Controlled-release fertilizers to increase efficiency of nutrient use and minimize environmental degradation: A review. *Fert. Res.* **1993**, *35*, 1–12. [CrossRef]

9. Dumroese, R.K.; Page-Dumroese, D.S.; Salifu, K.F.; Jacobs, D.F. Exponential fertilization of *Pinus monticola* seedlings: Nutrient uptake efficiency, leaching fractions, and early outplanting performance. *Can. J. For. Res.* **2005**, *35*, 2961–2967. [CrossRef]

10. Dumroese, R.K.; Page-Dumroese, D.S.; Wenny, D.L. Managing Pesticide and Fertilizer Leaching and Runoff in a Container Nursery. In Proceedings of the Intermountain Forest Nursery Association, Park City, UT, USA, 12–16 August 1991; Landis, T.D., Ed.; USDA Forest Service, Rocky Mountain Forest and Range Experiment Station: Fort Collins, CO, USA, 1992; pp. 27–33. Available online: https://www.researchgate.net/publication/272819594 (accessed on 15 March 2018).

11. Dumroese, R.K.; Wenny, D.L.; Page-Dumroese, D.S. Nursery Waste Water: The Problem and Possible Remedies. In Proceedings of the National Proceedings, Forest and Conservation Nursery Associations, Kearney, NE, USA, 7–11 August 1995; Landis, T.D., Cregg, B., Eds.; USDA Forest Service, Pacific Northwest Research Station: Portland, OR, USA, 1995; pp. 89–97. Available online: https://www.researchgate.net/publication/272819463 (accessed on 15 March 2018).

12. Juntunen, M.-L.; Hammar, T.; Rikala, R. Leaching of nitrogen and phosphorus during production of forest seedlings in containers. *J. Environ. Qual.* **2002**, *31*, 1868–1874. [CrossRef] [PubMed]

13. Apostol, K.G.; Dumroese, R.K.; Pinto, J.R.; Davis, A.S. Response of conifer species from three latitudinal populations to light spectra generated by light-emitting diodes and high-pressure sodium lamps. *Can. J. For. Res.* **2015**, *45*, 1711–1719. [CrossRef]

14. Riikonen, J.; Kettunen, N.; Gritsevich, M.; Hakala, T.; Särkkä, L.; Tahvonen, R. Growth and development of Norway spruce and Scots pine seedlings under different light spectra. *Environ. Exp. Bot.* **2016**, *121*, 112–120. [CrossRef]

15. Montagnoli, A.; Dumroese, R.K.; Terzaghi, M.; Pinto, J.R.; Fulgaro, N.; Scippa, G.S.; Chiatante, D. Tree seedling response to LED spectra: Implications for forest restoration. *Plant Biosyst.* **2018**, *152*, 515–523. [CrossRef]

16. Rose, R.; Haase, D.L. The use of coir as a containerized growing medium for Douglas-fir seedlings. *Nativ. Plants J.* **2000**, *1*, 107–111. [CrossRef]

17. Dumroese, R.K.; Page-Dumroese, D.S.; Brown, R.E. Allometry, nitrogen status, and carbon stable isotope composition of *Pinus ponderosa* seedlings in two growing media with contrasting nursery irrigation regimes. *Can. J. For. Res.* **2011**, *41*, 1091–1101. [CrossRef]

18. Heiskanen, J. Effects of compost additive in sphagnum peat growing medium on Norway spruce container seedlings. *New For.* **2013**, *44*, 101–118. [CrossRef]

19. Villa Castillo, J. Inoculating composted pine bark with beneficial organisms to make a disease suppressive compost for container production in Mexican forest nurseries. *Nativ. Plants J.* **2004**, *5*, 181–185. [CrossRef]

20. Caron, J.; Rochefort, L. Use of peat in growing media: State of the art on industrial and scientific efforts envisioning sustainability. *Acta Hortic.* **2013**, *982*, 15–22. [CrossRef]

21. Vaughn, S.F.; Eller, F.J.; Evangelista, R.L.; Moser, B.R.; Lee, E.; Wagner, R.E.; Peterson, S.C. Evaluation of biochar-anaerobic potato digestate mixtures as renewable components of horticultural potting media. *Ind. Crop Prod.* **2015**, *65*, 467–471. [CrossRef]

22. Landis, T.D.; Morgan, N. Growing Media Alternatives for Forest and Native Plant Nurseries. In Proceedings of the National Proceedings, Forest and Conservation Nursery Associations, Missoula, MT, USA, 23–25 June 2008; Dumroese, R.K., Riley, L.E., Eds.; USDA Forest Service, Rocky Mountain Research Station: Fort Collins, CO, USA, 2009; pp. 26–31. Available online: https://www.fs.usda.gov/treesearch/pubs/20894 (accessed on 15 March 2018).

23. Huber, G.W.; Iborra, S.; Corma, A. Synthesis of transportation fuels from biomass: Chemistry, catalysts, and engineering. *Chem. Rev.* **2006**, *106*, 4044–4098. [CrossRef] [PubMed]

24. Dumroese, R.K.; Heiskanen, J.; Englund, K.; Tervahauta, A. Pelleted biochar: Chemical and physical properties show potential use as a substrate in container nurseries. *Biomass Bioenergy* **2011**, *35*, 2018–2027. [CrossRef]

25. Tian, Y.; Sun, X.; Li, S.; Wang, H.; Wang, L.; Cao, J.; Zhang, L. Biochar made from green waste as peat substitute in growth media for *Calathea rotundifola cv. Fasciata*. *Sci. Hortic.* **2012**, *143*, 15–18. [CrossRef]

26. Vaughn, S.F.; Kenar, J.A.; Thompson, A.R.; Peterson, S.C. Comparison of biochars derived from wood pellets and pelletized wheat straw as replacements for peat in potting substrates. *Ind. Crop Prod.* **2013**, *51*, 437–443. [CrossRef]

27. Matt, C.P.; Keyes, C.R.; Dumroese, R.K. Biochar effects on the nursery propagation of 4 northern Rocky Mountain native plant species. *Nativ. Plants J.* **2018**, *19*, 14–25. [CrossRef]

28. Headlee, W.L.; Brewer, C.E.; Hall, R.B. Biochar as a substitute for vermiculite in potting mix for hybrid poplar. *Bioenergy Res.* **2014**, *7*, 120–131. [CrossRef]

29. Nemati, M.R.; Simard, F.; Fortin, J.-P.; Beaudoin, J. Potential use of biochar in growing media. *Vadose Zone J.* **2015**, *14*. [CrossRef]

30. Ennis, C.J.; Evans, A.G.; Islam, M.; Ralebitso-Senior, T.K.; Senior, E. Biochar: Carbon sequestration, land remediation, and impacts on soil microbiology. *Crit. Rev. Environ. Sci. Technol.* **2012**, *42*, 2311–2364. [CrossRef]

31. McCarl, B.A.; Peacocke, C.; Chrisman, R.; Kung, C.-C.; Sands, R.D. Chapter 19: Economics of Biochar Production, Utilization and Greenhouse Gas Offsets. In *Biochar for Environmental Management*; Lehmann, J., Joseph, S., Eds.; Earthscan: Stirling, VA, USA, 2009; pp. 341–358. ISBN 978-1-84407-658-1.

32. Page-Dumroese, D.S.; Anderson, N.M.; Windell, K.; Englund, K.; Jump, K. *Development and Use of a Commercial-Scale Biochar Spreader*; General Technical Report RMRS-GTR-354; USDA Forest Service, Rocky Mountain Research Station: Fort Collins, CO, USA, 2016; p. 10. Available online: https://www.fs.usda.gov/treesearch/pubs/52309 (accessed on 20 March 2018).

33. McElligott, K.M. Biochar Amendments to Forest Soils: Effects on Soil Properties and Tree Growth. Master of Science Thesis, University of Idaho, Moscow, ID, USA, 2011.

34. Blake, G.R.; Hartge, K.H. Bulk Density. In *Methods of Soil Analysis. Part 1. Physical and Mineralogical Methods*, 2nd ed.; Agronomy Monograph 9; Klute, A., Ed.; American Society of Agronomy and Academic Press: Madison, WI, USA, 1986; pp. 363–375.

35. Heiskanen, J. Comparison of three methods for determining the particle density of soil with liquid pycnometers. *Commun. Soil Sci. Plant Anal.* **1992**, *23*, 841–846. [CrossRef]

36. Ball, D.F. Loss-on-ignition as an estimate of organic matter and organic carbon in non-calcareous soils. *Eur. J. Soil Sci.* **1964**, *15*, 84–92. [CrossRef]

37. Klute, A. *Methods of Soil Analysis. Part 1. Physical and Mineralogical Methods*, 2nd ed.; Agronomy Monograph 9; American Society of Agronomy and Academic Press: Madison, WI, USA, 1986.

38. Heiskanen, J. Variation in water retention characteristics of peat growth media used in tree nurseries. *Silva Fenn.* **1993**, *27*, 77–97. [CrossRef]

39. White, J.W.; Mastalerz, J.W. Soil moisture as related to "container capacity". *Am. Soc. Hortic. Sci.* **1996**, *89*, 758–765.

40. Wilson, G.C.S. The physico-chemical and physical properties of horticultural substrates. *Acta Hortic.* **1983**, *150*, 19–32. [CrossRef]

41. Schindler, U. Ein Schnellverfahren zur Messung der Wasserleitfähigkeit im teilgesättigten Boden an Stechzylinderproben. *Arch. Acker Pflanzenbau Bodenkd.* **1980**, *24*, 1–7. (In German)

42. Schindler, U.; Müller, L. Simplifying the evaporation method for quantifying soil hydraulic properties. *J. Plant Nutr. Soil Sci.* **2006**, *169*, 623–629. [CrossRef]

43. Mahler, R.L.; Naylor, D.V.; Fredrickson, M.K. Hot water extraction of boron from soils using sealed plastic pouches. *Commun. Soil Sci. Plant Anal.* **1984**, *15*, 479–492. [CrossRef]

44. Dumroese, R.K.; Montville, M.E.; Pinto, J.R. Using container weights to determine irrigation needs: A simple method. *Nativ. Plants J.* **2015**, *16*, 67–71. Available online: https://www.fs.usda.gov/treesearch/pubs/48087 (accessed on 15 March 2018). [CrossRef]

45. Timmer, V.R.; Aidelbaum, A.S. *Manual for Exponential Nutrient Loading of Seedlings to Improve Outplanting Performance on Competitive Forest Sites*; NODA/NFP Technical Report TR-25; Natural Resources Canada, Canadian Forest Service, Great Lakes Forestry Center: Marie, ON, Canada, 1996. Available online: http://www.cfs.nrcan.gc.ca/pubwarehouse/pdfs/9567.pdf (accessed on 15 March 2018).

46. Hedges, L.V.; Gurevitch, J.; Curtis, P.S. The meta-analysis of response ratios in experimental ecology. *Ecology* **1999**, *80*, 1150–1156. [CrossRef]

47. Haase, D.L.; Rose, R. Vector analysis and its use for interpreting plant nutrient shifts in response to silvicultural treatments. *For. Sci.* **1995**, *41*, 54–66.

48. Landis, T.D.; Tinus, R.W.; McDonald, S.E.; Barnett, J.P. *Seedling Nutrition and Irrigation. The Container Tree Nursery Manual: Agriculture Handbook 674*; USDA Forest Service: Washington, DC, USA, 1989; Volume 4, 119p. Available online: https://rngr.net/publications/ctnm/volume-4 (accessed on 18 March 2018).

49. Landis, T.D.; Haase, D.L.; Dumroese, R.K. Plant Nutrient Testing and Analysis in Forest and Conservation Nurseries. In Proceedings of the National Proceedings, Forest and Conservation Nursery Associations—2004, Charleston, NC, USA, 12–15 July 2004; Dumroese, R.K., Riley, L.E., Landis, T.D., Eds.; USDA Forest Service, Rocky Mountain Research Station: Fort Collins, CO, USA, 2005; pp. 76–83. Available online: https://www.fs.usda.gov/treesearch/pubs/20894 (accessed on 15 March 2018).

50. Bunt, A.C. *Media and Mixes for Container-Grown Plants. A Manual on the Preparation and Use of Growing Media for Pot Plants*, 2nd ed.; Unwin Hyman: London, UK, 1988; ISBN 978-94-011-7904-1.

51. Rikala, R.; Jozefek, H.J. Effect of dolomite lime and wood ash on peat subtrate and develoment of tree seedlings. *Silva Fenn.* **1990**, *24*, 323–334. [CrossRef]

52. Schmidt, M.W.I.; Skjemstad, J.O.; Czimczik, C.I.; Glaser, B.; Prentic, K.M.; Gelina, Y.; Kuhlbusch, T.A. Comparative analysis of black carbon in soils. *Glob. Biogeochem. Cycles* **2001**, *15*, 163–167. [CrossRef]

53. Liang, B.; Lehmann, J.; Solomon, D.; Kinyangi, J.; Grossman, J.; O'neill, B.; Skjemstad, J.O.; Thies, J.; Luizao, F.J.; Petersen, J.; et al. Black carbon increases cation exchange capacity in soils. *Soil Sci. Soc. Am. J.* **2006**, *70*, 1719–1730. [CrossRef]

54. DeLuca, T.H.; Gundale, M.J.; MacKenzie, M.D.; Jones, D.L. Biochar Effects on Soil Nutrient Transformations. In *Biochar for Environmental Management: Science, Technology and Implementation*, 2nd ed.; Lehmann, J., Joseph, S., Eds.; Earthscan: London, UK; New York, NY, USA, 2015; pp. 421–454. ISBN 978-0-415-70415-1.

55. Jeffery, S.; Verheijen, F.G.A.; van der Velde, M.; Bastos, A.C.A. Quantitative review of the effects of biochar applications to soils on crop productivity using meta-analysis. *Agric. Ecosyst. Environ.* **2011**, *144*, 175–187. [CrossRef]

56. Altland, J.E.; Locke, J.C. Gasified rice hull biochar is a source of phosphorus and potassium for container-grown plants. *J. Environ. Hortic.* **2013**, *31*, 138–144. [CrossRef]

57. Wrobel-Tobiszewska, A.; Boersma, M.; Sargison, J.P.; Adams, P.; Singh, B.; Franks, S.; Birch, C.J.; Close, D.C. Nutrient changes in potting mix and *Eucalyptus nitens* leaf tissue under macadamia biochar amendments. *J. For. Res.* **2018**, *29*, 383–393. [CrossRef]

58. Zhang, L.; Sun, X.-Y.; Tian, Y.; Gong, X.-G. Biochar and humic acid amendments improve the quality of composted green waste as a growth medium for the ornamental plant *Calathea insignis. Sci. Hortic.* **2014**, *176*, 70–78. [CrossRef]

59. Beets, P.N.; Oliver, G.R.; Kimberley, M.O.; Pearce, S.H.; Rodgers, B. Genetic and soil factors associated with variation in visual magnesium deficiency symptoms in *Pinus radiata. For. Ecol. Manag.* **2004**, *189*, 263–279. [CrossRef]

60. Heiskanen, J. Favourable water and aeration conditions for growth media used in containerized tree seedling production: A review. *Scand. J. For. Res.* **1993**, *8*, 337–358. [CrossRef]

61. Conversa, G.; Bonasia, A.; Lazzizera, C.; Elia, A. Influence of biochar, mycorrhizal inoculation, and fertilizer rate on growth and flowering of *Pelargonium* (*Pelargonium zonale* L.) plants. *Front. Plant Sci.* **2015**, *6*, 429. [CrossRef] [PubMed]

62. Graber, E.R.; Harel, Y.M.; Kolton, M.; Cytryn, E.; Silber, A.; David, D.R.; Tsechansky, L.; Borenshtein, M.; Elad, Y. Biochar impact on development and productivity of pepper and tomato grown in fertigated soilless media. *Plant Soil* **2010**, *337*, 481–496. [CrossRef]

63. Wenny, D.L.; Dumroese, R.K. *A Growing Regime for Containerized Ponderosa Pine Seedlings*; University of Idaho, Idaho Forest, Wildlife and Range Experiment Station: Moscow, ID, USA, 1987; p. 9. Available online: https://www.researchgate.net/publication/272828130 (accessed on 16 April 2018).

64. Landis, T.D.; Tinus, R.W.; McDonald, S.E.; Barnett, J.P. *Seedling Propagation. The Container Tree Nursery Manual: Agriculture Handbook 674*; USDA Forest Service: Washington, DC, USA, 1998; Volume 6, 166p. Available online: https://rngr.net/publications/ctnm/volume-6 (accessed on 16 April 2018).

65. Frenkel, O.; Jaiswal, A.K.; Elad, Y.; Lew, B.; Kammann, C.; Graber, E.R. The effect of biochar on plant diseases: What should we learn while designing biochar substrates? *J. Environ. Eng. Landsc.* **2017**, *25*, 105–113. [CrossRef]

66. Kope, H.H.; Axelrood, P.E.; Sutherland, J.; Reddy, M.S. Prevalence and incidence of the root-inhabiting fungi, *Fusarium, Cylindrocarpon* and *Pythium*, on container-grown Douglas-fir and spruce seedlings in British Columbia. *New For.* **1996**, *12*, 55–67. [CrossRef]

67. Unestam, T.; Beyer-Ericson, L.; Strand, M. Involvement of *Cylindrocarpon destructans* in root death of *Pinus sylvestris* seedlings: Pathogenic behaviour and predisposing factors. *Scand. J. For. Res.* **1989**, *4*, 521–535. [CrossRef]

68. Sutherland, J.R.; Shrimpton, G.M.; Sturrock, R.N. *Diseases and Insects in British Columbian Forest Seedling Nurseries*; FRDA Report 065; Forestry Canada and British Columbia Ministry of Forests: Victoria, BC, Canada, 1989; p. 85. Available online: https://www2.gov.bc.ca/assets/gov/farming-natural-resources-and-industry/forestry/forest-health/diseases_and_insects_in_bc_forest_seedling_nurseries-complete.pdf (accessed on 18 March 2018).

69. Dumroese, R.K.; James, R.L. Root diseases in bareroot and container nurseries of the Pacific Northwest: Epidemiology, management, and effects on outplanting performance. *New For.* **2005**, *30*, 185–202. [CrossRef]

70. Heiskanen, J. Water status of sphagnum peat and a peat–perlite mixture in containers subjected to irrigation regimes. *HortScience* **1995**, *30*, 281–284.

71. Heiskanen, J. Irrigation regime affects water and aeration conditions in peat growth medium and the growth of containerized Scots pine seedings. *New For.* **1995**, *9*, 181–195. [CrossRef]

72. Pinto, J.R.; Dumroese, R.K.; Davis, A.S.; Landis, T.D. Conducting seedling stocktype trials: A new approach to an old question. *J. For.* **2011**, *109*, 293–299. Available online: https://www.fs.usda.gov/treesearch/pubs/38391 (accessed on 22 March 2018).

Review

Seedling Quality: History, Application, and Plant Attributes

Steven C. Grossnickle [1,*] and Joanne E. MacDonald [2]

[1] NurseryToForest Solutions, 1325 Readings Drive, Sidney, BC V8L 5K7, Canada
[2] Natural Resources Canada, Canadian Forest Service—Atlantic Forestry Centre, PO Box 4000,
    Fredericton, NB E3B 5P7, Canada; joanne.macdonald@canada.ca
*   Correspondence: sgrossnickle@shaw.ca; Tel.: +1-250-655-9155

Received: 1 April 2018; Accepted: 8 May 2018; Published: 22 May 2018

Abstract: Since the early 20th century, silviculturists have recognized the importance of planting seedlings with desirable attributes, and that these attributes are associated with successful seedling survival and growth after outplanting. Over the ensuing century, concepts on what is meant by a quality seedling have evolved to the point that these assessments now provide value to both the nursery practitioner growing seedlings and the forester planting seedlings. Various seedling quality assessment procedures that measure numerous morphological and physiological plant attributes have been designed and applied. This paper examines the historical development of the discipline of seedling quality, as well as where it is today. It also examines how seedling quality is employed in forest restoration programs and the attributes that are measured to define quality. The intent is to provide readers with an overall perspective on the field of seedling quality and the people who developed this discipline from an idea into an operational reality.

Keywords: seedling quality; historical perspective; morphological attributes; physiological attributes

1. Introduction

Forest restoration is a complex process that requires many steps to ensure successful forest establishment. These steps include choosing suitable tree species and provenance, applying nursery cultural practices to produce quality seedlings, ensuring proper seedling handling practices, and making site modifications to improve the physical environment of the restoration site [1,2]. Implicit within a seedling production program is the recognition that inherent species attributes [3] and phenotypic traits created during nursery culture [4] are both important in determining initial seedling field performance. Thus, seedling quality is a critical component in ensuring a successful forest restoration program.

This review summarizes the evolution of seedling quality from three perspectives. First, a historical perspective outlines a timeline for the evolution of this discipline over the past century. Second, the application of seedling quality within restoration programs is discussed from the perspectives of monitoring the process and monitoring the product. Third, various plant attributes that have been considered or are currently in operational use for defining seedling quality are discussed. The intent of this review is to provide nursery practitioners and foresters with a better understanding of seedling quality so they can effectively apply these assessment practices in their forest restoration program.

2. Historical Perspective on Seedling Quality

The focus on seedling quality in forest restoration programs goes back at least a century (Table 1). Since the early 20th century, silviculturists have recognized the importance of planting seedlings with desirable attributes, and that successful establishment was associated with these attributes [5]. Early on, foresters examined plantation failures in an attempt to discern causes of poor performance, because of the silvicultural investment needed to ensure seedling establishment (e.g., [6–9]). Often, poor performance was attributed to environmental stress, animal grazing, or damage from disease or insects. However, poor-quality seedlings [8] and the inability of planted seedlings to grow roots [9] were also suggested as probable causes of plantation failure. Thus, these early researchers began to ask questions as to how best to grow quality seedlings and what plant attributes influence seedling survival and growth (i.e., field performance) after planting on reforestation sites. Furthermore, studies initiated on southern pines in the 1930s [10,11] were groundbreaking, in that they showed that seedling attributes measured at the end of nursery culture were related to subsequent seedling field performance.

Table 1. A chronological list of references that discuss seedling quality, review seedling quality issues, or provide conceptual ideas related to seedling survival and/or growth after outplanting.

Author(s)/Date	Relevance to the Discipline of Seedling Quality
Toumey (1916) [5]	Desirable seedlings are selected for their "vigor and growing power"
Kittredge (1929) [8]	Poor-quality planting stock is defined as the reason for plantation failure.
Wakeley (1935) [12]	Higher morphological (i.e., shoot and root length, diameter) grades of seedlings showed "consistent superiority" over lower grades of seedlings.
Rudolf (1939) [9]	The inability of planted seedlings to grow roots is defined as the reason for plantation failure.
Wakeley (1948) [10]	"Grades applied to nursery stock can be useful only so far as they distinguish seedlings with a high capacity for survival and growth after planting from those with a low capacity" (i.e., physiological grade).
Wakeley (1954) [11]	Recognized importance of physiological quality for survival and growth. Seedlings within a defined height range and increasing stem diameter grew best.
Stone (1955) [13]	"If the root system did not increase in size at a fairly rapid rate … the seedlings would die of drought … "
Stone and Schubert (1959) [14]; Stone et al. (1962) [15]	Determined that periodicity of root regeneration potential was the basis for defining lifting and cold-storage schedules that avoided early plantation failures.
Rowe (1964) [16]	Proposed that preconditioning might be useful for acclimatizing seedlings to improve their field performance.
Lavender and Cleary (1974) [17]	" … seedlings must be produced in such a way as to be physiologically ready to outplant into the field environment"
Tinus (1974) [18]	Seedlings must be in the "proper physiological state" to survive in the field environment.
Lavender (1976) [19]	Recognized importance of seedling physiology for field performance; initial stages of articulating seedling quality.
van den Driessche (1976) [20]	Stated "physiological factors likely to influence survival and growth," but questioned whether they can be incorporated into "a grading system"
Cleary et al. (1978) [21]	Seedlings with appropriate morphological characteristics that are properly conditioned and vigorous positively "influence(s) reforestation success"
Sutton (1979) [22]	Morphological attributes related to seedling performance, but variability in field performance leads to conclusion it is " … not what a tree looks like but how it performs in the field"
Sutton (1980a) [23]	"The quality of planting stock is the degree to which that stock realises the objectives of management at minimum cost. Quality is fitness for purpose."
Sutton (1980b) [24]	"In stressful outplanting situations … morphology is an inadequate or misleading indicator of performance."
Timmis (1980) [25]	Physiological variables define seedling performance; seedling response to site conditions drives growth.
Chavasse (1980) [26]	Seedling appearance is not a good measure of field performance. All steps in regeneration silviculture affect field performance.
Schmidt-Vogt (1981) [27]	Stress tolerance of seedlings "holds a key position" in the establishment of forests.
Burdett (1983) [4]	First comprehensive list of seedling characteristics that "enhance early plantation performance"
Iverson (1984) [28]	The biological goal is to plant seedlings that have the desired genetic, morphological, and physiological characteristics to utilize site resources most fully.
Ritchie (1984) [29]	Morphological characteristics exert primary influence on performance when seedlings are physiologically sound.
Duryea (1985a) [30]	The first seedling quality compendium detailing application of many seedling attributes still commonly used in assessment programs.
Duryea (1985b) [31]	"Having a wide array of tests to choose from may soon enable us to predict a seedling's suitability to a particular planting site … "
Kramer and Rose (1986) [32]	Physiological processes are the "machinery" through which genetics and nursery culture determine seedling quality.
Glerum (1988) [33]	Attributes define a seedling's "performance potential", but sound silvicultural practices are required for "optimal field performance"

Table 1. *Cont.*

Author(s)/Date	Relevance to the Discipline of Seedling Quality
Lavender (1988) [34]	"At present there is no really effective method to measure seedling vigour."
Puttonen (1989) [35]	Morphological traits describe "overall suitability" and physiological traits predict "acclimatization" to the site.
Hawkins and Binder (1990) [36]	"...no one test will be able to predict stock quality...," rather an integration of tests is required to define "seedling fitness" for field performance.
Rose et al. (1990) [37]	The "target seedling concept" was developed to define specific morphological and physiological seedling attributes "that can be quantitatively linked to reforestation success"
Johnson and Cline (1991) [38]	No single test is best and a "battery of tests is required to consistently predict seedling quality"
Langerud (1991) [39]	The term "viability" is the best descriptor for tests assessing seedling quality.
Omi (1993) [40]	No single attribute can "solely predict outplanting success". However, a "wide array of seedling tests may be impractical"
Grossnickle and Folk (1993) [41]	A combination of tests simulating field conditions are required to forecast, not predict, growth.
Folk and Grossnickle (1997) [42]	The distinction between seedling quality testing for initial survival or growth potential is required for better decision making in forest restoration programs.
Mattsson (1997) [43]	Single morphological attributes cannot forecast performance. A combination of morphological and physiological attributes can possibly "predict field performance"
Mohammed (1997) [44]	Measurement of attributes is critical for defining viable seedlings that can survive in the field, although it is difficult to reliably forecast growth.
Puttonen (1997) [45]	Morphological attributes can be used to "predict field performance"
Grossnickle (2000) [2]	Attributes supply useful performance information, although there are forecasting limitations depending on timing of tests and field site conditions.
Colombo (2004) [46]; Wilson and Jacobs (2006) [47]	First reviews to focus on hardwoods; their unique characteristics mean alternative morphological attributes or timing of physiological measurements should be considered.
Haase (2008) [48]	Many morphological and physiological variables can be measured to track and assess seedling quality. Defined a list of most commonly used morphological and physiological measurements of forest seedlings.
Ritchie et al. (2010) [49]	Morphological attributes "seldom change" after lifting, thus they project to the field, whereas physiological attributes "provide only a momentary analysis of plant quality"
Villar-Salvador et al. (2010) [50]	Review focused on the uniqueness of Mediterranean woody species and that, although somewhat similar, seedling quality practices need modification for species of this geographic region.
Landis (2011) [51]	The "target seedling concept" expanded to the "target plant concept" thereby including all types of plant materials (e.g., trees, shrubs, grasses) and including seeds, cuttings, or wildlings, as well as traditional nursery stock.
Dumroese et al. (2016) [52]	Application of the "target plant concept" to the nursery manager-client partnership with the goal of meeting forest restoration objectives.

In the mid-20th century, researchers began to critically examine what it took to grow quality seedlings in nurseries and what plant attributes conferred improved field performance (Table 1). These programs initially focused on bareroot seedlings [10,11,13]. Many of these measurements were related to morphological attributes [11] or root growth [13]. However, physiological attributes [10] and periodicity of root growth [14,15] were recognized as important factors affecting field performance.

In the 1970s, the emergence of container nurseries with their inherent ability to have greater control of cultural practices [53] created a realization that seedling physiology could be manipulated to change seedling quality (e.g., [17,18,54]). This realization began with the idea, proposed by Rowe [16], that cultural practices could be applied to acclimatize seedlings and improve their field performance. At this time, selection of species and locally adapted genetic sources also became part of the seedling quality discussion [18]. Together, these changes gave researchers and practitioners an opportunity to produce quality container-grown seedlings that resulted in new standards of field performance [55]. This was the start of seedling quality programs based on the need for a better understanding of seedling performance capabilities in relation to forest restoration sites (Table 1).

In the late 1970s and early 1980s, forest scientists were discussing the morphological and physiological attributes of seedling quality (Table 1). At this time, Sutton [23] proposed defining seedling quality as "fitness for purpose", meaning that seedlings are grown not just for the sake of producing nursery stock, but rather to achieve some objective(s) of management [24]. Subsequently, it became the standard definition for seedling quality, and remains so to this day [49,52]. Interestingly, this definition also mirrors one of the basic tenets of quality-assurance programs in manufacturing, i.e., that the product should be "fit for purpose" [56] (see Section 3.1).

Sutton [24] suggested that improvements in seedling quality would only occur when both morphological and physiological attributes were considered. Jaramillo [57] was one the first to provide a brief list of measurement techniques to evaluate seedling quality. Burdett [4] proposed a more comprehensive list of morphological (e.g., bud, shoot, root) and physiological (e.g., carbohydrate reserves, dormancy, drought tolerance, freezing tolerance, nutrient status) attributes which, if present in seedlings within the proper range of values, would "enhance" seedling performance after planting. These measured attributes quantify a seedling's growth potential, with field performance dictated by how site conditions affect this potential [58]. Burdett [4] proposed that phenotypic traits created during nursery culture were necessary for matching seedlings to site conditions (i.e., that these traits "preadapted" seedlings). Furthermore, he considered these phenotypic traits to be just as important as genotypic traits in determining initial field performance [4].

Further refinement of what seedling attributes defined field performance occurred during the early to mid-1980s (Table 1). Ritchie [29] articulated seedling properties that describe material attributes (i.e., single measures of seedling parameters) and performance attributes (i.e., integrated measures of various material attributes to test conditions). Iverson [28] believed that seedling selection needed to be based on that genetic, morphological, and physiological attributes that would be best suited to the intended field site. Duryea [31] envisioned that choosing from a wide array of attributes would allow one " ... to predict a seedling's suitability for a particular planting site ... ", thereby ensuring successful forest establishment. Furthermore, she believed a testing approach defining seedling quality just before planting would be desirable [31]. Moreover, Navratil et al. [59] voiced the need for an integrated stock quality program that assessed seedlings through all facets of the forest restoration process to improve both nursery production and restoration success.

Between 1988 and 1999, various researchers concluded that seedling quality could not be determined by an individual morphological or physiological attribute in isolation from other attributes (Table 1). In addition, it was recognized that measured attributes had to define seedling growth in relation to anticipated site conditions [35,36,60]. At this time, the "target seedling concept" was proposed, which suggested that "numerous seedling traits must work together to produce the desired field response" [37] (see Section 3.1). However, Langerud [39] warned that any measured attribute is a just a point-in-time assessment. Furthermore, a performance potential index was proposed at this time [61]. The idea was to create a battery of measured attributes that defined seedling performance in relation to potential field conditions [41]. Simpson and Ritchie [62] felt that the ability of a measured attribute (i.e., root growth potential) to define field performance was a function of both the seedling's level of stress resistance and the field site conditions. It was suggested that if the desire was to come closer to forecasting seedling field performance, then testing conditions should simulate environmental conditions at the planting site [42].

The range of seedling quality testing approaches continued to expand through the 1990s [2,43,44], even though many practitioners desired a single test that could measure seedling quality (Table 1). In a provocative paper, Puttonen [45] addressed whether there was the single "silver bullet" test that could be used in seedling quality assessment programs. He suggested that grouping morphological attributes together showed the best evidence of having "predictive value" in defining field performance, because they retain their mark on seedling identity for an extended time after the seedlings are field-planted and start to grow. Thus, such a grouping was the best candidate to be the "silver bullet" test [45]. However, Puttonen [45] concluded that physiological status cannot be ignored. This was in agreement with what other researchers were stating: that individual quality assessments should not be done in isolation [34], and that a combination of morphological and physiological attributes are required to describe seedling quality (e.g., [41,43]).

As the field of seedling quality expanded to hardwoods, it was recognized that, although some conifer attributes were applicable to hardwoods, these genera had unique attributes when it came to quality assessment procedures (Table 1). Variation in hardwood phenology and ecology requires that sampling periods and sampled tissues need to be carefully considered when devising a quality assessment program [47]. Species-specific variation also creates a need to modify quality-assessment approaches [50]. Thus, refinement of conifer procedures was needed to effectively measure the quality of hardwood seedlings.

In conclusion, from the realization that establishment success was associated with seedling attributes [5], through recognizing that seedling attributes were related to seedling performance [11], to defining these measurements as being related to a seedling's "fitness for purpose" [23], these perspectives have focused on defining seedling attributes that define their field performance. Moreover, this view was the main premise of the "target seedling concept" [37]. Use of this concept within an operational setting [63] was viewed as an effective way to create a nursery–client partnership that would permit open dialogue leading to a successful restoration outcome [51,64–66]. Finally, the idea that this concept be expanded to include native plant (i.e., woody and non-woody forest and range species) material (e.g., seedlings, cuttings) used in restoration programs has been proposed, and is known as the "target plant concept" [51,52,64,66].

3. Application of Seedling Quality within a Forest Restoration Program

Seedling quality assessment procedures occur in the nursery both during culture (see Section 3.1) and at lifting (see Section 3.2). The following is a review of the conceptual approach used to assess seedling quality from these two perspectives.

3.1. Monitoring the Process

In Monitoring the Process, the nursery manager creates a system for monitoring culture practices and crop development, which allows them to grow seedlings to the desired specifications. The proper application of nursery practices to produce quality seedlings is a key component of successful restoration programs using both bareroot [67,68] and container-grown [2,55,69] seedlings. To develop an effective seedling quality program that monitors the process, one needs to understand how the crop responds to cultural conditions. A crop's physiological response to the environment and its subsequent developmental response ultimately determine its growth performance in the nursery [70]. If nursery staff understand a species' physiological capability in relation to environmental conditions, then these detailed cultural practices can become standard operating procedures (SOPs). Various authors have suggested that SOPs need to be integrated into crop plans to consistently produce high-quality seedlings each year, whether they are bareroot [11,67,71] or container grown [55,72].

A conceptual model for monitoring a nursery production system that consistently produces high-quality seedlings is outlined in Figure 1. As mentioned, to create SOPs, one needs to fully understand each species' performance attributes, which entails understanding the ecophysiological and growth characteristics that define seedling development. Furthermore, SOPs for a given species can vary significantly with seedlot and/or target morphological and physiological specifications needed for a given outplanting site. In addition, SOPs for every phase of nursery culture need to be created because seedling development changes throughout culture. Furthermore, SOPs are the 'knowledge tools' nursery practitioners develop and subsequently use to guide them through each crop production cycle.

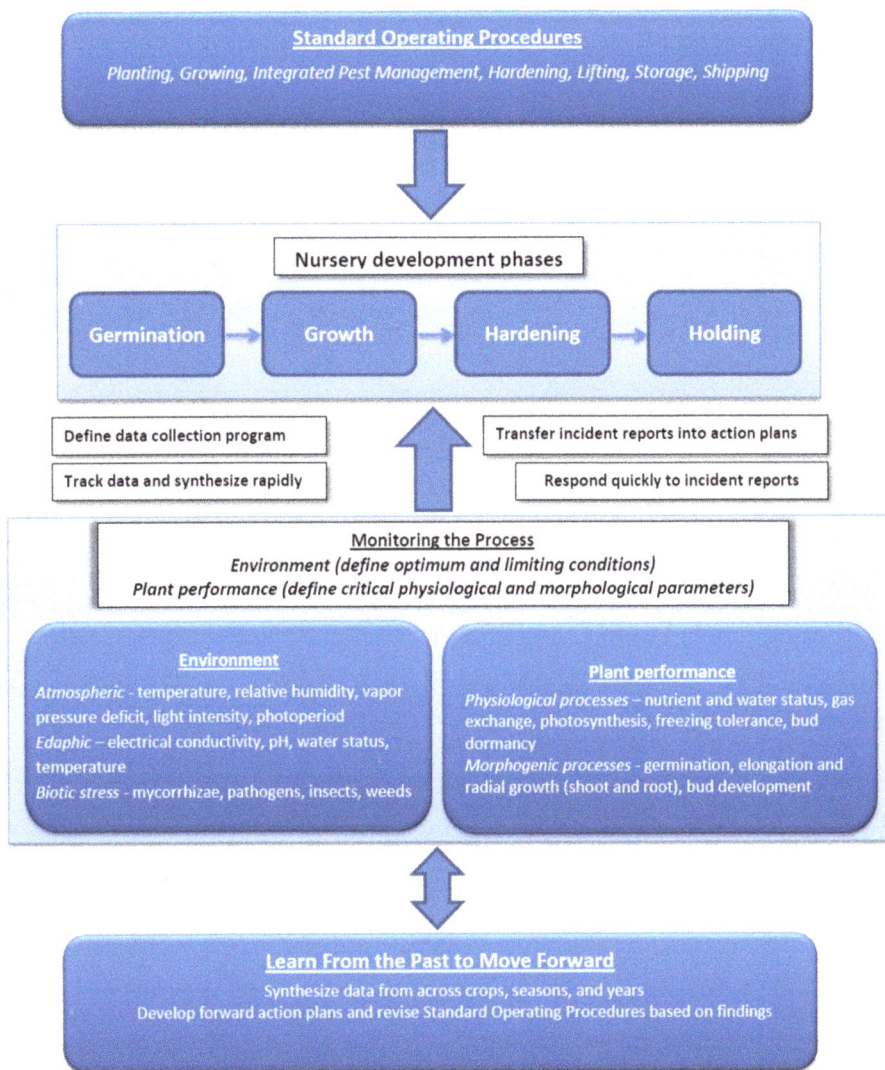

Figure 1. Conceptual model for monitoring seedling quality during crop development in the nursery.

Once the crop plan with SOPs has been developed, a tracking system is needed to ensure cultural guidelines are being followed and the crop is growing according to the plan (Figure 1). Such a system involves tracking both the nursery environment and crop performance [73]. Environmental conditions are tracked to define both optimum and limiting conditions for crop performance. Atmospheric conditions (e.g., air and plant temperature, relative humidity or vapor pressure deficit, light intensity, carbon dioxide level) and edaphic parameters (e.g., substrate temperature, substrate water content) can be monitored continuously with automated environmental sensors. Fertigation parameters (e.g., pH, electrical conductivity) can be monitored by handheld devices or semipermanent substrate probes. Automation of environmental monitoring provides rapid data synthesis that allows one to quickly understand how various parameters are affecting crop performance.

Crop performance is tracked by selecting morphological and physiological parameters that both mark important stages in seedling development [39] and can be easily measured. Alternatively, new technologies (e.g., fluorescence-imaging systems [74]) are becoming available that measure crop performance at a large scale and provide staff with the ability to understand how cultural practices are affecting seedling ecophysiological response. Such technology can be integrated with the irrigation system so that irrigation/fertigation automatically occurs at the first sign of stress. However, one also needs to "walk the crop" on a regular basis, thereby ensuring that the measured data corresponds with actual crop development. Furthermore, continued monitoring of the crop for pests is a critical part of maintaining crop quality during nursery development.

A crop performance database, together with a database for operational and cultural adjustments to the crop plan, is needed for such a monitoring system. Data collection and entry need to be efficient, as ongoing data analysis alerts staff when an incident that takes the crop away from the intended plan has occurred. In addition, one needs to understand seedling development in relation to planned cultural practices and use assessments to discern if corrective actions are needed to ensure the development of quality seedlings. Then, remedial action can quickly be taken to return to the crop plan. Deviations from the crop plan are recorded, so that after crop lifting, a crop review allows nursery staff to develop an understanding of what worked, what didn't, and where improvements in the crop plan can be made (Figure 1). In addition, deviations are compared with crops across a number of years to gain a perspective on crop performance under a range of conditions. Both retrospections allow the nursery practitioner to make adjustments to cultural practices, thereby further refining SOPs to improve future crop performance. In this way, a quality assurance program develops, and becomes a system of positive change and continued improvement in crop cultural practices.

This approach is part of the "target seedling concept", in which attention to the crop plan, as proposed above, is important to achieving the desired seedling product [37,75]. This approach is also similar to ISO quality assurance programs that monitor the production process to ensure achievement of planned results [56,76]. Furthermore, Grossnickle [73] described a quality-assurance program designed and operated at ten nurseries across North America that produced tens of millions of high-quality somatic loblolly pine (*Pinus taeda* L.) seedlings, which were planted throughout the southeastern United States. Creating and running this quality program demonstrated that, when designed to monitor the process, quality seedlings were the final output [73].

3.2. Monitoring the Product

In Monitoring the Product, an information database is created that allows dialogue between nursery and client on seedling performance capabilities. When nursery staff and silviculturists consider using a quality-assurance program to assess their seedlings, two questions are commonly asked. How to select stock that ensures the best field performance after planting? How to select tests that are useful in culling seedlings that do not meet desired quality standards? These questions are addressed in the paragraphs below.

A conceptual model for modeling seedling quality at the end of nursery practice is presented in Figure 2. This model provides a perspective on how one applies various assessment procedures when measuring seedling attributes that define field survival and/or field growth potential. Ritchie [29] discussed seedling quality in terms of material and performance attributes. Material attributes are single-point measures of individual parameters representing specific plant subsystems (e.g., morphology, osmotic potential, root electrolyte leakage, nutrient content/concentration, individual gas exchange measurements). In contrast, performance attributes reflect an integration of various material attributes, are environmentally sensitive plant properties, and are measured under specific testing conditions (e.g., root growth potential, freezing tolerance, 14-day gas exchange integrals). Both attribute types provide information on initial survival and field performance potential. Nursery staff and silviculturists need to define specific objectives before selecting testing procedures within a seedling quality program. In this way, they will achieve one of the basic principles of the "target seedling concept", which is for nurseries to deliver seedlings with morphological and physiological attributes that meet targets set by land managers for their restoration program [37,52] (see Section 2).

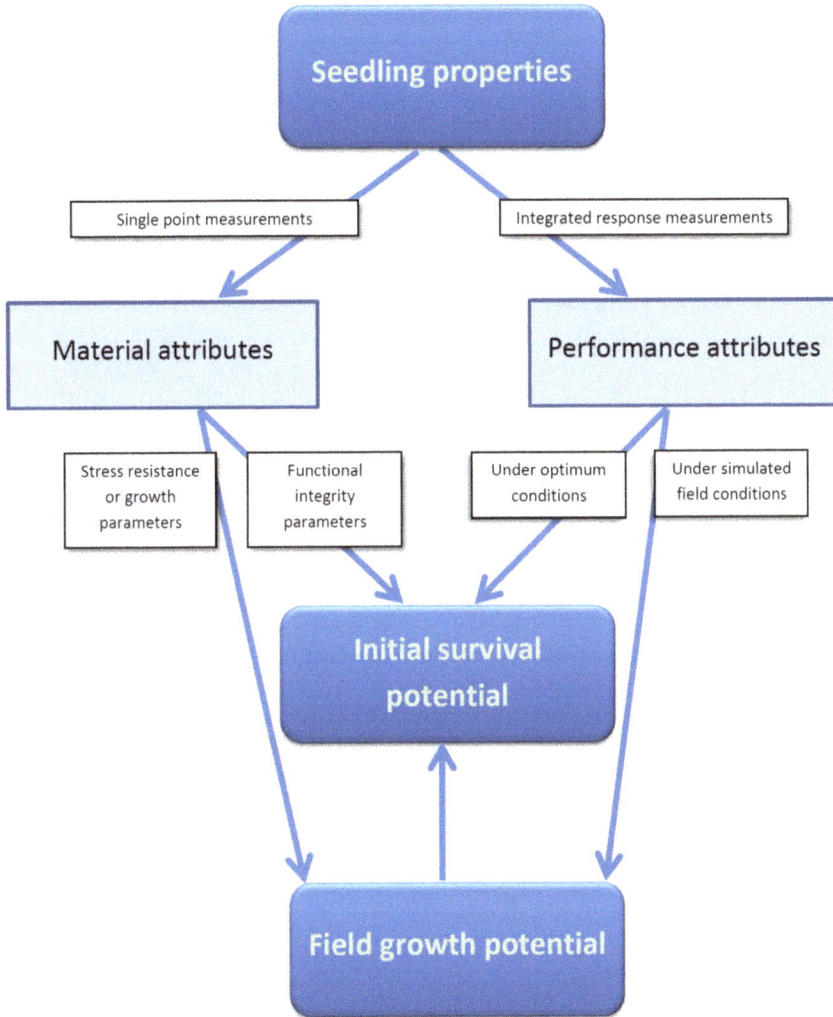

Figure 2. Conceptual model for monitoring seedling quality at the end of nursery culture (adapted from Folk and Grossnickle [42]).

One can never assume that planting high-quality seedlings "predicts" good field performance, as success is also influenced by appropriate silvicultural treatments before planting, as well as site conditions after planting [39,41,77]. After seedlings are planted, they may undergo various transplanting stresses before they can initiate growth and become "coupled" with the forest ecosystem [78]). Furthermore, if these environmental stresses are excessive [78], or seedlings have "too low a viability for the planting site" [39], then mortality [79] or a lack of proper growth [80] can occur. This is why seedling growth just after planting is critical to seedling survival and establishment [81,82]. Furthermore, once seedlings are established, seedling performance depends on inherent growth potential (which is related to their morphological and physiological attributes), together with their ecophysiological response to site environmental conditions that limit or enhance that potential [2]. The degree to which seedlings are suited to site conditions has the greatest influence on their performance immediately after planting [4,37]. Finally, as part of a comprehensive forest restoration program, measurement of seedling quality provides the silviculturist with information to "forecast" future plantation performance.

In planning a seedling quality program, one needs to choose attributes that assess seedling potential both to survive initially and to grow after field planting. The following paragraphs discuss attributes that measure initial survival potential and growth potential.

Initial survival potential is a measure of seedling "functional integrity" [41]. Functional integrity indicates whether seedlings are, or are not, damaged to the point of limiting primary physiological processes. Indeed, seedlings with reduced functional integrity can have poor field survival [2,79]. That said, seedlings meeting minimum standards typically have the capability of surviving in all but the most severe field site conditions [60]. Testing for functional integrity can be used at lifting to cull seedlings that do not meet minimum physiological performance standards, and includes assessment techniques such as root growth potential, root electrolyte leakage, and chlorophyll fluorescence. Root growth potential [13,62,83–87] and root electrolyte leakage [88,89] indicate root system integrity. Shoot system integrity is indicated directly by chlorophyll fluorescence [90–92] and indirectly by root growth potential [29]. Morphological attributes such as shoot height, stem diameter, root mass, and shoot-to-root ratio, together with physiological attributes such as drought resistance, mineral nutrient status, freezing tolerance, and root growth, have been shown, in some instances, to forecast survival after planting (reviewed by [2,79]). However, there is no guarantee that testing for initial survival potential provides information on field growth under limiting environmental conditions.

Plant attributes forecasting field growth need to define the intrinsic growth potential of seedlings with regard to site conditions [60]. A number of plant attributes measured at lifting (e.g., height, diameter, shoot-to-root ratio, root growth potential, nutrient status, drought resistance, freezing tolerance) have been reviewed for their capability to forecast growth [80]. When considering a more detailed assessment of seedling performance potential, it is important to select plant attributes that characterize performance in relation to the anticipated field site environmental conditions [31,35,36,41,42,60] (Figure 2). However, field conditions can only be roughly simulated. Furthermore, these are single-point assessments within a seasonal performance pattern [41] that changes as seedlings go through their phenological cycle [70]. Therefore, this approach forecasts, but is not able to predict, field performance. With these caveats in mind, multiple plant attributes have been combined that characterize seedling performance relative to stress events typically encountered on restoration sites (e.g., performance potential index [61], covariate morphological attributes [93], multivariate analysis [94], multiple variable models [95]) and provide forecasting models.

4. Plant Attributes that Define Seedling Quality

Plant attributes have been assessed at the morphological and physiological levels (Tables 2 and 3). However, in reality, only a limited number of these attributes are used within operational programs [44], because an "ideal operational measure" needs to be rapid, simple, cheap, reliable, nondestructive, quantitative, and diagnostic [96]. Indeed, researchers have agreed that only a subset of the most easily measured attributes listed in Tables 2 and 3 [48–50,97] be considered for seedling quality programs in nurseries [2,4,36,41,43–45]. However, each researcher has his/her preferred attributes. Furthermore, the "ideal operational measure" filter has also limited the operational use of comprehensive tests that combine multiple morphological and physiological attributes [36,43,45,97].

Table 2. Morphological attributes commonly used in seedling quality assessment programs to monitor either the process during nursery culture or the product at the end of culture.

Attribute	Application	Monitor the Process	Monitor the Product	References [1]
Bud development	Growth		×	[98–100]
Dry weight fraction	Lift/store	×		[101–103]
Height and diameter	Crop development	×		[11,21,37,55,75,104]
Height and diameter	Survival, growth		×	[4,5,11,21,22,26,27,29,104,105]
Morphological ratios	Survival, growth		×	[4,5,11,25,29,35,98]
Root system	Crop development	×		[11,21,37,75,104,106]
Root system	Survival, growth		×	[4,5,11,19,27,32,33,35,98]
Shoot and root weight	Survival, growth		×	[11,98]
Shoot system dimensions	Growth		×	[2,107]
Qualitative shoot trait [2]	Survival, growth		×	[5,11,48–50,98]
Qualitative root trait [3]	Survival, growth		×	[5,11,48–50,97,98]

[1] References are either the initial research conducted on an attribute and/or citations that initially recognized the attribute for inclusion in seedling quality programs at nurseries; [2] Examples: lack of terminal bud, multiple stems, stem curvature; [3] Examples: deformed root, poor plug development.

Table 3. Physiological attributes commonly used in seedling quality assessment programs to monitor either the process during nursery culture or the product at the end of culture.

Attribute	Application	Monitor the Process	Monitor the Product	References [1]
Chlorophyll fluorescence	Lift/store, viability	×		[90–92,108,109]
Chlorophyll fluorescence	Survival, growth		×	[48,49,110]
Freezing tolerance	Lift/store	×		[25,29,33,35,111,112]
Freezing tolerance	Survival, growth		×	[29,33,35]
Nutrient status	Crop development	×		[11,17,21,55,67,71,113–115]
Nutrient status	Survival, growth		×	[4,11,18,35,116–119]
Pest status	Crop development	×		[38,55,120–122]
Pest status	Survival, growth		×	[11,97]
Plant water status	Crop development	×		[21,115,123]
Plant water status	Survival, growth		×	[38,124–126]
Root electrolyte leakage	Crop development	×		[49,127]
Root electrolyte leakage	Survival, growth		×	[49,88,89,126]
Root growth potential	Survival, growth		×	[4,13,21,29,33,35,83–87]

[1] References are either the initial research conducted on an attribute and/or citations that initially recognized the attribute for inclusion in seedling quality programs at nurseries.

Despite these challenges, assessment programs for nurseries have been developed by selecting a set of attributes whose intended purpose is to ensure quality control, enhance consumer confidence, avoid planting damaged stock, and improve nursery cultural practices [50,97,128–130]. In addition, there have been a number of published discussions describing measurement procedures for the most common attributes (e.g., [48,49,97]). As mentioned, the field of seedling quality has evolved to the point that nursery practitioners and silviculturists now have a range of plant attributes that they can measure to understand the quality of their seedlings. The following discussion briefly examines the application of commonly used morphological (Table 2) and physiological (Table 3) attributes in forest restoration programs.

4.1. Commonly Used Plant Attributes

Morphological and physiological attributes are used to measure crop development in the nursery (See Section 3.1). Commonly measured morphological attributes include height, diameter, and root development for bareroot (e.g., [11,104]) and container-grown (e.g., [55,75]) seedlings. Typically, height and diameter are compared with standardized growth curves defined for each species, seedlot, and stocktype, which allows the adjustment of the nursery environment and cultural practices in order to keep seedlings on the crop plan. Root development is also monitored in container-grown seedlings to determine plug integrity, which is critical at lifting [131]. Physiological attributes commonly measured during crop development include nutrient status and plant water status. These attributes provide information for tracking crop performance, thereby supporting cultural adjustments to the crop plan. However, root electrolyte leakage is measured during crop development if there is a concern about damage. Furthermore, measuring chlorophyll fluorescence, electrolyte leakage, or whole-plant freezing during crop development in the autumn provides an assessment of freezing tolerance, with the goal of determining the proper lift/store date to develop sufficient stress resistance so high quality seedlings are stored (reviewed by [109,132,133]). Finally, at lifting, various morphological attributes, together with the physiological attributes of plant water relations, freezing tolerance, mineral nutrient status, root growth potential, and root electrolyte leakage are commonly assessed (See Section 3.2).

Morphological attributes are also used to relate seedling quality at lifting to subsequent field performance (See Section 3.2). Commonly measured morphological attributes include height, stem diameter, root systems, and shoot-to-root ratios [134]. These attributes are easy to measure in operational settings, ensuring their use in small-scale nurseries in developing countries [135] and large, commercial nurseries in first-world countries [2,136,137]. Morphological attributes influence seedling survival and growth after planting on forest restoration sites, because they retain their mark on seedling attributes for extended timeframes (reviewed by [79,80]). Greater stem diameter and root system size confer a higher chance of survival and growth, because they limit susceptibility to planting stress by improving water uptake and transport to foliage. Interestingly, South [138] revisited the morphological criteria defined by Wakeley [11] and found that root collar diameter was still the attribute that best forecast field growth potential. Greater height provides a competitive advantage (i.e., access to light) on sites with competing vegetation. However, where potential site environmental conditions are limiting (e.g., dry soils, high evaporative demand), seedlings with smaller shoot systems or lower shoot-to-root ratios are better adapted. Finally, morphological attributes are only measures that help define overall seedling size, growth potential, and balance [98,105], whereas seedling physiological attributes also have a major influence on field performance.

Other morphological attributes have been used in seedling quality programs, but with limited acceptance (Table 2). Bud development has been used in Ontario, Canada as a measure of potential seedling shoot growth [97]. Dry weight fraction has been used in Scandinavia to assess the development of stress resistance in the fall (c.f. [102]). Shoot dimensions (i.e., phyllotaxy of needles on shoots and arrangement of shoots along the leading stem) can be an important measure of seedling development for some (e.g., spruce [139]) but not all species.

Physiological attributes are also used to relate seedling quality at lifting to field performance after planting (See Section 3.2). Drought resistance, mineral nutrient status, root growth potential, root electrolyte leakage, and freezing tolerance have been used to assess seedling quality in relation to field survival (reviewed by [79]) and growth (reviewed by [80]) after planting. Improved survival is to the result of greater drought resistance and improved seedling nutrition at planting, which increases the speed with which seedlings can overcome planting stress, become established, and grow on the forest restoration site. Shoot water potential and root electrolyte leakage provide critical information on whether seedlings are damaged to the point of limiting physiological function; planting undamaged seedlings improves their survival and growth. Additional measurement of seedling functional integrity (e.g., root growth potential) is recommended if earlier tests detect a level of damage that could potentially limit field performance. Root growth potential on its own is valuable in many instances in forecasting field performance, because improved survival and growth due to greater root growth immediately after planting (reviewed by [79,80]) confers improved seedling survival and subsequent establishment within the first few months after planting.

In conclusion, it is important to emphasize that no single attribute can assess all seedling quality issues [43,45]. Morphological attributes cannot be used in isolation to assess seedling quality because morphology does not describe physiological vigor [105,134]). Furthermore, seedling quality cannot be determined by individual physiological attributes in isolation from other physiological and morphological attributes [34]. Thus, a seedling quality program needs to combine morphological and physiological attributes to provide the information necessary for making both sound nursery cultural decisions and restoration site decisions. Furthermore, a combination of desirable morphological and physiological attributes forecasts greater chances of survival and increased growth after establishment.

4.2. Novel Attributes and Tests for Plant Attributes

As the field of seedling quality assessment has developed, "novel" attributes and measurement techniques have been examined for their usefulness. The following paragraphs briefly outline novel physiological attributes or novel measurement techniques for traditional physiological attributes (Table 4), and novel biochemical, biophysical, and molecular techniques (Table 5).

Table 4. Novel physiological attributes or novel measurement techniques for traditional physiological attributes, proposed for use in seedling quality-assessment programs to monitor either the process during nursery culture or the product at the end of culture, which were not adopted.

Attribute or Technique	Application	Monitor the Process	Monitor the Product	References [1]
Auto-fluorescence	Viability		×	[44,140]
Bud dormancy	Lift/store, viability	×		[29,112,141,142]
Carbohydrate status	Survival, growth		×	[143–146]
Chlorophyll content, foliage color	Crop development	×		[147]
Chlorophyll content, foliage color	Growth		×	[24,49,98]
Crop-level chlorophyll fluorescence	Crop development	×		[74]
Drought avoidance	Survival, growth		×	[148]
Drought tolerance	Survival, growth		×	[4,11,19,25,27,29]
Electrical impedance	Lift/store, viability	×		[111,149,150]
Gas exchange [2]	Survival, growth		×	[107,151,152]
Heat tolerance	Survival		×	[153]
Infrared thermography	Lift/store, viability	×		[154–156]
Mycorrhizal status	Growth		×	[157–161]
Nuclear magnetic resonance	Survival		×	[162]
OSU [3] vigor test	Survival		×	[34,125,163]
Performance under stress	Growth		×	[42,61]
Root hydraulic conductivity	Survival, growth		×	[164–166]
Stress-induced volatile emissions	Survival		×	[167–170]
Xylem cavitation	Survival		×	[171–173]

[1] References are the initial work conducted on an attribute or a measurement technique; [2] Examples: needle conductance, photosynthesis, transpiration; [3] Oregon State University.

Table 5. Novel measurement techniques at the biochemical, biophysical, and molecular levels, proposed for use in seedling quality-assessment programs to monitor either the process during nursery culture or the product at the end of culture, which were not adopted or were recently reported in the literature.

Technique	Application	Monitor the Process	Monitor the Product	References [1]
Biochemical				
Enzymatic activity	Survival		×	[35,174]
Fluorescein diacetate staining	Viability		×	[175,176]
Triphenyl tetrazolium chloride staining	Survival		×	[36,177]
Vegetative storage proteins	Lift/store, viability	×		[103]
Biophysical				
Extracellular electropotential	Viability	×		[178–180]
Root electrical impedance	Lift/store	×		[181]
Molecular				
Gene expression	Lift/store	×		[182–186]
Molecular markers	Survival, growth		×	[187]

[1] References are the initial research conducted on a measurement technique in the context of seedling quality.

Some physiological attributes and measurement techniques were categorized as "novel" (Table 4), because other than in the articles describing them or in subsequent review articles, there is scant information that nursery practitioners are operationally using them. Indeed, when these attributes and techniques were compared against the criteria for "ideal operational measures" of seedling quality [96], many failed for one reason or another. Some fail the criterion of being rapid (e.g., bud dormancy, OSU vigor test). Others fail the criteria of simple and cheap because they require technically trained staff to run relatively expensive instruments for the analysis (e.g., drought tolerance, chlorophyll content, electrical impedance, infrared thermography, gas exchange, crop-level chlorophyll fluorescence, nuclear magnetic resonance, root hydraulic conductivity, stress-induced volatile emissions, xylem cavitation). Furthermore, whether the information is a reliable assessment of seedling quality (e.g., drought avoidance, foliage color, mycorrhizal status) plays a role in whether a nursery would spend the time to conduct the test.

Most of the biochemical, biophysical, and molecular techniques (Table 5), which were developed during the late 1980s and early 1990s have yet to be applied in nurseries. In general, molecular testing has not fulfilled the expectation voiced over 20 years ago that they would offer rapid measures of seedling quality [45]. However, more recent gene-expression analysis on freezing tolerance [188] has the potential to replace other tests (e.g., whole-plant freezing, electrolyte leakage, chlorophyll fluorescence [109,111]) used to make lift/store decisions. Genes involved in freezing tolerance in Scot's pine [188], Norway spruce [188], and Douglas-fir [183] have been identified, and then correlated with results from shoot electrolyte leakage tests to develop an assay that measures gene activity during freezing tolerance acquisition [188]. Furthermore, a related spin-off company (nsure®) has commercialized the assay. Clients sample, stabilize, and ship shoot tips to the lab, which conducts the test; level of freezing tolerance is e-mailed to clients within 2 days of sample arrival at the lab. It is yet to be determined whether this assay will replace the traditional measures of freezing tolerance used by nurseries.

5. Summary

Seedling quality is an important component of any successful forest restoration program. Over the past century, the concept of what is meant by seedling quality has evolved to the point that these plant attributes are used to improve seedling nursery culture and to forecast seedling survival and growth after outplanting. Such seedling quality information can now be used within the "target forest or plant seedling" concept to enable nursery practitioners and foresters to have an effective dialogue on how seedlings with certain attributes will meet forest restoration objectives. Even though planting seedlings with desirable plant attributes does not guarantee high survival and good growth after planting, planting seedlings with desirable attributes increases chances for a successful forest restoration program.

Author Contributions: S.C.G. and J.E.M. shared equally in the development and writing of this paper.

Acknowledgments: J.E.M. acknowledges Natural Resources Canada, Canadian Forest Service operating funds used in publishing this article in open access.

Conflicts of Interest: The authors declare no conflict of interest. Mention of any company name implies no endorsement of the company's product by the authors' respective organizations.

References

1. Gladstone, W.T.; Ledig, F.T. Reducing pressure on natural forests through high-yield forestry. *For. Ecol. Manag.* **1990**, *35*, 69–78. [CrossRef]
2. Grossnickle, S.C. *Ecophysiology of Northern Spruce Species. The Performance of Planted Seedlings*; NRC Research Press: Ottawa, ON, Canada, 2000.
3. Zobel, B.J.; Talbert, J.T. *Applied Forest Tree Improvement*; John Wiley & Sons, Inc.: New York, NY, USA, 1984.

4. Burdett, A.N. Quality control in the production of forest planting stock. *For. Chron.* **1983**, *59*, 132–138. [CrossRef]
5. Toumey, J.W. *Seeding and Planting, a Manual for the Guidance of Forestry Students, Foresters, Nurserymen, Forest Owners, and Farmers*, 1st ed.; John Wiley & Sons, Inc.: New York, NY, USA, 1916.
6. Tillotson, C.R. *Forest Planting in the Eastern United States*; Bulletin No. 153; U.S. Department of Agriculture: Washington, DC, USA, 1915.
7. Young, L.J. Forest planting in southern Michigan. *J. For.* **1921**, *19*, 1–8.
8. Kittredge, J. *Forest Planting in the Lake States*; Bulletin No. 1497; U.S. Department of Agriculture: Washington, DC, USA, 1929.
9. Rudolf, P.O. Why forest plantations fail. *J. For.* **1939**, *37*, 377–383.
10. Wakeley, P.C. Physiological grades of southern pine nursery stock. In *Proceedings Society of American Foresters 1948 Meeting*; Society of American Foresters: Washington, DC, USA, 1948; pp. 311–322.
11. Wakeley, P.C. *Planting the Southern Pines*; Agriculture Monograph No. 18; U.S. Department of Agriculture, Forest Service: Washington, DC, USA, 1954.
12. Wakeley, P.C. *Artificial Reforestation in the Southern Pine Region*; Technical Bulletin 492; U.S. Department of Agriculture, Forest Service: Washington, DC, USA, 1935.
13. Stone, E.C. Poor survival and the physiological condition of planting stock. *For. Sci.* **1955**, *1*, 89–94.
14. Stone, E.C.; Schubert, G.H. Root regeneration by ponderosa pine seedlings lifted at different times of the year. *For. Sci.* **1959**, *5*, 322–332.
15. Stone, E.C.; Jenkinson, J.L.; Krugman, S.L. Root-regenerating potential of Douglas-fir seedlings lifted at different times of the year. *For. Sci.* **1962**, *8*, 288–297.
16. Rowe, J.S. Environmental preconditioning with special reference to forestry. *Ecology* **1964**, *45*, 399–403. [CrossRef]
17. Lavender, D.P.; Cleary, B.D. Coniferous seedling production techniques to improve seedling establishment. In Proceedings of the North American Containerized Forest Tree Seedling Symposium, Denver, CO, USA, 26–29 August 1974; Tinus, R.W., Stein, W.I., Balmer, W.E., Eds.; Great Plains Agricultural Council Publication No. 68: Lincoln, NE, USA, 1974; pp. 177–180.
18. Tinus, R.W. Characteristics of seedlings with high survival potential. In Proceedings of the North American Containerized Forest Tree Seedling Symposium, Denver, CO, USA, 26–29 August 1974; Tinus, R.W., Stein, W.I., Balmer, W.E., Eds.; Great Plains Agricultural Council Publication No. 68: Lincoln, NE, USA, 1974; pp. 276–282.
19. Lavender, D.P. Role of forest tree physiology in producing planting stock and establishing plantations. In Proceedings of the XVI IUFRO World Congress, Oslo, Norway, 20 June–2 July 1976; Norwegian IUFRO Congress Committee: Ås, Norway, 1976; pp. 34–45.
20. Van den Driessche, R. How far do seedling standards reflect seedling quality? In Proceedings of the XVI IUFRO World Congress, Oslo, Norway, 20 June–2 July 1976; Norwegian IUFRO Congress Committee: Ås, Norway, 1976; pp. 50–52.
21. Cleary, B.D.; Greaves, R.D.; Owsten, P.W. Seedlings. In *Regenerating Oregon's Forests: A Guide for the Regeneration Forester*; Cleary, B.D., Greaves, R.D., Hermann, R.K., Eds.; Oregon State University, Extension Service, Extension Manual 7: Corvallis, OR, USA, 1978; pp. 63–97.
22. Sutton, R.F. Planting stock quality and grading. *For. Ecol. Manag.* **1979**, *2*, 123–132. [CrossRef]
23. Sutton, R.F. Evaluation of planting stock quality. *N. Z. J. For. Sci.* **1980**, *10*, 293–300.
24. Sutton, R.F. Techniques for evaluating planting stock quality. *For. Chron.* **1980**, *56*, 116–120. [CrossRef]
25. Timmis, R. Stress resistance and quality criteria for tree seedlings: Analysis, measurement and use. *N. Z. J. For. Sci.* **1980**, *10*, 21–53.
26. Chavasse, C.G.R. Planting stock quality: A review of factors affecting performance. *N. Z. J. For. Sci.* **1980**, *25*, 144–171.
27. Schmidt-Vogt, H. Morphological and physiological characteristics of planting stock: Present state of research and research tasks for the future. In Proceedings of the XVII IUFRO World Congress, Kyoto, Japan, 6–17 September 1981; Japanese IUFRO Congress Committee: Ibaraki, Japan, 1981; pp. 433–446.
28. Iverson, R.D. Planting stock selection: Meeting biological needs and operational realities. In *Forest Nursery Manual: Production of Bareroot Seedlings*; Duryea, M.L., Landis, T.D., Eds.; Martinus Nijhoff/Dr. W. Junk Publishers: The Hague, The Netherlands, 1984; pp. 261–268.

29. Ritchie, G.A. Assessing seedling quality. In *Forest Nursery Manual: Production of Bareroot Seedlings*; Duryea, M.L., Landis, T.D., Eds.; Martinus Nijhoff/Dr. W. Junk Publishers: The Hague, The Netherlands, 1984; pp. 243–266.

30. Duryea, M.L. *Evaluating Seedling Quality: Principles, Procedures, and Predictive Abilities of Major Tests*; Forest Research Laboratory, Oregon State University: Corvallis, OR, USA, 1985.

31. Duryea, M.L. Evaluating seedling quality: Importance to reforestation. In *Evaluating Seedling Quality: Principles, Procedures, and Predictive Abilities of Major Tests*; Duryea, M.L., Ed.; Forest Research Laboratory, Oregon State University: Corvallis, OR, USA, 1985; pp. 1–4.

32. Kramer, P.J.; Rose, R.R. Physiological characteristics of loblolly pine seedlings in relation to field performance. In Proceedings of the International Symposium on Nursery Management Practices for the Southern Pines, Montgomery, AL, USA, 4–9 August 1985; South, D.B., Ed.; School of Forestry, Auburn University and IUFRO Subject Group "Nursery Operations": Auburn, AL, USA, 1986; pp. 416–440.

33. Glerum, C. Evaluation of planting stock quality. In *Taking Stock: The Role of Nursery Practice in Forest Renewal, Proceedings of a Symposium under the Auspices of the Ontario Forestry Research Committee, Kirkland Lake, ON, Canada, 14–17 September 1987*; OFRC Proceedings O-P-16; Smith, C.R., Reffle, R.J., Eds.; Canadian Forestry Service, Great Lakes Forestry Centre: Sault Ste. Marie, ON, Canada, 1988; pp. 44–49.

34. Lavender, D.P. Characterization and manipulation of the physiological quality of planting stock. Proceeding of the 10th North American Forest Biology Workshop, Physiology and Genetics of Reforestation, Vancouver, BC, Canada, 20–22 July 1988; Worrall, J., Loo-Dinkins, J., Lester, D., Eds.; UBC Press: Vancouver, BC, Canada, 1988; pp. 32–57.

35. Puttonen, P. Criteria for using seedling performance potential tests. *New For.* **1989**, *3*, 67–87. [CrossRef]

36. Hawkins, C.D.B.; Binder, W.D. State of the art stock quality tests based on seedling physiology. In Proceedings of the Target Seedling Symposium: Proceedings, Combined Meeting of the Western Forest Nursery Associations, Roseburg, OR, USA, 13–17 August 1990; RM-GTR-200; Rose, R., Campbell, S.J., Landis, T.D., Eds.; U.S. Department of Agriculture, Forest Service: Fort Collins, CO, USA, 1990; pp. 91–122.

37. Rose, R.; Carlson, W.C.; Morgan, P. The target seedling concept. In *Target Seedling Symposium, Proceedings of the Combined Meeting of the Western Forest Nursery Associations, Roseburg, OR, USA, 13–17 August 1990*; RM-GTR-200; Rose, R., Campbell, S.J., Landis, T.D., Eds.; U.S. Department of Agriculture, Forest Service: Fort Collins, CO, USA, 1990; pp. 1–8.

38. Johnson, J.D.; Cline, M.L. Seedling quality of southern pines. In *Forest Regeneration Manual*; Duryea, M.L., Dougherty, P.M., Eds.; Kluwer Academic Publishers: Dordrecht, The Netherlands, 1991; pp. 143–162.

39. Langerud, B.R. 'Planting stock quality': A proposal for better terminology. *Scand. J. For. Res.* **1991**, *6*, 49–51. [CrossRef]

40. Omi, S.K. The target seedling and how to produce it. In Proceedings of the Nursery Management Workshop, Alexandria, LA, USA, 10–12 September 1991; Publication 148. Texas Forest Service: College Station, TX, USA, 1993; pp. 88–118.

41. Grossnickle, S.C.; Folk, R.S. Stock quality assessment: Forecasting survival or performance on a reforestation site. *Tree Plant. Notes* **1993**, *44*, 113–121.

42. Folk, R.S.; Grossnickle, S.C. Determining field performance potential with the use of limiting environmental conditions. *New For.* **1997**, *13*, 121–138. [CrossRef]

43. Mattsson, A. Predicting field performance using seedling quality assessment. *New For.* **1997**, *13*, 227–252. [CrossRef]

44. Mohammed, G.H. The status and future of stock quality testing. *New For.* **1997**, *13*, 491–514. [CrossRef]

45. Puttonen, P. Looking for the "silver" bullet—Can one test do it all? *New For.* **1997**, *13*, 9–27. [CrossRef]

46. Colombo, S.J. How to improve the quality of broadleaved seedlings produced in tree nurseries. In Proceedings of the Conference, Nursery Production and Stand Establishment of Broad-Leaves to Promote Sustainable Forest Management, Rome, Italy, 7–10 May 2001; Ciccarese, L., Lucci, S., Mattsson, A., Eds.; Italian Republic, Agency for the Protection of the Environment and for Technical Services: Rome, Italy, 2004; pp. 41–53.

47. Wilson, B.C.; Jacobs, D.F. Quality assessment of temperate zone deciduous hardwood seedlings. *New For.* **2006**, *31*, 417–433. [CrossRef]

48. Haase, D.L. Understanding forest seedling quality: Measurements and interpretation. *Tree Plant. Notes* **2008**, *52*, 24–30.

49. Ritchie, G.A.; Landis, T.D.; Dumroese, R.K.; Haase, D.L. Assessing plant quality, Chapter 2. In *The Container Tree Nursery Manual, Volume 7, Seedling Processing, Storage, and Outplanting*; Agriculture Handbook 674; Landis, T.D., Dumroese, R.K., Haase, D.L., Eds.; U.S. Department of Agriculture, Forest Service: Washington, DC, USA, 2010; pp. 17–82.

50. Villar-Salvador, P.; Puértolas, J.; Penuelas, J.L. Assessing morphological and physiological plant quality for Mediterranean woodland restoration projects. In *Land Restoration to Combat Desertification: Innovative Approaches, Quality Control and Project Evaluation*; Bautista, S., Aronson, J., Ramón Vallejo, V.J., Eds.; Fundación Centro de Estudios Ambientales del Mediterráneo—CEAM: Paterna, Valencia, Spain, 2010; pp. 103–120.

51. Landis, T.D. The target plant concept—A history and brief overview. In *National Proceedings: Forest and Conservation Nursery Associations—2010*; RMRS-P-65; U.S. Department of Agriculture, Forest Service: Fort Collins, CO, USA, 2011; pp. 61–66.

52. Dumroese, R.K.; Landis, T.D.; Pinto, J.R.; Haase, D.L.; Wilkinson, K.W.; Davis, A.S. Meeting forest restoration challenges: Using the target plant concept. *Reforesta* **2016**, *1*, 37–52. [CrossRef]

53. Tinus, R.W. A greenhouse nursery system for rapid production of container planting stock. In Proceedings of the 1971 Annual Meeting of the American Society of Agricultural Engineers, Pullman, WA, USA, 27–30 June 1971; American Society of Agricultural Engineers: St. Joseph, MI, USA, 1971.

54. Tanaka, Y.; Timmis, R. Effects of container density on the growth and cold hardiness of Douglas-Fir seedlings. In Proceedings of the North American Containerized Forest Tree Seedling Symposium, Denver, CO, USA, 26–29 August 1974; Tinus, R.W., Stein, W.I., Balmer, W.E., Eds.; Great Plains Agricultural Council Publication No. 68: Lincoln, NE, USA, 1974; pp. 181–186.

55. Tinus, R.W.; McDonald, S.E. *How to Grow Tree Seedlings in Containers in Greenhouses*; GTR-RM-60; U.S. Department of Agriculture, Forest Service: Fort Collins, CO, USA, 1979.

56. Stebbing, L. *Quality Assurance: The Route to Efficiency and Competitiveness*, 3rd ed.; Ellis Horwood Series in Applied Science and Industrial Technology; Prentice Hall: New York, NY, USA, 1993.

57. Jaramillo, A. Review of techniques used to evaluate seedling quality. Proceedings of Intermountain Nurseryman's Association and Western Forest Nursery Association Combined Meeting, Boise, ID, USA, 12–18 August 1980; INT-GTR-109; USDA Forest Service, Intermountain Forest and Range Experiment Station: Ogden, UT, USA, 1981; pp. 84–95.

58. Sutton, R.F. Plantation establishment with bareroot stock: Some critical factors. In Proceedings of the Artificial Regeneration of Conifers in the Upper Great Lakes Region, Green Bay, WI, USA, 26–28 October 1982; Mroz, G.D., Berner, J.F., Eds.; Michigan Technical University: Houghton, MI, USA, 1982; pp. 304–321.

59. Navratil, S.; Brace, L.G.; Edwards, I.K. *Planting Stock Quality Monitoring*; Information Report NOR-X-279; Canadian Forestry Service, Northern Forestry Centre: Edmonton, AB, Canada, 1986.

60. Sutton, R.F. Planting stock quality is fitness for purpose. In *Taking Stock: The Role of Nursery Practice in Forest Renewal, Proceedings of a Symposium under the Auspices of the Ontario Forestry Research Committee, Kirkland Lake, ON, Canada, 14–17 September 1987*; OFRC Proceedings O-P-16; Smith, C.R., Reffle, R.J., Eds.; Canadian Forestry Service, Great Lakes Forestry Centre: Sault Ste. Marie, ON, Canada, 1988; pp. 39–43.

61. Grossnickle, S.C.; Major, J.E.; Arnott, J.T.; Lemay, V.M. Stock quality assessment through an integrated approach. *New For.* **1991**, *5*, 77–91. [CrossRef]

62. Simpson, D.G.; Ritchie, G.A. Does RGP predict field performance? A debate. *New For.* **1997**, *13*, 253–277. [CrossRef]

63. Rose, R.; Haase, D.L. The target seedling concept: Implementing a program. In *National Proceedings: Forest and Nursery Conservation Associations—1995*; PNW-GTR-365; U.S. Department of Agriculture, Forest Service: Portland, OR, USA, 1995; pp. 124–130.

64. Landis, T.D. The target seedling concept: The first step in growing or ordering native plants. In Proceedings of the Conference Native Plant Propagation and Restoration Strategies, Eugene, OR, USA, 12–13 December 2001; Haase, D.L., Rose, R., Eds.; Nursery Tech Cooperative, Oregon State University: Corvallis, OR, USA; Western Forestry and Conservation Association: Portland, OR, USA, 2001; pp. 71–79.

65. Landis, T.D.; Dumroese, R.K. Applying the target plant concept to nursery stock quality. In *Plant Quality—A Key to Success in Forest Establishment, Proceedings of the COFORD Conference, Tulow, County Curlow, Ireland, 20–21 September 2006*; MacLennan, L., Fennessy, J., Eds.; COFORD, National Council for Forest Research and Development: Dublin, Ireland, 2006; pp. 1–10.

66. Landis, T.D.; Wilkinson, K.M. Defining the target plant. In *Tropical Nursery Manual: A Guide to Starting and Operating a Nursery for Native and Traditional Plants*; Agriculture Handbook 732; Wilkinson, K.M., Landis, T.D., Haase, D.L., Daley, B.F., Dumroese, R.K., Eds.; U.S. Department of Agriculture, Forest Service: Washington, DC, USA, 2014; pp. 44–65.

67. Duryea, M.L. Nursery cultural practices: Impacts on seedling quality. In *Forest Nursery Manual: Production of Bareroot Seedlings*; Duryea, M.L., Landis, T.D., Eds.; Martinus Nijhoff/Dr. W. Junk Publishers: The Hague, The Netherlands, 1984; pp. 143–164.

68. Mexal, J.G.; South, D.B. Bareroot seedling culture. In *Forest Regeneration Manual*; Duryea, M.L., Dougherty, P.M., Eds.; Kluwer Academic Publishers: Dordrecht, The Netherlands, 1991; pp. 89–115.

69. Landis, T.D.; Dumroese, R.K.; Haase, D.L. *The Container Tree Nursery Manual, Volume 7, Seedling Processing, Storage, and Outplanting*; Agriculture Handbook 674; U.S. Department of Agriculture, Forest Service: Washington, DC, USA, 2010.

70. Grossnickle, S.C. Restoration silviculture: An ecophysiological perspective—Lessons learned across 40 years. *Reforesta* **2016**, *1*, 1–36. [CrossRef]

71. May, J.T. Seedling growth and development. In *Southern Pine Nursery Handbook*; Lantz, C.W., Ed.; U.S. Department of Agriculture, Forest Service: Southern Region: Washington, DC, USA, 1985; Chapter 7; pp. 7-1–7-18.

72. Landis, T.D.; Tinus, R.W.; Barnett, J.P. *The Container Tree Nursery Manual, Volume 6, Seedling Propagation*; Agriculture Handbook 674; U.S. Forest Service: Washington, DC, USA, 1999.

73. Grossnickle, S.C. Tissue culture of conifer seedlings—Twenty years on: Viewed through the lens of seedling quality. In *National Proceedings of the Forest and Conservation Nursery Associations—2010*; RMRS-P-65; U.S. Department of Agriculture, Forest Service: Fort Collins, CO, USA, 2011; pp. 139–146.

74. Wang, H.; Qian, X.; Zhang, L.; Xu, S.; Li, H.; Xia, X.; Dai, L.; Xu, L.; Yu, J.; Liu, X. Detecting crop population growth based on chlorophyll fluorescence imaging. *Appl. Opt.* **2017**, *56*, 9762–9769. [CrossRef] [PubMed]

75. Landis, T.D.; Tinus, R.W.; McDonald, S.E.; Barnett, J.P. *The Container Tree Nursery Manual, Volume 1, Nursery Planning, Development, and Management*; Agriculture Handbook 674; U.S. Department of Agriculture, Forest Service: Washington, DC, USA, 1995.

76. Anonymous. *ISO 9000:2000, Quality Management Systems—Fundamentals and Vocabulary*, 2nd ed.; International Organization for Standardization: Geneva, Switzerland, 2000.

77. Schultz, R.I. *Loblolly Pine: The Ecology and Culture of Loblolly Pine (Pinus taeda L.)*; Agriculture Handbook 713; U.S. Department of Agriculture, Forest Service: Washington, DC, USA, 1997.

78. Grossnickle, S.C. Importance of root growth in overcoming planting stress. *New For.* **2005**, *30*, 273–294. [CrossRef]

79. Grossnickle, S.C. Why seedlings survive: Importance of plant attributes. *New For.* **2012**, *43*, 711–738. [CrossRef]

80. Grossnickle, S.C.; MacDonald, J.E. Why seedlings grow: Influence of plant attributes. *New For.* **2018**, *49*, 1–34. [CrossRef]

81. Burdett, A.N. Physiological processes in plantation establishment and the development of specifications for forest planting stock. *Can. J. For. Res.* **1990**, *20*, 415–427. [CrossRef]

82. Margolis, H.A.; Brand, D.G. An ecophysiological basis for understanding plantation establishment. *Can. J. For. Res.* **1990**, *20*, 375–390. [CrossRef]

83. Ritchie, G.A.; Dunlap, J.R. Root growth potential: Its development and expression in forest tree seedlings. *N. Z. J. For. Sci.* **1980**, *10*, 218–248.

84. Ritchie, G.A. Root growth potential: Principles, procedures, and predictive ability. In *Evaluating Seedling Quality: Principles, Procedures, and Predictive Abilities of Major Tests*; Duryea, M.L., Ed.; Forest Research Laboratory, Oregon State University: Corvallis, OR, USA, 1985; pp. 93–106.

85. Burdett, A.N. Understanding root growth capacity: Theoretical considerations in assessing planting stock quality by means of root growth tests. *Can. J. For. Res.* **1987**, *17*, 768–775. [CrossRef]

86. Ritchie, G.A.; Tanaka, Y. Root growth potential and the target seedling. In *Target Seedling Symposium, Proceedings of the Combined Meeting of the Western Forest Nursery Associations, Roseburg, OR, USA, 13–17 August 1990*; RM-GTR-200; Rose, R., Campbell, S.J., Landis, T.D., Eds.; U.S. Department of Agriculture, Forest Service: Fort Collins, CO, USA, 1990; pp. 37–51.

87. Sutton, R.F. Root growth capacity in coniferous forest trees. *HortScience* **1990**, *25*, 259–266.

88. McKay, H.M. Electrolyte leakage from fine roots of conifer seedlings: A rapid index for plant vitality following cold storage. *Can. J. For. Res.* **1992**, *22*, 1371–1377. [CrossRef]

89. Bigras, F.J.; Calmé, S. Viability tests for estimating root cold tolerance of black spruce seedlings. *Can. J. For. Res.* **1994**, *24*, 1039–1048. [CrossRef]

90. Vidaver, W.; Binder, W.; Brooke, R.C.; Lister, G.R.; Toivonen, P.M.A. Assessment of photosynthetic activity of nursery-grown *Picea glauca* seedlings using an integrating fluorometer to monitor variable chlorophyll fluorescence. *Can. J. For. Res.* **1989**, *19*, 1478–1482. [CrossRef]

91. Vidaver, W.; Lister, G.R.; Brooke, R.C.; Binder, W.D. *A Manual for the Use of Variable Chlorophyll Fluorescence in the Assessment of the Ecophysiology of Conifer Seedlings*; FRDA Report 163, Co-Published with the British Columbia Ministry of Forests; Forestry Canada, Pacific Forestry Centre: Victoria, BC, Canada, 1991.

92. Binder, W.D.; Fielder, P.; Mohammed, G.H.; L'Hirondelle, S.J. Applications of chlorophyll fluorescence for stock quality assessment with different types of fluorometers. *New For.* **1997**, *13*, 63–89. [CrossRef]

93. Kaczmarek, D.J.; Pope, P.E. Covariate analysis of northern red oak seedling growth. In Proceedings of the Seventh Biennial Southern Silvicultural Research Conference, Mobile, AL, USA, 17–19 November 1992; GTR-SO-93. Brissette, J.C., Ed.; U.S. Department of Agriculture, Forest Service: New Orleans, LA, USA, 1993; pp. 351–356.

94. D'Aoust, A.L.; Delisle, C.; Giouard, R.; Gonzales, A.; Bernier-Cardou, M. *Containerized Spruce Seedlings: Relative Importance of Measured Morphological and Physiological Variables in Characterizing Seedlings for Reforestation*; Information Report, LAU-X-110E; Natural Resources Canada, Canadian Forest Service, Quebec Region: Sainte-Foy, QC, Canada, 1994.

95. Jacobs, D.F.; Salifu, K.F.; Seifert, J.R. Relative contribution of initial root and shoot morphology in predicting field performance of hardwood seedlings. *New For.* **2005**, *30*, 235–251. [CrossRef]

96. Zaerr, J.B. The role of biochemical measurements in evaluating vigor. In *Evaluating Seedling Quality: Principles, Procedures, and Predictive Abilities of Major Tests*; Duryea, M.L., Ed.; Forest Research Laboratory, Oregon State University: Corvallis, OR, USA, 1985; pp. 137–141.

97. Colombo, S.J.; Sampson, P.H.; Templeton, C.W.G.; McDonough, T.C.; Menes, P.A.; DeYoe, D.; Grossnickle, S.C. Nursery stock quality assessment in Ontario. In *Regenerating the Canadian Forest: Principles and Practice for Ontario*; Wagner, R.G., Colombo, S.J., Eds.; Fitzhenry & Whiteside Ltd.: Markham, ON, Canada, 2001; pp. 307–324.

98. Thompson, B.E. Seedling morphological evaluation: What you can tell by looking. In *Evaluating Seedling Quality: Principles, Procedures, and Predictive Abilities of Major Tests*; Duryea, M.L., Ed.; Forest Research Laboratory, Oregon State University: Corvallis, OR, USA, 1985; pp. 59–72.

99. Colombo, S.J. Second-year shoot development in black spruce *Picea mariana* (Mill.) B.S.P. container seedlings. *Can. J. For. Res.* **1986**, *16*, 68–73. [CrossRef]

100. Templeton, C.W.G.; Odlum, K.D.; Colombo, S.J. How to identify bud initiation and count needle primordia in first-year spruce seedlings. *For. Chron.* **1993**, *69*, 431–437. [CrossRef]

101. Colombo, S.J. Bud dormancy status, frost hardiness, shoot moisture content, and readiness of black spruce container seedlings for frozen storage. *J. Am. Soc. Hortic. Sci.* **1990**, *115*, 302–307.

102. Calmé, S.; Margolis, H.A.; Bigras, F.J. Influence of cultural practices on the relationship between frost tolerance and water content of containerized black spruce, white spruce, and jack pine seedlings. *Can. J. For. Res.* **1993**, *23*, 503–511. [CrossRef]

103. Binnie, S.C.; Grossnickle, S.C.; Roberts, D.R. Fall acclimation patterns of interior spruce seedlings and their relationship to changes in vegetative storage proteins. *Tree Physiol.* **1994**, *14*, 1107–1120. [CrossRef] [PubMed]

104. Armson, K.A.; Sadreika, V. *Forest Tree Nursery Soil Management and Related Practices*; Ontario Ministry of Natural Resources: Toronto, ON, Canada, 1979.

105. Mexal, J.G.; Landis, T.D. Target seedling concepts: Height and diameter. In *Target Seedling Symposium, Proceedings of the Combined Meeting of the Western Forest Nursery Associations, Roseburg, OR, USA, 13–17 August 1990*; RM-GTR-200; Rose, R., Campbell, S.J., Landis, T.D., Eds.; U.S. Department of Agriculture, Forest Service: Fort Collins, CO, USA, 1990; pp. 17–36.

106. Davis, S.D.; Jacobs, D.F. Quantifying root system quality of nursery seedlings and relationship to outplanting performance. *New For.* **2005**, *30*, 295–311. [CrossRef]

107. Grossnickle, S.C.; Arnott, J.T.; Major, J.E.; Tschaplinski, T.J. Influence of dormancy induction treatment on western hemlock seedlings. I. Seedling development and stock quality assessment. *Can. J. For. Res.* **1991**, *21*, 164–174. [CrossRef]

108. Binder, W.D.; Fielder, P. Chlorophyll fluorescence as an indicator of frost hardiness in white spruce seedlings from different latitudes. *New For.* **1996**, *11*, 233–253.

109. Burr, K.E.; Hawkins, C.D.B.; L'Hirondelle, S.J.; Binder, W.D.; George, M.F.; Repo, T. Methods for measuring cold hardiness of conifers. In *Conifer Cold Hardiness*; Bigras, F., Colombo, S.J., Eds.; Kluwer Academic Publishers: Dordrecht, The Netherlands, 2001; pp. 369–401.

110. L'Hirondelle, S.J.; Simpson, D.G.; Binder, W.D. Chlorophyll fluorescence, root growth potential, and stomatal conductance as estimates of field performance potential in conifer seedlings. *New For.* **2007**, *34*, 235–251. [CrossRef]

111. Glerum, C. Frost hardiness of coniferous seedlings: Principles and applications. In *Evaluating Seedling Quality: Principles, Procedures, and Predictive Abilities of Major Tests*; Duryea, M.L., Ed.; Forest Research Laboratory, Oregon State University: Corvallis, OR, USA, 1985; pp. 107–123.

112. Burr, K.E. The target seedling concept: Bud dormancy and cold-hardiness. In Proceedings of the Target Seedling Symposium: Proceedings, Combined Meeting of the Western Forest Nursery Associations, Roseburg, OR, USA, 13–17 August 1990; RM-GTR-200; Rose, R., Campbell, S.J., Landis, T.D., Eds.; U.S. Department of Agriculture, Forest Service: Fort Collins, CO, USA, 1990; pp. 79–90.

113. Youngberg, C.T. Soil and tissue analysis: Tools for maintaining soil fertility. In *Forest Nursery Manual: Production of Bareroot Seedlings*; Duryea, M.L., Landis, T.D., Eds.; Martinus Nijhoff/Dr. W. Junk Publishers: The Hague, The Netherlands, 1984; pp. 75–80.

114. Landis, T.D. Mineral nutrition as an index of seedling quality: Principles and applications. In *Evaluating Seedling Quality: Principles, Procedures, and Predictive Abilities of Major Tests*; Duryea, M.L., Ed.; Forest Research Laboratory, Oregon State University: Corvallis, OR, USA, 1985; pp. 29–48.

115. Landis, T.D.; Tinus, R.W.; McDonald, S.E.; Barnett, J.P. *The Container Tree Nursery Manual, Volume 4, Seedling Nutrition and Irrigation*; Agriculture Handbook 674; U.S. Department of Agriculture, Forest Service: Washington, DC, USA, 1989.

116. Benzian, B.; Brown, R.M.; Freeman, S.C.R. Effect of late-season topdressing of N (and K) applied to conifer transplants in the nursery on their survival and growth on British forest sites. *Forestry* **1974**, *47*, 153–184. [CrossRef]

117. Brix, H.; van den Driessche, R. Mineral nutrition of container-grown tree seedlings. In Proceedings of the North American Containerized Forest Tree Seedling Symposium, Denver, CO, USA, 26–29 August 1974; Tinus, R.W., Stein, W.I., Balmer, W.E., Eds.; Great Plains Agricultural Council Publication No. 68: Lincoln, NE, USA, 1974; pp. 77–84.

118. Van den Driessche, R. Effects of nutrients on stock performance in the forest. In *Mineral Nutrition of Conifer Seedlings*; van den Driessche, R., Ed.; CRC Press: Boca Raton, FL, USA, 1991; pp. 229–260.

119. Timmer, V.R. Exponential nutrient loading: A new fertilization technique to improve seedling performance on competitive sites. *New For.* **1997**, *13*, 279–299. [CrossRef]

120. Sutherland, J.R. Pest management in Northwest bareroot nurseries. In *Forest Nursery Manual: Production of Bareroot Seedlings*; Duryea, M.L., Landis, T.D., Eds.; Martinus Nijhoff/Dr. W. Junk Publishers: The Hague, The Netherlands, 1984; pp. 203–210.

121. Landis, T.D. Disease and pest management. In *The Container Tree Nursery Manual, Volume 5, the Biological Component: Nursery Pests and Mycorrhizae Seedling Propagation*; Agriculture Handbook, 674, Landis, T.D., Tinus, R.W., McDonald, S.E., Barnett, J.P., Eds.; U.S. Forest Service: Washington, DC, USA, 1989; pp. 1–99.

122. Brissette, J.C.; Barnett, J.P.; Landis, T.D. Container seedlings. In *Forest Regeneration Manual*; Duryea, M.L., Dougherty, P.M., Eds.; Kluwer Academic Publishers: Dordrecht, The Netherlands, 1991; pp. 117–142.

123. Day, R.J.; Bunting, W.R.; Glerum, C.; Harvey, E.M.; Polhill, B.; Reese, K.H.; Wynia, A. *Evaluating the Quality of Bareroot Forest Nursery Stock*; Ontario Ministry of Natural Resources Report: Toronto, ON, Canada, 1987.

124. Joly, R.J. Techniques for determining seedling water status and their effectiveness in assessing stress. In *Evaluating Seedling Quality: Principles, Procedures, and Predictive Abilities of Major Tests*; Duryea, M.L., Ed.; Forest Research Laboratory, Oregon State University: Corvallis, OR, USA, 1985; pp. 17–28.

125. McCreary, D.D.; Duryea, M.L. Predicting field performance of Douglas-fir seedlings: Comparison of root growth potential, vigor, and plant moisture stress. *New For.* **1987**, *1*, 153–169. [CrossRef]

126. McKay, H.M.; White, I.M.S. Fine root electrolyte leakage and moisture content and indices of Sitka spruce and Douglas-fir seedling performance after desiccation. *New For.* **1997**, *13*, 139–162. [CrossRef]

127. Ritchie, G.A.; Landis, T.D. Seedling quality tests: Root electrolyte leakage. In *Forest Nursery Notes, Winter 2006*; R6-CP-TP-08-05; Landis, T.D., Ed.; U.S. Department of Agriculture, Forest Service, Pacific Northwest Region: Fort Collins, CO, USA, 2006; pp. 6–10.

128. Dunsworth, G.B. Plant quality assessment: An industrial perspective. *New For.* **1997**, *13*, 439–448. [CrossRef]

129. Sampson, P.H.; Templeton, C.W.G.; Colombo, S.J. An overview of Ontario's stock quality assessment program. *New For.* **1997**, *13*, 469–487. [CrossRef]

130. Tanaka, Y.; Brotherton, P.; Hostetter, S.; Chapman, D.; Dyce, S.; Belander, J.; Johnson, B.; Duke, S. The operational planting stock quality testing program at Weyerhaeuser. *New For.* **1997**, *13*, 423–437. [CrossRef]

131. Grossnickle, S.C.; El-Kassaby, Y. Bareroot versus container stocktypes: A performance comparison. *New For.* **2016**, *47*, 1–51. [CrossRef]

132. Colombo, S.J.; Menzies, M.I.; O'Reilly, C. Influence of nursery cultural practices on cold hardiness of coniferous forest tree seedlings. In *Conifer Cold Hardiness*; Bigras, F., Colombo, S.J., Eds.; Kluwer Academic Publishers: Dordrecht, The Netherlands, 2001; pp. 223–252.

133. Grossnickle, S.C.; South, D.B. Fall acclimation and the lift/store pathway: Effect on reforestation. *Open For. Sci. J.* **2014**, *7*, 1–20. [CrossRef]

134. Pinto, J.R. Morphology targets: What do seedling morphological attributes tell us. In *National Proceedings: Forest and Conservation Nursery Associations—2010*; RMRS-P-65; U.S. Department of Agriculture, Forest Service: Fort Collins, CO, USA, 2011; pp. 74–79.

135. Takoutsing, B.; Tchoundjeu, Z.; Degrande, A.; Asaah, E.; Gyau, A.; Nkeumoe, F.; Tsobeng, A. Assessing the quality of seedlings in small-scale nurseries in the highlands of Cameroon: The use of growth characteristics and quality thresholds as indicators. *Small Scale For.* **2013**, *13*, 65–77. [CrossRef]

136. South, D.B. *Planting Morphologically Improved Pine Seedlings to Increase Survival and Growth*; Forestry and Wildlife Research Series No. 1; Auburn University, Alabama Agricultural Experiment Station: Auburn, AL, USA, 2000.

137. Grossnickle, S.C.; South, D.B. Seedling quality of southern pines: Influence of plant attributes. *Tree Plant. Notes* **2017**, *60*, 29–40.

138. South, D.B. A re-evaluation of Wakeley's "critical tests" of morphological grades of southern pine nursery stock. *S. Afr. For. J.* **1987**, *142*, 56–59. [CrossRef]

139. Grossnickle, S.C. Seedling size and reforestation success. How big is big enough. In *The Thin Green Line: A Symposium on the State-of-the-Art in Reforestation, Proceedings, Thunder Bay, ON, Canada, 26–28 July 2005*; Forest Research Information Paper No. 160; Colombo, S.J., Compiler, Eds.; Queen's Printer for Ontario, Ontario Forest Research Institute, Ontario Ministry of Natural Resources: Sault Ste. Marie, ON, Canada, 2005; pp. 138–144.

140. Adams, G.T.; Lintilhac, P.M. Fluorescence microscopy of fresh tissue as a rapid technique for assessing early injury to mesophyll. *Biotech. Histochem.* **1993**, *68*, 3–7. [CrossRef] [PubMed]

141. Carlson, W.C.; Binder, W.D.; Feenan, C.O.; Preisig, C.L. Changes in mitotic index during onset of dormancy in Douglas-fir seedlings. *Can. J. For. Res.* **1980**, *10*, 371–378. [CrossRef]

142. Lavender, D.P. Bud dormancy. In *Evaluating Seedling Quality: Principles, Procedures, and Predictive Abilities of Major Tests*; Duryea, M.L., Ed.; Forest Research Laboratory, Oregon State University: Corvallis, OR, USA, 1985; pp. 7–15.

143. Ritchie, G.A. Carbohydrate reserves and root growth potential in Douglas-fir seedlings before and after cold storage. *Can. J. For. Res.* **1982**, *12*, 905–912. [CrossRef]

144. Marshall, J.D. Carbohydrate status as a measure of seedling quality. In *Evaluating Seedling Quality: Principles, Procedures, and Predictive Abilities of Major Tests*; Duryea, M.L., Ed.; Forest Research Laboratory, Oregon State University: Corvallis, OR, USA, 1985; pp. 49–58.

145. Puttonen, P. Carbohydrate reserves in *Pinus sylvestris* seedling needles as an attribute of seedling vigor. *Scand. J. For. Res.* **1986**, *1*, 181–193. [CrossRef]

146. Villar-Salvador, P.; Uscola, M.; Jacobs, D.F. The role of stored carbohydrates and nitrogen in the growth and stress tolerance of planted forest trees. *New For.* **2015**, *46*, 813–839. [CrossRef]

147. Van den Driessche, R. Relationship between spacing and nitrogen fertilization of seedlings in the nursery, seedling mineral and nutrition, and outplanting performance. *Can. J. For. Res.* **1984**, *14*, 431–436. [CrossRef]

148. Vanhinsberg, N.B.; Colombo, S.J. Effect of temperature on needle anatomy and transpiration of *Picea mariana* after bud initiation. *Can. J. For. Res.* **1990**, *20*, 598–601. [CrossRef]

149. Van den Driessche, R.; Cheung, K.W. Relationship of stem electrical impedance and water potential of Douglas-fir seedlings to survival and cold storage. *For. Sci.* **1979**, *25*, 507–517.

150. Glerum, C. Electrical impedance techniques in physiological studies. *N. Z. J. For. Sci.* **1980**, *10*, 196–207.

151. Örlander, G.; Rosvall-Ahnebrink, G. Evaluating seedling quality by determining their water status. *Scand. J. For. Res.* **1987**, *2*, 167–177. [CrossRef]

152. Langerud, B.R.; Puttonen, P.; Troeng, E. Viability of *Picea abies* seedlings with damaged roots and shoots. *Scand. J. For. Res.* **1991**, *6*, 59–72. [CrossRef]

153. Colombo, S.J.; Colclough, M.L.; Timmer, V.R.; Blumwald, E. Clonal variation in heat tolerance and heat shock protein expression in black spruce. *Silvae Genet.* **1992**, *41*, 234–239.

154. Weatherspoon, C.P.; Laacke, R.J. Infrared thermography for assessing seedling condition—Rationale and preliminary observations. In *Evaluating Seedling Quality: Principles, Procedures, and Predictive Abilities of Major Tests*; Duryea, M.L., Ed.; Forest Research Laboratory, Oregon State University: Corvallis, OR, USA, 1985; pp. 127–135.

155. Örlander, G.; Egnell, G.; Forsén, S. Infrared thermography as a means of assessing seedling quality. *Scand. J. For. Res.* **1989**, *4*, 215–222. [CrossRef]

156. Egnell, G.; Örlander, G. Using infrared thermography to assess viability of *Pinus sylvestris* and *Picea abies* seedlings before planting. *Can. J. For. Res.* **1993**, *23*, 1737–1743. [CrossRef]

157. Cordell, C.E.; Marx, D.H. Ectomycorrhizae: Benefits and practical application in forest tree nurseries and field plantings. In Proceedings of the North American Forest Tree Nursery Soils Workshop, Syracuse, NY, USA, 28 July–1 August 1980; Abrahamson, L.P., Bickelhaupt, D.H., Eds.; U.S. Department of Agriculture, Forest Service: Washington, DC, USA, 1980; pp. 217–224.

158. Molina, R.; Trappe, J.M. Mycorrhiza management in bareroot nurseries. In *Forest Nursery Manual: Production of Bareroot Seedlings*; Duryea, M.L., Landis, T.D., Eds.; Martinus Nijhoff/Dr. W. Junk Publishers: The Hague, The Netherlands, 1984; pp. 211–226.

159. Cordell, C.E.; Omdal, D.W.; Marx, D.H. Operational ectomycorrhizal fungus inoculations in forest tree nurseries: 1989. In Proceedings of the Intermountain Forest Nursery Association, Bismarck, ND, USA, 14–18 August 1989; GTR-RM-184. Landis, T.D., Technical Coordinator, Eds.; U.S. Department of Agriculture, Forest Service: Fort Collins, CO, USA, 1989; pp. 86–92.

160. Davey, C.B. Mycorrhizae and realistic nursery management. In Proceedings of the Target Seedling Symposium: Proceedings, Combined Meeting of the Western Forest Nursery Associations, Roseburg, OR, USA, RM-GTR-200; 13–17 August 1990; Rose, R., Campbell, S.J., Landis, T.D., Eds.; U.S. Department of Agriculture, Forest Service: Fort Collins, CO, USA, 1990; pp. 67–77.

161. Castellano, M.A.; Molina, R. Mycorrhizae, Chapter 2. In *The Container Tree Nursery Manual, Volume 5, the Biological Component: Nursery Pests and Mycorrhizae Seedling Propagation*; Agriculture Handbook 674; Landis, T.D., Tinus, R.W., McDonald, S.E., Barnett, J.P., Eds.; U.S. Forest Service: Washington, DC, USA, 1989; pp. 101–167.

162. Southon, T.E.; Mattsson, A.; Jones, R.A. NMR imaging of roots: Effects after root freezing of containerised conifer seedlings. *Physiol. Plant.* **1992**, *86*, 329–334. [CrossRef]

163. McCreary, D.D.; Duryea, M.L. OSU vigor test: Principles, procedures, and predictive ability. In *Evaluating Seedling Quality: Principles, Procedures, and Predictive Abilities of Major Tests*; Duryea, M.L., Ed.; Forest Research Laboratory, Oregon State University: Corvallis, OR, USA, 1985; pp. 85–92.

164. Colombo, S.J.; Asselstine, M.F. Root hydraulic conductivity and root growth capacity of black spruce (*Picea mariana*) seedlings. *Tree Physiol.* **1989**, *5*, 73–81. [CrossRef] [PubMed]

165. Carlson, W.C.; Miller, D.E. Target seedling root system size, hydraulic conductivity, and water use during seedling establishment. In Proceedings of the Target Seedling Symposium: Proceedings, Combined Meeting of the Western Forest Nursery Associations, Roseburg, OR, USA, 13–17 August 1990; RM-GTR-200; Rose, R., Campbell, S.J., Landis, T.D., Eds.; U.S. Department of Agriculture, Forest Service: Fort Collins, CO, USA, 1990; pp. 53–67.

166. Ritchie, G.A. A rapid method for detecting cold injury in conifer seedling root systems. *Can. J. For. Res.* **1990**, *20*, 26–30. [CrossRef]

167. DeYoe, D.R.; Drakeford, D.R. Assessing seedling response to stress—An operational approach. *Plant Physiol.* **1989**, *89*, 88.

168. Drakeford, D.R.; Hawkins, C.D.B. *The Stress-Induced Volatile Emissions (SIVE) Technique for Measuring Levels of Stress in Conifer Seedlings*; FRDA Report 084, Co-Published by the British Columbia Ministry of Forests; Forestry Canada, Pacific Forestry Centre: Victoria, BC, Canada, 1989.

169. Hawkins, C.B.D.; DeYoe, D.R. *SIVE, a New Stock Quality Test: The First Approximation*; FRDA Report 175, Co-Published by the British Columbia Ministry of Forests; Forestry Canada, Pacific Forestry Centre: Victoria, BC, Canada, 1992.

170. Templeton, C.W.G.; Colombo, S.J. A portable system to quantify seedling damage using stress—Induced volatile emission. *Can. J. For. Res.* **1995**, *25*, 682–686. [CrossRef]

171. Tyree, M.T.; Sperry, J.S. Do woody plants operate near the point of catastrophic xylem dysfunction caused by dynamic water stress? Answers from a model. *Plant Physiol.* **1988**, *88*, 547–580. [CrossRef]

172. Tyree, M.T.; Yang, S. Hydraulic conductivity recovery versus water pressure in xylem of *Acer saccharum*. *Plant Physiol.* **1992**, *100*, 669–676. [CrossRef] [PubMed]

173. Kavanagh, K.L.; Zaerr, J.B. Xylem cavitation and loss of hydraulic conductance in western hemlock following planting. *Tree Physiol.* **1997**, *17*, 59–63. [CrossRef] [PubMed]

174. Lindström, A.; Nyström, C. Seasonal variation in root hardiness of container grown Scots pine, Norway spruce, and lodgepole pine seedlings. *Can. J. For. Res.* **1987**, *17*, 787–793. [CrossRef]

175. Huang, C.-N.; Cornejo, M.J.; Bush, D.S.; Jones, R.L. Estimating viability of plant protoplasts using double and single staining. *Protoplasma* **1986**, *135*, 80–87. [CrossRef]

176. Kuoksa, T.; Hohtola, A. Freeze-preservation of buds from Scots pine trees. *Plant Cell Tissue Organ Cult.* **1991**, *27*, 89–93. [CrossRef]

177. Steponkus, P.L.; Lanphear, F.O. Refinement of the triphenyl tetrazolium chloride method of determining cold injury. *Plant Physiol.* **1967**, *42*, 1423–1426. [CrossRef] [PubMed]

178. Gensler, W.G. An electrochemical instrumentation system for agriculture and the plant sciences. *J. Electrochem. Soc.* **1980**, *127*, 2365–2370. [CrossRef]

179. Gensler, W.G. Stem diameter and electrochemical measurements. In *Advanced Agriculture Instrumentation. Design and Use*; Gensler, W.G., Ed.; Martinus Nijhoff Publishers: Dordrecht, The Netherland, 1986; pp. 457–476.

180. Gensler, W.G. The phytogram technique. In Proceedings of the XIX IUFRO World Congress, Montreal, QC, Canada, 5–11 August 1990; Canadian IUFRO World Congress Organizing Committee: Hull, QC, Canada, 1990; pp. 78–87.

181. Di, B.; Luoranen, J.; Lehto, T.; Himanen, K.; Silvennoinen, M.; Silvennoinen, R.; Repo, T. Biophysical changes in the roots of Scots pine seedlings during cold acclimation and after frost damage. *For. Ecol. Manag.* **2018**. [CrossRef]

182. Joosen, R.V.; Lammers, M.; Balk, P.A.; Brønnum, P.; Konings, M.C.; Perks, M.; Stattin, E.; van Wordragen, M.F.; van der Geest, A.L.H. Correlating gene expression to physiological parameters and environmental conditions during cold acclimation of Pinus sylvestris, identification of molecular markers using cDNA microarrays. *Tree Physiol.* **2006**, *26*, 1297–1313. [CrossRef] [PubMed]

183. Van Wordragen, M.F.; Balk, P.; Haase, D. Successful trial with innovative cold NSure test on Douglas-fir seedlings. In *Forest Nursery Notes, Summer 2007*; R6-CP-TP-04-2007; Landis, T.D., Ed.; U.S. Department of Agriculture, Forest Service, Pacific Northwest Region: Fort Collins, CO, USA, 2007; pp. 21–23.

184. Balk, P.A.; Haase, D.L.; van Wordragen, M.F. Gene activity test determines cold tolerance in Douglas-fir seedlings. In *National Proceedings: Forest and Conservation Nursery Associations—2007*; RMRS-P-57; U.S. Department of Agriculture, Forest Service: Fort Collins, CO, USA, 2008; pp. 140–148.

185. Stattin, E.; Verhoef, N.; Balk, P.; van Wordragen, M.; Lindström, A. Development of a molecular test to determine the vitality status of Norway spruce (*Picea abies*) seedlings during frozen storage. *New For.* **2012**, *43*, 665–678. [CrossRef]

186. Wallin, E.; Gräns, D.; Jacobs, D.F.; Lindström, A.; Verhoef, N. Short-day photoperiods affect expression of genes related to dormancy and freezing tolerance in Norway spruce seedlings. *Ann. For. Sci.* **2017**, *74*. [CrossRef]

187. Mayne, M.; Coleman, J.R.; Blumwald, E. Identification and characterization of two drought-induced cDNA clones in jack pine seedlings (*Pinus banksiana*) Lamb. In Proceedings of the Making the Grade: An International Symposium on Planting Stock Performance and Quality Assessment, Sault Ste. Marie, ON, Canada, 11–15 September 1994; Maki, D.S., McDonough, T.M., Noland, T.L., Compilers, Eds.; Ontario Forest Research Institute: Sault Ste. Marie, ON, Canada, 1994; p. 62.
188. Landis, T.D.; van Wordragen, M.F. Seedling Quality Testing at the Gene Level. In *Forest Nursery Notes, Summer 2006*; R6-CP-TP-04-2006; Landis, T.D., Ed.; U.S. Department of Agriculture, Forest Service, Pacific Northwest Region: Fort Collins, CO, USA, 2006; pp. 4–5.

forests

MDPI

Article

Development of Somatic Embryo Maturation and Growing Techniques of Norway Spruce Emblings towards Large-Scale Field Testing

Mikko Tikkinen *, Saila Varis and Tuija Aronen

Natural Resources Institute Finland, FI-58450 Punkaharju, Finland; saila.varis@luke.fi (S.V.); tuija.aronen@luke.fi (T.A.)
* Correspondence: mikko.tikkinen@luke.fi; Tel.: +358-29-532-8475

Received: 23 March 2018; Accepted: 12 May 2018; Published: 4 June 2018

Abstract: The possibility to utilize non-additive genetic gain in planting stock has increased the interest towards vegetative propagation. In Finland, the increased planting of Norway spruce combined with fluctuant seed yields has resulted in shortages of improved regeneration material. Somatic embryogenesis is an attractive method to rapidly facilitate breeding results, not in the least, because juvenile propagation material can be cryostored for decades. Further development of technology for the somatic embryogenesis of Norway spruce is essential, as the high cost of somatic embryo plants (emblings) limits deployment. We examined the effects of maturation media varying in abscisic acid (20, 30 or 60 µM) and polyethylene glycol 4000 (PEG) concentrations, as well as the effect of cryopreservation cycles on embryo production, and the effects of two growing techniques on embling survival and growth. Embryo production and nursery performance of 712 genotypes from 12 full-sib families were evaluated. Most embryos per gram of fresh embryogenic mass (296 ± 31) were obtained by using 30 µM abscisic acid without PEG in the maturation media. Transplanting the emblings into nursery after one-week in vitro germination resulted in 77% survival and the tallest emblings after the first growing season. Genotypes with good production properties were found in all families.

Keywords: Norway spruce; *Picea abies* L. Karst.; somatic embryogenesis; forest biotechnology; forest regeneration material; cryopreservation; maturation; embling production

1. Introduction

In Finland, the increased planting of Norway spruce (*Picea abies* L. Karst.) seedlings and difficulties in seed production has resulted in intermittent shortages of regeneration material of a high breeding value [1]. One solution to this problem is to use vegetative propagation, e.g., somatic embryogenesis (SE), which was observed in Norway spruce for the first time in 1985 [2,3]. Vegetative propagation enables more efficient tree improvement e.g., by capturing non-additive genetic gain [4,5].

Additionally, the cryopreservation of embryogenic tissue (ET) in liquid nitrogen (LN) enables long-term storage of regeneration material in its juvenile state [6–8]. Cryopreservation techniques are available for several conifer species, based on either applying cryoprotectant before freezing, or either on drying embryos or embryogenic tissues in different developmental stages [9–14]. Additionally, for Norway spruce, reliable cryopreservation protocols applicable for large number of samples have been developed [8,15]. As a result, acceptable recovery rates together with high morphological and genetic fidelity have been observed [8,15].

The commercial scale production of conifer emblings, i.e., somatic embryo plants, has been achieved in Denmark, Ireland and France (*Abies*, *Picea* and *Pinus* species) and is being piloted in Sweden (*Picea abies*). Companies producing conifers (e.g., *Pseudotsuga menziesii*, *Picea glauca engelmannii complex*,

Picea glauca (Moench) Voss and *Picea sitchensis* (Bong.) Carr., *Pinus radiata* D. Don, *Pinus taeda* L. and *Pinus elliottii* Engelm.) with SE for planting stock exists in North America and New Zealand [6,16,17].

The main reason limiting the commercial application of SE in forestry is the high cost of emblings compared to seedlings [17,18]. Several efforts to reduce costs have been made e.g., producing emblings as donor plants for rooted shoot cuttings, thus fragmenting the high cost of emblings to several hundred rooted shoot cuttings [17,19]. Despite all the efforts, emblings are still rather expensive compared to seedlings, which limits their deployment especially in Nordic conditions [20]. Additionally, the loss of genetic material during SE is a major challenge [21]. However, this could be mitigated by improving production methods in the post cryopreservation phases and in the laboratory-nursery interface [20–23].

Abscisic acid (ABA) is a relatively expensive plant hormone, widely used in conifer SE to promote embryo maturation [24]. It has a positive effect in promoting the maturation of embryogenic tissues, but it can also inhibit the germination and height growth of emblings for several growing seasons after exposure [7,25–27]. The type and amount of ABA concentration significantly affects the maturation results, and the optimal concentration varies between species and genotypes [27–29].

Similar, species and genotype specific, responses in embryo maturation have been reported when various amounts and types of polyethylene glycol has been added to the maturation medium in different conifer species [30–33]. Polyethylene glycol is added to the maturation media to reduce the moisture content of somatic embryos, thus increasing the content of the storage materials in later phases of maturation compared to ABA [31,32,34]. In Norway spruce, polyethylene glycol 4000 (PEG), when added to the maturation media, is known to increase the number of somatic embryos but is also known to have a negative effect on the later growth and development of the embryos [25,35,36]. PEG has also been found to speed up somatic embryo maturation by several weeks [31,36].

The aim of this work was to improve the efficiency of Norway spruce embling production in order to enable large scale testing of numerous SE lines. To achieve this we studied, (I) the effects of different levels of ABA and PEG in maturation media and the effect of an additional cryopreservation cycle on embryo production capacity; and (II) the effects of two different growing techniques on embling yield and early performance. Furthermore, the embryo production capacity and survival rate in the nursery were tested for a wide range of genotypes (712) originating from 12 full-sib families, with the aim to initiate field testing with rooted cuttings. This was done to improve the properties and yield of emblings (I and II) and to evaluate embling production schemes needed for large-scale field testing and variation among full-sib families affecting them (III).

2. Materials and Methods

2.1. Origin of Embryogenic Lines

The embryogenic lines used in this study were initiated in 2014 and 2015 from immature zygotic embryos of full-sibling families of progeny tested plus trees from Southern Finland. The medium and methods developed by Klimaszewska et al. [37], as described by Varis et al. [8] were used for culture establishment. Zygotic embryos without megametophytes, were placed on a modified Litvay's medium (mLM) containing half-strength macroelements [37,38], 10 μM 2,4-dicholophenoxyacetic acid (2,4-D), and 5 μM 6-benzyladenine (BA). The sucrose concentration of the medium was 1% (*w/v*) and the pH was adjusted to 5.8 prior to adding gellan gum (4 g/L, Phytagel™, Sigma-Aldrich, Saint Louis, MO, USA) and autoclaving. The cultures were kept in the dark (at 24 °C) for two to eight weeks without subculturing, until embryogenic tissue (ET) started to grow. Established ETs were subcultured bi-weekly, on a fresh Petri dish of the same medium.

Cryopreservation of ETs was done according to Varis et al. [8]. From each genotype one to four samples were cryopreserved right after initiation. Slowly growing ETs were not cryopreserved. The number of samples per genotype was kept low to increase the number of genotypes in cryostorage. For maturation in all experiments, from 150 to 200 milligrams of fresh embryogenic mass was weighed and absorbed on filter paper (Whatman # 2), using a Buchner funnel as done by Varis et al. [8].

2.2. Experiment I

The effect of the ABA concentration on embryo production and plant viability was studied using three different trials (1, 2 and 3). In Trial 1, six continuously subcultured Norway spruce embryogenic lines from five families initiated in 2014 were matured in December 2015, and seven lines from four families initiated in 2015 were matured in February 2016. Each line was matured on six filter papers (Whatman # 2) which were placed in petri dishes filled with 28 mL of LM-media containing 60 or 20 µM ABA (Later referred to 60ABA and 20ABA), and three maturations of each treatment (Table 1). Filter papers on the latter media were moved to fresh media two times at one-week intervals at the beginning of maturation. Cotyledonary embryos with visible initial shoot and root meristems and at least four cotyledons were manually counted after eight weeks maturation in the dark at 24 °C room temperature, as was done in previous studies [39,40].

Table 1. Schematic description of experiments and treatments used in different phases of SE production in Experiments I to III.

Exp./Trial	Treatment	Maturation	Germination	1st Growing Period/Growing Season	2nd Growing Season
I/1; 2; 3	60ABA	8 weeks			
I/1	3* 20ABA	8 weeks			
I/2	60ABA + PEG	8 weeks			
I/2	30ABA + PEG	8 weeks			
I/2; 3	30ABA	8 weeks			
	GT-I	8 weeks	18:6 Day-night	Transplanted to Miniplugs	Grown outside since
		60 µM	1 week [(1)]	Controlled environment	June 2017
II		30 µM ABA		Transplanted to Plantek 81f	
				Winterized and cold stored	
	GT-II	8 weeks	18:6 Day-night	Transplanted to Plantek 81f	
		60 µM	1 week [(1)]	Nursery greenhouse in March 2017	
		30 µM ABA		Grown outside since June 2017	
	Thawing lots A to D	8 weeks	18:6 Day-night	Transplanted to Miniplugs	Grown outside since
		60 µM ABA	1 week [(1)]	Controlled environment	June 2017
				Winterized and cold stored	
	Thawing lot E	8 weeks	18:6 Day-night	Transplanted to Miniplugs	
		30 µM ABA	1 week [(1)]	Controlled environment	
III				Transplanted to Plantek 81f in March 2017	
				Grown outside since June 2017	
	Thawing lot F	8 weeks	18:6 Day-night	Transplanted to Plantek 81f	
		30 µM ABA	1 week [(1)]	Nursery greenhouse in March 2017	
				Grown outside since June 2017	

In Experiment I, the effect of different concentrations/combinations of abscisic acid (ABA) and polyethylene glycol 4000 (PEG) on the yield of cotyledonary embryos was tested. 3* means a transfer of ET twice to fresh media. In Experiment II, two ex vitro growing techniques for emblings were tested (GT-I and GT-II). In Experiment III, large number of samples from 12 full-sib families was thawed from cryopreservation in lots A to F, to produce emblings (cutting donors) for clone testing. [(1)] three days in five $\mu mol/m^{-2}/s^{-1}$, two days in 50 $\mu mol/m^{-2}/s^{-1}$ and two days in 150 $\mu mol/m^{-2}/s^{-1}$.

Trial 2 consisted of seven genotypes (from different full-sib families) which were matured in May 2016. Two of the lines were initiated in 2015 and maintained in a subculture. Five lines initiated in 2014 were thawed from LN. Four different ABA (60 and 30 µM) and PEG (4.75% concentration in media) combinations were used in mLM-media (later referred to 60ABA, 30ABA, 60ABA + PEG and 30ABA + PEG) (Table 1). The filters were kept in the original petri dishes for eight weeks under the same conditions as in Trial 1, after which cotyledonary embryos were counted.

Trial 3 consisted of 120 cryopreserved genotypes from 12 families (10 genotypes per family), which were thawed from LN and matured using LM media with two different ABA concentrations (Table 1). The first lot of 120 genotypes was thawed and matured in 2016. From these genotypes, samples were cryopreserved again and one sample per genotype was thawed and matured in 2017. From the first lot, samples from 65 genotypes were matured on media containing 60 µM of ABA, and 55 genotypes were matured on media with a 30 μM ABA concentration. In the second lot, all 120 genotypes were matured on media containing 30 μM ABA. The filters were kept in the original Petri dishes for eight weeks in the same conditions as Trial 1, after which cotyledonary embryos were counted.

2.3. Experiment II

Two different growing techniques (later referred to GT-I and GT-II) in a nursery were evaluated by germinating cotyledonary embryos from 18 genotypes (9 families) according to the 1w-filter protocol described by Tikkinen et al. [23]. In short: cotyledonary embryos, cold stored at +3 °C (for at least four weeks in a large refrigerator unit) on the same filter papers in Petri dishes where the maturation was carried out, were germinated one-week in vitro under LED (Light emitting diode) lights (at a temperature from 20 to 23 °C inside the Petri dishes). The emblings were transplanted to a peat-based growth media after germination in vitro, using the 'pricking out' method as described by Landis et al. [23,41], in which forceps were used to transfer the emblings and to place them in peat. Peat was gently compressed around the embling to provide sufficient edaphic conditions for developing roots, as demonstrated by Landis et al. [23,41].

In GT-I, 36 emblings from each genotype were transplanted in small containers (Preforma 126/JIF, ViVi Pak, ViVi, Burgh Haamstede, Netherlands) with 126 plugs per container (plug volume 3.4 mL) (Miniplugs), and grown in a controlled environment for 50 days, until the temperature sum reached 1300 degree days (d.d.) (later referred to as the growing period). The controlled environment refers to a growth room, where the light period, temperature and humidity were adjusted to obey suggested levels for the different stages of growth of Norway spruce seedlings [42]. After this artificial growing season, the emblings were transplanted into Plantek 81f containers, winterized and cold stored in a large cooler unit. After cold storage the emblings were transferred outside together with a large lot of seedlings (Table 1).

In GT-II, 81 emblings from each genotype were transplanted straight into Plantek 81f containers (81 separate ventilated compartments of 85 cm^3 size) and were grown in a greenhouse as described by Tikkinen et al. [23]. These emblings were grown together with the emblings from thawing lots A to D of Experiment III (Table 1).

2.4. Experiment III

To initiate the field testing of SE-lines, emblings were produced for donor plants for shoot cuttings. This was done with cryopreserved genotypes from 12 full-sib families, initiated in 2014 (Table 2). The aim was to produce emblings from 20 genotypes from each full-sib family to initiate field testing with rooted shoot cuttings.

ETs were thawed from 712 genotypes at six different times, in thawing lots A to F. Thawings were carried out during 2016 (A to E) and 2017 (F) (Table 2). ETs were thawed, subcultured bi-weekly, cryopreserved again and matured (three Petri dishes each) in a five or ten weeks production cycle [8,37,38]. If the ET did not proliferate enough for cryopreservation and maturation in ten weeks it was discarded.

The ETs were cryopreserved again to increase the number of samples from each genotype for future use. Cryopreservation was prioritized, so that maturation was delayed for five weeks, if necessary. In the cases of poorly proliferating ETs, they were matured first and cryopreserved only if enough embryogenic tissue was available after 10 weeks of proliferation. This was done to increase the number of genotypes available for field testing.

Table 2. Description of material included in Experiment III.

Crossing	Explants	Initiated	Cryostored	Thawing	Thawed	Cryo + Maturation	E/gFW	Over 200 E/gFW	Surv. % A to D	Surv. % E to F
E1551 × E2229	99	67	61	III, IV and VI	48	45	94 (±13)	7	28 (±2)	72 (±2)
E162 × E81	200	101	66	II and VI	48	42	89 (±12)	5	14 (±2)	87 (±2)
E18 × E436	200	133	127	III and VI	47	30	110 (±18)	6	62 (±2)	85 (±2)
E207 × E1373	200	145	140	III, IV and VI	50	36	81 (±11)	3	54 (±2)	70 (±5)
E207 × E252	200	60	52	II, IV and VI	51	43	124 (±14)	6	30 (±2)	78 (±2)
E2105c × E2283	200	114	98	II, IV and VI	50	45	40 (±9)	1	14 (±2)	66 (±5)
E212 × E54	200	127	92	I, IV and V	92	61	45 (±7)	2	14 (±2)	80 (±2)
E242 × E222	400	187	106	I, IV, V and VI	106	49	36 (±7)	2	10 (±2)	82 (±2)
E46 × E3222	400	244	209	III and VI	78	53	89 (±12)	5	25 (±2)	88 (±2)
E462 × E64	200	160	137	I and VI	48	41	81 (±13)	5	54 (±2)	81 (±1)
E799 × E1366	45	42	36	II and IV	34	32	144 (±21)	7	24 (±2)	
E9 × E3231	400	132	123	I and VI	60	52	68 (±10)	5	5 (±1)	71 (±2)
Overall	2744	1512	1247	I to VI	712	529	79 (±4)	54	30 (±1)	80 (±1)

Number of explants, genotypes initiated and cryopreserved in 2014. Thawing lot, number of genotypes thawed, cryopreserved again and matured. Yield of cotyledonary embryos (E/gFM) from three maturation dishes (mean ± standard error), as well as the number of genotypes producing over 200 E/gFW and the survival rates (Surv. %) of emblings (mean ± standard error) from different crossings from thawing lots A to D and E to F, in Experiment III.

2.4.1. Thawing, Proliferation Maturation and Germination

After eight weeks of maturation under the same conditions as Experiment I, cotyledonary embryos were manually counted and cold stored on filter papers, as described by Tikkinen et al. [23]. In the case of thawing lot B, the maturation dishes were moved to cold storage before counting. The maturation medium contained 60 µM ABA, in thawing lots A to D, and 30 µM ABA in thawing lots E and F (Table 1). Maturation media was changed between thawing lots D and E, because of the higher yield of cotyledonary embryos in media with a lowered ABA concentration observed in Experiment I.

2.4.2. Embling Production

A one-week in vitro germination protocol was used in all cases following the methods described by Tikkinen et al. [23]. Because of limited resources, e.g., work force and growing space, for thawing lots A to E, only up to 36 cotyledonary embryos from each genotype were selected for cultivation depending on the availability of cotyledonary embryos. The cotyledonary embryos were grown as in GT-1 (Table 1). The emblings from thawing lot E were grown according to GT-I, with the exception that the emblings were transferred straight to the nursery in spring 2017, while they were still growing height (Table 1).

Germinated emblings from thawing lot F were grown as in GT-II (Table 1). From thawing lots E and F, up to 81 cotyledonary embryos were selected for cultivation depending on the availability of cotyledonary embryos.

2.5. Measurements and Data Analysis

Cotyledonary embryos were counted and the embryo productivity was calculated per one gram of fresh cell mass (E/gFM) in all experiments. Mean values are presented with their standard errors ($\pm$). All measurements and inventory results were analyzed using the IBM SPSS Statistics 22 software package (International Business Machines Corporation, Ammonk, NY, USA). The level of confidence used was 5%.

In Experiment I, nonparametric tests were used because normal distribution could not be assumed. In Trials 1 and 3 of Experiment I, differences between treatments were analyzed with Mann–Whitney U-test. In Trial 2 of Experiment I, the Kruskal–Wallis test (one-way ANOVA on ranks) was used to analyze differences between treatments.

To compare the two growing techniques in Experiment II, the survival of the emblings was inventoried from GT-I after the first growing period in a controlled environment, i.e., before cold storage. The survival and height measurements for the 2017 growing season, were obtained for both growing techniques. Logistic regression was used to examine the differences in survival between growing techniques, after the first growing period and after the growing season of 2017. A non-parametric test (Mann–Whitney U) was used to test the differences in embling height between the two growing techniques, because a normal distribution could not be assumed.

In Experiment III, the differences in embling survival between families were examined with logistic regression after the first growing period and after the growing season of 2017. A crossing covariate was used to investigate a possible parental effect. A thawing covariate was used to distinguish differences between different thawing lots. Among thawing lots, variation occurs in the date of thawing, the ABA concentration in the maturation media, the growing method and the lenght of the cold storage period. The effect of the thawing lot had no effect on the percentage of cases predicted correctly and was left out from the final models. The effect of location inside the containers was studied by using row and column covariates, which defined the location of a single embling inside a container. This effect was significant, but explained only 0.1% of the correctly predicted cases; hence row and column covariates were excluded from the final models. Embryo production between full-sib families was analyzed with Kruskal-Wallis and Mann–Whitney U nonparametric tests, because normal distributions could not be assumed.

3. Results

3.1. Effect of ABA Concentration and PEG in the Maturation Media

In Trial 1, with the first group of genotypes (Experiment 1), the mean yield of cotyledonary embryos was 80 ± 23 E/gFM when the media contained 60 µM ABA, and 185 ± 24 E/gFM when the ETs on filters were twice transferred to fresh media containing 20 µM ABA (Figure 1). Reducing the ABA concentration in the maturation media enhanced embryo production by 131% ($p < 0.001$).

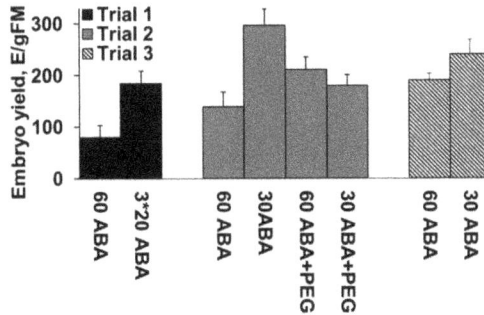

Figure 1. Yield of cotyledonary embryos in different treatments in three Trials in Experiment I. Mean values are presented with standard error bars. In Trial 3, values are presented for genotypes for which maturation with both treatments was available.

In Trial 2, reducing the amount of ABA enhanced the productivity even though the cell mass was not transferred to fresh media. The mean productivity on 30 µM ABA was 296 ± 31 E/gFM while on 60 µM ABA it was 139 ± 28 E/gFM, with the increase being 113% ($p = 0.001$) (Figure 1). Adding PEG increased the productivity only when combined with 60 µM ABA (211 ± 24 E/gFM). With 30 µM ABA the number of cotyledonary embryos decreased to 179 ± 22 E/gFM when PEG was included in the media (Figure 1).

In Trial 3, the average yield of cotyledonary embryos among genotypes in the first lot was 180 ± 9 E/gFM. For the 65 genotypes matured on 60 µM ABA, the average embryo yield was 191 ± 1 E/gFM. For the 55 genotypes matured on 30 µM ABA, the embryo yield was 166 ± 14 E/gFM. In the second lot, following cryopreservation, with 30 µM ABA used in the maturation media for all genotypes, the overall embryo yield was 206 ± 19 E/gFM. For the genotypes matured with 60 or 30 µM ABA in the first lot, the average embryo yields in the second lot were 240 ± 28 E/gFM and 167 ± 24 E/gFM, respectively. In the second lot, 116 genotypes were successfully regenerated from cryostorage and 114 were matured. No significant change in the average embryo production was found between lots when separately examining the genotypes matured on media containing 30 µM ABA in both lots, or between genotypes matured with 60 µM ABA in the first lot.

3.2. In Vitro Germination, Survival and Height Growth

In Experiment II, all selected cotyledonary embryos germinated and were transplanted ex vitro in both growing techniques. In GT-I, the average survival rate was 93 ± 1%, after the growing period, before cold storage. After cold storage and the growing season of 2017 in the nursery, the average survival rate and height were 53 ± 2% and 5.8 ± 0.3 cm, respectively. In GT-II, the average survival rate and height of the emblings after the first growing season (2017) in the nursery were 77 ± 1% and 10.9 ± 0.2 cm, respectively.

The difference in the embling survival rates between GT-I (after first growing period) and GT-II (after the first growing season of 2017) was significant ($p < 0.001$) (Table 3). The survival and height of the emblings obtained from different growing techniques varied after the growing season of 2017 ($p < 0.001$, in both) (Table 3).

Table 3. Logistic regression models used for analyzing binary response (living or dead) in different growing techniques (GT-I and GT-II), after first growing season in the nursery or artificial growing season in growth room (GP1) and after the growing season of 2017 in the nursery, for Experiments II and III.

Experiment	Measurement	Model	Variable	Sig.	Odds Ratio (95% Confidence Interval)	Correct, %
Experiment II	After GP1	$log(p/1−p) =$ $2.839 − 1.372g_1 − 0.076c_2 − 0.437c_4 − 0.748c_6 − 0.820c_7 + 0.470c_8 − 0.748c_9 − 0.163c_{10} − 0.546c_{11}$	Growing technique	<0.001	GT-I 1; GT-II 0.254 (0.177–0.363)	81
			Family	<0.001	E1551 × E2229 1; E18 × E436 0.927 (0.591–1.453); E207 × E252 0.623 (0.341–1.138); E212 × E54 0.473 (0.293–0.765); E242 × E222 0.44 (0.25–0.775); E462 × E64 0.473 (0.293–0.765); E46 × E3222 1.6 (0.956–2.676); E799 × E1366 0.849 (0.509–1.419); E9 × E3231 0.579 (0.324–1.035)	
Experiment II	Autumn 2017	$log(p/1−p) =$ $0.386 + 1.210g_1 + 0.150c_2 − 0.437c_4 − 0.588c_6 − 0.654c_7 + 0.491c_8 − 1.248c_9 − 0.899c_{10} − 1.095c_{11}$	Growing technique	<0.001	GT-I 1; GT-II 3.354 (2.684–4.191)	74.1
			Family	<0.001	E1551 × E2229 1; E18 × E436 1.162 (0.785–1.72); E207 × E252 0.646 (0.368–1.134); E212 × E54 0.556 (0.361–0.855); E242 × E222 0.52 (0.31–0.873); E462 × E64 0.287 (0.188–0.437); E46 × E3222 1.634 (1.06–2.52); E799 × E1366 0.407 (0.266–0.622); E9 × E3231 0.335 (0.203–0.55)	
Experiment III	After GP1	$log(p/1−p) =$ $0.859 + 0.238c_1 + 0.945c_2 + 0.421c_3 + 0.018c_4 + 0.104c_5 + 0.405c_6 + 0.419c_7 + 1.206c_8 + 0.651c_9 + 0.574c_{10} − 0.003c_{11}$	Family	<0.001	E1551 × E2229 1; E162 × E81 1.269 (1.019–1.58); E18 × E436 2.573 (2.035–3.252); E207 × E1373 1.523 (1.188–1.954); E207 × E252 1.018 (0.83–1.25); E2105c × E2283 1.11 (0.832–1.48); E212 × E54 1.499 (1.183–1.899); E242 × E222 1.52 (1.197–1.931); E462 × E64 3.34 (2.601–4.289); E46 × E3222 1.918 (1.575–2.336); E799 × E1366 1.775 (1.371–2.298); E9 × E3231 0.997 (0.812–1.225)	78

Table 3. *Cont.*

Experiment	Measurement	Model	Variable	Sig.	Odds Ratio (95% Confidence Interval)		Correct, %
Experiment III	Autumn 2017	$log(p/1-p) =$	$-2.652 + 0.613b_1 + 1.186b_2 + 0.915b_3 + 3.753b_4 + 3.334b_5 + 0.632c_1 + 1.186c_2 + 0.886c_3 + 0.912c_4 + 0.108c_5 + 0.365c_6 + 0.401c_7 + 1.494c_8 + 0.831c_9 + 0.840c_{10} + 0.126c_{11}$				
			Thawing	<0.001	1	A	76.5
					1.846 (1.326–2.571)	B	
					6.66 (4.935–8.988)	C	
					2.497 (1.714–3.638)	D	
					42.655 (29.934–60.782)	E	
					28.049 (21.653–36.334)	F	
			Family	<0.001	1	E1551 × E2229	
					1.881 (1.427–2.479)	E162 × E81	
					3.275 (2.659–4.034)	E18 × E436	
					2.425 (1.931–3.046)	E207 × E1373	
					2.49 (1.907–3.25)	E207 × E252	
					1.114 (0.776–1.601)	E2105c × E2283	
					1.441 (0.973–2.135)	E212 × E54	
					1.493 (1.049–2.123)	E242 × E222	
					4.456 (3.266–6.08)	E462 × E64	
					2.297 (1.905–2.769)	E46 × E3222	
					2.316 (1.659–3.232)	E799 × E1366	
					1.134 (0.88–1.462)	E9 × E3231	

In the models g_1 is a design variable for growing technique II, c_1 to c_{11} are design variables for full-sib families (Family) and b_1 to b_5 are design variables for thawing lots (Thawing).

3.3. Donor Plant Production

In Experiment III, 76% (51% to 94% variation among full-sib families) of the thawed genotypes produced enough ET to be matured. Cotyledonary embryos were produced from 67% of the thawed genotypes, varying from 43% to 91% of the thawed genotypes among the full-sib families. The average yield of cotyledonary embryos was 79 ± 4 E/gFW (Table 2).

In thawing lots A to D, the average yield of cotyledonary embryos was 80 ± 4 E/gFW. The embryo yield varied between families ($p < 0.001$). In thawing lot E, in which all the remaining cryopreserved genotypes were thawed from two full-sib families with the lowest number of genotypes producing cotyledonary embryos, on average 46 ± 9 E/gFW were produced. In thawing lot F, the mean yield of cotyledonary embryos was 92 ± 10 E/gFW. In thawing lots E and F, the embryo yield varied among full-sib families ($p = 0.002$).

Overall, 12,910 germinated emblings from thawing lots A to F were transplanted into peat-based growth media. From thawing lots A to D, 8904 germinated emblings from 340 genotypes (varying from 21 to 34 genotypes per family) were transplanted into small growing containers. From thawing lots E and F, 4013 emblings were transplanted straight into Plantek 81f containers. In thawing lots A to F, the embling survival rate after the first growing period or season varied between full-sib families ($p < 0.001$).

From thawing lots A to D, 3837 growing emblings (43%) were recorded after the growing period, varying from 21 to 34 genotypes per family. After cold storage and the 2017 growing season in the nursery, only 1481 (17% overall) of the transplanted emblings from 12 to 28 genotypes per family were alive.

In thawing lots E and F, 4006 emblings from 127 genotypes, from two to 24 genotypes per family, were transplanted into Plantek 81f containers in May 2017. In autumn 2017, the number of vital emblings was 3196 (80% survival), consisting of 121 genotypes, from two to 23 genotypes per family.

In October 2017, a total of 4677 plants were vital from 356 genotypes (50% of all thawed genotypes), from 42 to 18 genotypes among full-sib families. The embling survival rate varied between thawing lots and full-sib families ($p < 0.001$, in both). When the inventories from thawing lots A to D after the first growing period were combined with the inventory from thawing lots E and F after the 2017 growing season, the average survival rate of transplanted emblings was 54%. Survival varied from 50% to 89% among full-sib families. Numbers include emblings from 439 genotypes (62% of the thawed genotypes) varying from 26 to 48 genotypes between full-sib families.

4. Discussion

The embryo yield increased significantly in different groups of genotypes when the ABA concentration in the maturation media was reduced. The increase in embryo yield was over two-fold in Trials 1 and 2 of Experiment I when the reduced amount of ABA was compared to the control 60 µM ABA treatment. Adding PEG to the semi-solid maturation media increased the embryo yield when compared to 60 µM ABA content. However, the highest embryo yield was gained without PEG. Using maturation media with 20 µM ABA and subculturing embryogenic tissues with filter paper resulted in the highest embryo yield. However, the subculturing is labour intensive when used with a large number of samples. For this reason, a maturation protocol, including media containing 30 µM ABA without subculturing was introduced as standard protocol.

When the ABA content in the maturation media was lowered, the average embryo production increased slightly among a large number of samples in a group of different genotypes (Trial 3 in Experiment I) which had been cryopreserved and thawed at least once before. This happened although more variation between individual genotypes was observed. Additionally, in previous studies, cryopreservation has not systematically reduced embryo production (see e.g., [8,25]). The loss of genetic material by cryopreservation accounted for 5% of the thawed genotypes distributed between five full-sib families. The possible loss of genotypes due to cryopreservation is a potential risk when a small number of samples are initially cryopreserved (see e.g., [8,25]).

The highest survival rate was recorded when germinated emblings were transplanted straight into the nursery (GT-II), although the average survival rate was significantly higher among emblings grown in small containers for the first growing period (GT-I) before cold storage. The decline in embling survival among emblings in GT-I was most likely due to improper cold storage conditions and fungicide treatments, which resulted in a severe infestation of mold (possibly *Botrytis* sp.).

The plants grown in treatment GT-II were significantly taller, although they were grown for one growing period less. This is most likely due to insufficient lighting in the growth room in treatment GT-I combined with the damage caused to initial meristems during cold storage. In treatment GT-I, the growth room light intensity in the photosynthetically active radiation (PAR) region in full light was set to 150 $\mu mol/m^{-2}/s^{-1}$. In the greenhouse, emblings were subjected to sunlight filtered through the plastic outer wall of the greenhouse. The suggested minimum light intensity to support height growth and prevent premature terminal bud formation on seedlings is 250 $\mu mol/m^{-2}/s^{-1}$ according to Landis et al. [41,42]. Emblings survived well in greenhouse conditions after one-week in vitro germination in treatment GT-II. This implies that a higher light intensity is not harmful for small emblings, as long as the air temperature is favourable. This was also observed previously by Tikkinen et al. [23].

Growing emblings in small containers in artificial conditions is plausible, although special attention needs to be paid to the cold storage conditions, similar to seedlings. Improvements in environmental control are still needed to match the properties of the emblings or seedlings obtained from conventional nurseries. Although, growing space can be saved by using small containers, the method requires additional labour and supplies (i.e., containers, substrate etc.) compared to the method where emblings are transplanted straight into the nursery (see e.g., [23,41]). Additionally, in Nordic conditions supplementary lighting is necessary during a short photoperiod. All the previous matters have an effect on the cost of the emblings. A detailed cost analysis is needed to define the threshold for embling vigor in containers, to determine whether using small containers is profitable.

The embling survival of 77% for cotyledonary embryos selected for in vitro germination is close to the germination and early vigour percentages of unimproved Norway spruce seeds and seedlings reported in earlier studies (see e.g., [43]). The observed mean height of 10.9 cm indicates that the emblings will reach sufficient height standards for planting during the second growing season. By selecting well acclimatizing genotypes for production, survival can be further increased, as suggested by Tikkinen et al. [23].

Genotypes with a high embryo production capacity (over 200 E/gFW, with lowered ABA content) were found, and vigorous emblings were successfully produced from several genotypes from all full-sib families (Figure 1; Table 2). Not all of the cryostored genotypes from the families used were tested, although the number of samples was fairly large varying from 34 to 106 per family. Högberg [44] found that the loss of genotypes during the overall SE process can be vast. According to our results, genotypes with a good embryo production capacity can be found in all families. We thawed a sample of 57% of the 1247 cryopreserved genotypes, representing 12 full-sib families. With the current rates of genotypes successfully initiated (55%) and cryopreserved (45%) of the total number of explants (2744) from selected families, the sample of 712 thawed genotypes accounts for 1292 to 1567 explants (Table 2). In the autumn of 2017, vital emblings were recorded from 356 genotypes, i.e., 23% to 28% of the estimated explants. Without the loss of emblings due to improper storage, emblings from 28% to 34% of the estimated explants could have been obtained. These genotypes could possibly be used for the large-scale propagation of emblings.

Plant loss during the cold storage of emblings from thawing lots A to D was severe, as 61% of emblings which had formed initial buds after the first growing period were lost during cold storage after the first growing period. This was most likely due to the same reasons as those found in the GT-I Experiment II, as the emblings were kept in the same cold storage unit. It is unacceptable to lose emblings at this magnitude after the most critical steps of embling production have been endured.

Hence, controlling the storage conditions and applying chemical treatments against mold must not be overlooked.

The genetic entry and early phases in the SE process, set limits for the later phases of the production (see e.g., [20]). Our results show, that the final output, e.g., the yield and quality of the emblings, together with the genetic diversity of the output of the SE process can be significantly improved by developing the later phases of production.

When a maturation media with a lower ABA concentration was used, the embryo yield increased to such an extent that transplanting germinated emblings in vitro straight into the nursery may be considered the best growing method. This is in line with the findings of [23,45]. Clone testing can be initiated significantly faster by using 12 emblings as cutting donors from each genotype, compared to using seedlings as cutting donors (see e.g., [46]). To succeed, on average 16 cotyledonary embryos need to be germinated for one-week in vitro and transplanted straight into a nursery for each genotype, with a 77% survival rate (Experiment II).

To initiate clone testing directly with emblings, 24 to 32 ramets per genotype are needed in the Finnish spruce breeding programme [47]. To produce this number of emblings, an average of 32 to 42 cotyledonary embryos are needed, with an expected survival rate of 77% from Experiment II. From the poorest family in Experiment III, 37 to 49 cotyledonary embryos are needed with an expected survival rate of 77%.

From the current data, covering 12 families in Experiment III, 394 genotypes reached the embryo production limit of 16 cotyledonary embryos from three maturation dishes (25% to 30% of estimated explants, 27 genotypes in the poorest family) and 336 produced over 32 cotyledonary embryos from three maturation dishes (21% to 26% of the estimated explants, 17 genotypes in poorest family). The current results indicate that large-scale field testing can be initiated from cryostored ETs with the current production protocol. Protocol includes short maturation on semisolid media containing 30 µM ABA, cold storing cotyledonary embryos on maturation dishes, and transplanting emblings into a nursery after one-week in vitro germination.

5. Conclusions

Reducing the ABA concentration in the maturation media increased the yield of cotyledonary embryos. When combined with state-of-the-art embryo storage and in vitro germination protocols, emblings can be grown and large-scale field testing can be initiated with current nursery protocols with material from a wide genetic background. Intensive planning, considering production in the laboratory and in the nursery, is essential to achieve good results. Despite large variations in embryo/embling production, genotypes with good production properties were found in all families in this study. Furthermore, automation is needed to further increase the production numbers and cost effectiveness, especially in the laboratory-nursery interface, where the embryos and emblings are individually handled.

Author Contributions: M.T. had the main responsibility for the planning and experimental set up for the study (Experiments I (Trial 3), II, III) and had the main responsibility for data measurements and analysis, and the writing of the manuscript. S.V. had the main responsibility for planning, the experimental set up, taking measurements and data analysis of Experiment I (Trials 1 and 2), and participated in the writing of the manuscript. T.A. participated in the planning of the experimental set up for the study in all experiments (including data measurements and analyses) and in the writing of the manuscript.

Funding: The writers would like to thank the European Regional Development Fund, South Savo Regional Council, Savonlinna Business Services Ltd. and Savonlinna municipality for funding this research.

Conflicts of Interest: The authors declare no conflict of interest.

References

1. Haapanen, M.; Leinonen, H.; Leinonen, K. Männyn ja kuusen siemenviljelyssiemenen taimitarhakäytön kehitys 2006–2016: Alueellinen tarkastelu. *Metsätieteen Aikakauskirja* **2017**. [CrossRef]

2. Chalupa, V. Somatic embryogenesis and plantlet regeneration from cultured immature and mature embryos of *Picea abies* (L.) Karst. *Commun. Inst. For. Czech Repub.* **1985**, *14*, 57–63.

3. Hakman, I.; Fowke, L.C.; von Arnold, S.; Eriksson, T. The development of somatic embryos in tissue cultures initiated from immature embryos of *Picea abies* (Norway spruce). *Plant Sci.* **1985**, *38*, 53–59. [CrossRef]

4. Bonga, J.M. Conifer clonal propagation in tree improvement programs. In *Vegetative Propagation of Forest Trees*; Park, Y.-S., Bonga, J.M., Moon, H.-K., Eds.; Korea Forest Research Institute: Seoul, Korea, 2016; pp. 3–31.

5. Burdon, R.D.; Aimers-Halliday, Y.J. Risk management for clonal forestry with *Pinus radiate*–Analysis and review 1: Strategic issues and risk spread. *N. Z. J. For. Sci.* **2003**, *33*, 156–180.

6. Grossnickle, S.C.; Cyr, D.; Polonenko, D.R. Somatic embryogenesis tissue culture for the propagation of conifer seedlings: A technology comes of age. *Tree Plant. Notes* **1996**, *47*, 48–57.

7. Högberg, K.-A.; Bozhkov, P.V.; Grönroos, R.; von Arnold, S. Critical Factors Affecting *Ex Vitro* Performance of Somatic Embryo Plants of *Picea abies*. *Scand. J. For. Res.* **2001**, *16*, 295–304. [CrossRef]

8. Varis, S.; Ahola, S.; Jaakola, L.; Aronen, T. Reliable and practical methods for cryopreservation of embryogenic cultures and cold storage of somatic embryos of Norway spruce. *Cryobiology* **2017**, *76*, 8–17. [CrossRef] [PubMed]

9. Cyr, D.R.; Lazaroff, W.R.; Grimes, S.M.; Quan, G.; Bethune, T.D.; Dunstan, D.I.; Roberts, D.R. Cryopreservation of interior spruce (*Picea glauca engelmannii complex*) embryogenic cultures. *Plant Cell Rep.* **1994**, *13*, 574–577. [CrossRef] [PubMed]

10. Find, J.; Floto, F.; Krogstrup, P.; Møller, J.D.; Nørgaard, J.V.; Kristensen, M.M.H. Cryopreservation of an embryogenic suspension culture of Picea sitchensis and subsequent plant regeneration. *Scand. J. For. Res.* **1993**, *8*, 156–162. [CrossRef]

11. Hazubska-Przybył, T.; Chmielarz, P.; Michalak, M.; Bojarczuk, K. Cryopreservation of embryogenic tissues of *Picea omorika* (Serbian spruce). *Plant Cell Tissue Organ Cult.* **2010**, *102*, 35–44. [CrossRef]

12. Häggman, H.M.; Ryynänen, L.A.; Aronen, T.S.; Krajnakova, J. Cryopreservation of embryogenic cultures of Scots pine. *Plant Cell Tissue Organ Cult.* **1998**, *54*, 45–53. [CrossRef]

13. Klimaszewska, K.; Ward, C.; Cheliak, W. Cryopreservation and plant regeneration from embryogenic cultures of larch (*larix X eurolepis*) and black spruce (*Picea mariana*). *J. Exp. Bot.* **1992**, *43*, 73–79. [CrossRef]

14. Kong, L.; von Aderkas, P. A novel method of cryopreservation without a cryoprotectant for immature somatic embryos of conifer. *Plant Cell Tissue Organ Cult.* **2011**, *106*, 115–125. [CrossRef]

15. Nørgaard, J.V.; Duran, V.; Johnsen, Ø.; Krogstrup, P.; Baldursson, S.; von Arnold, S. Variations in cryotolerance of embryogenic *Picea abies* cell lines and the association to genetic, morphological, and physiological factors. *Can. J. For. Res.* **1993**, *23*, 2560–2567. [CrossRef]

16. Adams, G.W.; Kunze, H.A.; McCartney, A.; Millican, S.; Park, Y.-S. An industrial perspective on the use of advanced reforestation stock technologies. In *Vegetative Propagation of Forest Trees*; Park, Y.-S., Bonga, J.M., Moon, H.-K., Eds.; Korea Forest Research Institute: Seoul, Korea, 2016; pp. 323–334.

17. Lelu-Walter, M.-A.; Thompson, D.; Harvengt, L.; Sanchez, L.; Toribio, M.; Pâques, L.E. Somatic embryogenesis in forestry with a focus on Europe: State-of-the-art, benefits, challenges and future direction. *Tree Genet. Genomes* **2013**, *9*, 883–899. [CrossRef]

18. Bonga, J.M. A comparative evaluation of the application of somatic embryogenesis, rooting of cuttings, and organogenesis of conifers. *Can. J. For. Res.* **2015**, *45*, 379–383. [CrossRef]

19. Carson, M.; Carson, S.; Te Riini, C. Successful Varietal Forestry with Radiata Pine in New Zealand. *N. Z. J. For.* **2015**, *60*, 8–11.

20. Thompson, D. Challenges for the large-scale propagation of forest trees by somatic embryogenesis—A review. In Proceedings of the 3rd International Conference of the IUFRO Unit 2.09.02 on "Woody Plant Production Integrating Genetic and Vegetative Propagation Technologies", Vitoria-Gasteiz, Spain, 8–12 September 2014; pp. 81–91.

21. Högberg, K.-A.; Ekberg, I.; Norell, L.; von Arnold, S. Integration of somatic embryogenesis in a tree breeding programme: A case study with *Picea abies*. *Can. J. For. Res.* **1998**, *28*, 1536–1545. [CrossRef]

22. Majada, J.P.; Sierra, M.I.; Sanchez-Tames, R. Air exchange rate affects the in vitro developed leaf cuticle of carnation. *Sci. Hortic.* **2001**, *87*, 121–130. [CrossRef]

23. Tikkinen, M.; Varis, S.; Peltola, H.; Aronen, T. Improved germination conditions for Norway spruce somatic cotyledonary embryos increased survival and height growth of emblings. *Trees*. submitted.

24. Lelu, M.-A.; Bastien, C.; Klimaszewska, K.; Ward, C.; Charest, P.J. An improved method for somatic plantlet production in hybrid larch (*Larix x leptoeuropaea*). Part 1. Somatic embryo maturation. *Plant Cell Tissue Organ Cult.* **1994**, *36*, 107–115. [CrossRef]

25. Find, J. Changes in endogenous ABA levels in developing somatic embryos of Norway spruce (*Picea abies* (L.) Karst.) in relation to maturation medium, desiccation and germination. *Plant Sci.* **1997**, *128*, 78–83. [CrossRef]

26. Von Arnold, S.; Hakman, I. Regulation of somatic embryo development in *Picea abies* by abscisic acid (ABA). *J. Plant Physiol.* **1988**, *132*, 164–169. [CrossRef]

27. Gutmann, M.; von Aderkas, P.; Label, P.; Lelu, M.-A. Effects of abscisic acid on somatic embryo maturation of hybrid larch. *J. Exp. Bot.* **1996**, *47*, 1905–1917. [CrossRef]

28. Lelu, M.-A.; Bastien, C.; Klimaszewska, K.; Ward, C.; Charest, P.J. An improved method for somatic plantlet production in hybrid larch (*Larix x leptoeuropaea*). Part 2. Control of germination and plantlet development. *Plant Cell Tissue Organ Cult.* **1994**, *36*, 117–127. [CrossRef]

29. Lelu, M.-A.; Label, P. Changes in the levels of abscisic acid and its glucose ester conjugate during maturation of hybrid larch (*Larix x leptoeuropaea*) somatic embryos, in relation to germination and plantlet recovery. *Physiol. Plant.* **1994**, *92*, 53–60. [CrossRef]

30. Attree, S.M.; Fowke, L.C. Embryogeny of gymnosperms: Advances in synthetic seed technology of conifer plants. *Plant Cell Tissue Organ Cult.* **1993**, *35*, 1–35. [CrossRef]

31. Attree, S.M.; Pomeroy, M.K.; Fowke, L.C. Development of white spruce (Picea glauca (Moench.) Voss) somatic embryos during culture with abscisic acid and osmoticum, and their tolerance to drying and frozen storage. *J. Exp. Bot.* **1995**, *285*, 433–439. [CrossRef]

32. Kong, L.; Yeung, E.C. Effects of silver nitrate and polyethylene glycol on white spruce (*Picea glauca*) somatic embryo development: Enhancing cotyledonary embryo formation and endogenous ABA content. *Physiol. Plant.* **1995**, *93*, 298–304. [CrossRef]

33. Merkle, S.A.; Montello, P.M.; Reece, H.M.; Kong, L. Somatic embryogenesis and cryostorage of eastern hemlock and Carolina hemlock for conservation and restoration. *Trees* **1994**, *28*, 1767–1776. [CrossRef]

34. Attree, S.M.; Pomeroy, M.K.; Fowke, L.C. Manipulation of conditions for the culture of somatic embryos of white spruce for improved triacylglycerol biosynthesis and desiccation tolerance. *Planta* **1992**, *198*, 395–404. [CrossRef] [PubMed]

35. Bozhkov, P.V.; von Arnold, S. Polyethylene glycol promotes maturation but inhibits further development of Picea abies somatic embryos. *Physiol. Plant.* **1998**, *104*, 211–224. [CrossRef]

36. Svobodová, H.; Albrechtová, J.; Kumstýřová, L.; Lipavská, H.; Vágner, M.; Vondráková, Z. Somatic embryogenesis in Norway spruce: Anatomical study of embryo development and influence of polyethylene glycol on maturation process. *Plant Physiol. Biochem.* **1999**, *37*, 209–221. [CrossRef]

37. Klimaszewska, K.; Lachance, D.; Pelletier, G.; Lelu, A.-M.; Seguin, A. Regeneration of transgenic *Picea glauca*, *P. mariana*, and *P. abies* after cocultivation of embryogenic tissue with Agrobacterium tumefaciens. *In Vitro Cell. Dev. Biol. Plant* **2001**, *37*, 748–755.

38. Litvay, J.D.; Verma, D.C.; Johnson, M.A. Influence of loblolly pine (*Pinus taeda* L.) culture medium and its components on growth and somatic embryogenesis of wild carrot (*Daucus carota* L.). *Plant Cell Rep.* **1985**, *4*, 325–328. [CrossRef] [PubMed]

39. Belmonte, M.F.; Yeung, E.C. The effects of reduced and oxidized glutathione on white spruce somatic embryogenesis. *In Vitro Cell. Dev. Biol. Plant* **2004**, *40*, 61–66. [CrossRef]

40. Klimaszewska, K.; Park, Y.-S.; Overton, C.; Maceacheron, I.; Bonga, J.M. Optimized somatic embryogenesis in *Pinus strobus* L. *In Vitro Cell. Dev. Biol. Plant* **2001**, *37*, 392–399. [CrossRef]

41. Landis, T.D.; Dumroese, R.K.; Haase, D. *The Container Tree Nursery Manual. Vol. 6. Seedling Propagation*; Agricultural Handbook 674; U.S. Department of Agriculture, Forest Service: Washington, DC, USA, 2010; pp. 3–164.

42. Rikala, R. *Metsäpuiden Paakkutaimien Kasvatusopas (Container Seedling Growing Manual for Forest Trees)*; The Finnish Forest Research Institute: Suonenjoki, Finland, 2012; 247 p.

43. Nygren, M. *Metsäpuiden Sisemenopas (Seedling Guide for Forest Trees)*; Metsäntutkimuslaitoksen Tiedonantoja: Suonenjoki, Finland, 2003; Volume 882, 144 p.

44. Högberg, K.-A. Possibilities and limitations of vegetative propagation of Norway spruce. In *Acta Universitatis Agriculturae Sueciae, Silvestria 294*; Swedish University of Agricultural Sciences: Uppsala, Switzerland, 2003.

45. Von Aderkas, P.; Kong, L.; Prior, N.A. In Vitro techniques for conifer embryogenesis. In *Vegetative Propagation of Forest Trees*; Park, Y.-S., Bonga, J.M., Moon, H.-K., Eds.; Korea Forest Research Institute: Seoul, Korea, 2016; pp. 335–350.

46. Tikkinen, M.; Varis, S.; Peltola, H.; Aronen, T. Norway spruce Emblings as Cutting Donors for Tree Breeding and Production. *Scand. J. For. Res.* **2017**, *33*, 207–214. [CrossRef]

47. Haapanen, M. Clones in Finnish tree breeding. In *Proceedings of the Nordic Meeting Held in September 10th–11th 2008 at Punkaharju, Finland*; Working papers of the Finnish Forest Research Institute; Finnish Forest Research Institute: Vantaa, Finland, 2009; Volume 114, pp. 16–19.

forests

Article

Effects of Seedling Quality and Family on Performance of Northern Red Oak Seedlings on a Xeric Upland Site

Cornelia C. Pinchot [1,*], Thomas J. Hall [2], Arnold M. Saxton [3], Scott E. Schlarbaum [4] and James K. Bailey [5]

[1] USDA Forest Service, Northern Research Station, Delaware, OH 43015, USA
[2] Pennsylvania Bureau of Forestry, Harrisburg, PA 17105, USA; thall@pa.gov
[3] Department of Animal Science, The University of Tennessee, Knoxville, TN 37996, USA; asaxton@utk.edu
[4] Department of Forestry, Wildlife, and Fisheries, The University of Tennessee, Knoxville, TN 37996, USA; tenntip@utk.edu
[5] Pennsylvania Bureau of Forestry, Harrisburg, PA 17105, USA; acornsbailey@verizon.net
* Correspondence: corneliapinchot@fs.fed.us; Tel.: +1-740-368-0039

Received: 2 May 2018; Accepted: 11 June 2018; Published: 13 June 2018

Abstract: Cultural practices to develop larger, more robust oak seedlings have been developed, however, the potential improvement conferred by these larger seedlings has received limited testing in the Northeast. We evaluated the effect of seedling size and pedigree on the survival, growth, and competitive ability of northern red oak (*Quercus rubra* L.) seedlings planted on a xeric site in northeastern Pennsylvania. We planted seedlings from a state tree nursery that represented locally available seedling stock, as well as high-quality seedlings from seven half-sibling families grown following improved nursery protocol. Half-sibling families were split into three size classes based on their root collar diameter and height; large, average, and poor. Eleven years after planting, survival across seedling treatments ranged from 45 percent for locally available seedlings, to 96 percent for one half-sibling family. Two families showed superior growth, survival, and competitive ability compared with the others. Seedling size class conferred moderate height and diameter advantage in four and three of the families, respectively. Initial seedling size was an important variable in models predicting survival, diameter, and dominance (competitive ability). Over time, the relationship between initial diameter and height diminished.

Keywords: northern red oak; *Quercus*; *Quercus rubra*; artificial regeneration; seedling quality

1. Introduction

Oak (*Quercus* spp. L.) is an important genus throughout much of the Northern Hemisphere, though its abundance has declined in recent years, largely due to regeneration failures [1,2] caused by changes to disturbance regimes [3,4], browsing by white tailed deer (*Odocoileus virginianus* Zimmerman) [5], fire suppression [6,7] and interference from invasive plant species [8]. A robust field of research focusing on silvicultural methods for increasing oak advance regeneration has developed to address these declines, e.g., [9,10]. In the central Appalachians, shelterwood and midstory removal treatments, prescribed fire, and control of interfering vegetation, often in tandem with control of browsing, have been found to successfully encourage establishment of oak regeneration [11,12].

The presence of competitive oak seedlings, saplings and/or sprouts before a harvest is a critical requirement for successful regeneration of oak post-treatment [13,14]. Where adequate advance reproduction of oak is lacking, artificial regeneration can be a useful tool to meet stocking goals. Establishment success of planted oak into forested stands, however, has been variable, with failures caused by poor seedling quality [10], competition from fast growing species [4,15,16], moisture stress associated with transplant shock [17], browsing by deer [18,19], and lack of genetically improved planting stock [20].

The importance of seedling size to early survival and growth in northern red oak (*Quercus rubra* L.) has been well studied [10,21–28]. Larger oak seedlings tend to have larger root systems, which provide carbohydrates, water, and nutrients necessary to promote rapid growth, a key trait for competing with fast-growing seedlings and sprouts [10,29]. Cultural practices to develop larger, more robust oak seedlings have been developed [30], however, improvement conferred by these larger seedlings has received limited testing, most occurring on mesic sites in the southeastern U.S. [24,31,32]. Early testing (<8 years) suggests that using high-quality seedlings (sensu [30]) produced through advanced irrigation and fertilization protocols yields improved growth and competitive ability of northern red oak when planted in previously harvested stands [24,31,32]. However, the use of seedlings produced using these protocols has not been tested in xeric stands in the Northeast. While oak is easier to naturally regenerate on drier as compared with more productive sites, the presence of fast growing red maple (*Acer rubrum* L.) and birch (*Betula* spp. L.) often limits successful oak recruitment on xeric sites [33]. High-quality seedlings may be able to compete with fast-growing advance reproduction on such sites, offering a management alternative where oak regeneration is lacking. To test this, we compared the survival, growth, and competitive ability among varying size classes and families of northern red oak seedlings planted on a xeric site in northeastern Pennsylvania. Because there are a limited number of studies using high-quality northern red oak seedlings that also include an assessment of family-origin effects on establishment success [34], we also included a family treatment.

2. Materials and Methods

2.1. Experimental Material

Acorns from seven open-pollinated northern red oak mother trees, kept separate by pedigree, and a bulked collection were used in this study. The acorns from the half-sibling families were harvested from mother trees located in natural forest stands at the United States Military Academy Reservation, West Point, NY, and proximal area in the fall of 2003. Mother trees were located at least 0.40 km apart to avoid collecting closely related material. Acorns were sown in separate family seed lots in one nursery bed at the Georgia Forestry Commission's Flint River Nursery in Byromville, GA in December, 2003 at a density of 65 seeds per m^2. Families were not replicated within the nursery bed. Fertilization and irrigation schedules followed guidelines developed by Kormanik and others [30] and were designed to produce high-quality seedlings. The seedlings were grown without root-pruning or top-clipping protocols. The one year (1–0) half-sibling seedlings were lifted in late January, 2005, transported to Knoxville, TN, USA, and placed in cold storage (~1 °C). Total height and root collar diameter of each seedling were measured, and an individual identification tag was attached to each seedling. All half-sibling seedlings were visually sorted into three size classes within each family; small, average, and large; according to height and root collar diameter (sensu [35]); (Tables 1 and 2). We used the minimum root collar diameter (8–10 mm, [36]) recommended for northern red oak seedlings as the standard for our average seedling size class. Acorns for the treatment representing locally available seedlings were collected from multiple mother trees across the Ridge and Valley region of Pennsylvania and planted at a density of approximately 190 acorns per m^2 at the Penn Nursery in Spring Mills, PA, USA. Seedlings were grown for one year and 1–0 bare-root seedlings were lifted in the spring of 2005 and kept in cold storage until they were transported to the study site for planting. These seedlings are termed "locally available" throughout the paper, represent seedlings available to landowners and forest managers in the Northeast at the time of the study, and are compared with the high-quality seedlings produced using advanced nursery protocol (sensu [30]).

Table 1. Mean height (± standard error) of families and seedling size classes within family at planting and eleven years after planting. LA indicates locally available seedlings. Means among families for family treatment and within family for size class (family) treatment with the same letters do not differ statistically (α = 0.05).

Family	Family Height at Planting (cm)	Family Height after 11 Years (cm)	Size Class within Family	Size Class Height at Planting (cm)	Size Class Height after 11 Years (cm)
8	50 ± 2 C	406 ± 13 A	Small	38 ± 4 C	401 ± 25 A
			Average	51 ± 3 B	391 ± 18 A
			Large	62 ± 3 A	426 ± 45 A
9	42 ± 1 E	322 ± 10 B	Small	37 ± 2 B	302 ± 12 B
			Average	44 ± 2 A	303 ± 13 B
			Large	44 ± 3 A	360 ± 18 A
10	44 ± 1 DE	319 ± 10 BC	Small	40 ± 2 B	282 ± 12 C
			Average	43 ± 2 B	318 ± 14 B
			Large	50 ± 2 A	358 ± 14 A
11	57 ± 1 B	320 ± 10 BC	Small	48 ± 2 B	236 ± 16 C
			Average	52 ± 2 B	329 ± 15 B
			Large	72 ± 2 A	394 ± 13 A
12	44 ± 1 DE	306 ± 12 BC	Small	36 ± 2 B	302 ± 17 A
			Average	46 ± 2 A	289 ± 17 A
			Large	50 ± 3 A	328 ± 17 A
14	46 ± 2 CD	295 ± 13 CD	Small	39 ± 3 B	275 ± 20 B
			Average	43 ± 2 B	212 ± 17 C
			Large	56 ± 4 A	397 ± 22 A
16	64 ± 2 A	377 ± 14 A	Small	45 ± 4 C	348 ± 25 A
			Average	61 ± 3 B	408 ± 21 A
			Large	87 ± 3 A	373 ± 21 A
LA	30 ± 1 F	232 ± 9 E	LA	30 ± 1	232 ± 9

Table 2. Basal diameter (BD, mm) at planting, diameter at breast height (DBH, mm) after eleven years—with trees too short to record DBH (1.27 m) given a DBH of 0 mm—for family and for seedling size class within family. LA indicates locally available seedlings. Means among families for family treatment and within family for size class (family) treatment with the same letters do not differ statistically (α = 0.05).

Family	Family BD at Planting	Family 11th Year DBH	Size Class within Family	Size Class BD at Planting	% of Seedlings Large Enough to Record DBH	Size Class 11th Year DBH—All Living Trees
8	8.5 ± 0.2 AB	28.0 ± 4.9 A	Small	6.8 ± 0.5 C	90	27.4 ± 6 A
			Average	8.6 ± 0.3 B	95	26.4 ± 5 A
			Large	10.0 ± 0.3 A	95	29.2 ± 5 A
9	8.0 ± 0.1 B	18.6 ± 4.7 BC	Small	7.0 ± 0.2 C	90	15.5 ± 5 A
			Average	8.0 ± 0.2 B	75	19.8 ± 5 A
			Large	9.1 ± 0.3 A	90	18.8 ± 5 A
10	8.3 ± 0.1 B	19.8 ± 4.6 B	Small	7.1 ± 0.1 C	78	18.8 ± 5 B
			Average	8.3 ± 0.2 B	82	16.8 ± 5 B
			Large	9.4 ± 0.2 A	100	22.4 ± 5 A
11	8.4 ± 0.1 AB	19.1 ± 4.7 BC	Small	7.0 ± 0.3 C	77	12.7 ± 5 B
			Average	8.0 ± 0.3 B	96	18.2 ± 5 B
			Large	10.2 ± 0.2 A	95	25.1 ± 5 A
12	8.4 ± 0.2 AB	20.3 ± 4.8 BC	Small	6.7 ± 0.3 C	94	17.4 ± 5 A
			Average	8.6 ± 0.3 B	88	18.4 ± 5 A
			Large	9.8 ± 0.3 A	100	23.8 ± 5 A
14	8.0 ± 0.2 B	17.6 ± 4.9 BC	Small	6.8 ± 0.3 C	83	12.2 ± 6 B
			Average	7.8 ± 0.3 B	63	12.7 ± 5 B
			Large	9.6 ± 0.4 A	100	26.0 ± 6 A
16	8.9 ± 0.2 A	26.8 ± 5 A	Small	7.4 ± 0.5 C	89	22.5 ± 6 A
			Average	8.6 ± 0.4 B	100	28.5 ± 6 A
			Large	10.7 ± 0.4 A	100	28.3 ± 6 A
LA	6.7 ± 0.1 C	16.5 ± 4.7 C	LA	6.7 ± 0.01	70	15.8 ± 5

2.2. Study Area

This study was established in April, 2005 on the Delaware State Forest in Blooming Grove, PA, USA (41°25′ N, 75°03′ W, elevation 420 m). This area represents the glaciated low plateau section province of northeastern Pennsylvania, and is dominated by oaks; primarily white oak (*Quercus alba* L.), northern red oak and chestnut oak (*Q. prinus* L.); hickory (*Carya* spp. Nutt), white pine (*Pinus strobus* L.), pitch pine (*P. rigida* Mill.), and red maple. The soils at the site are of the Manlius Series and are characterized as strongly acidic, rocky-silt loam with low soil moisture retention. The stand was clearcut in 1975 as part of a commercial harvest to regenerate oak and other economically desirable hardwood species. Preferential browsing by overabundant white tailed deer inhibited hardwood seedling regeneration and facilitated the establishment of a thick understory of sweet fern (*Comptonia peregrina* (L.) J.M. Coult) and ericaceous shrubs (primarily *Vaccinium* spp. L.). An 8-hectare 2.4 m tall woven-wire deer fence was erected on the site in 2005, prior to planting to protect the experimental material from deer browsing. The fence was removed in 2014 once the majority of planted and naturally regenerating seedlings had surpassed deer browse height. Aside from fencing, no other management measures were implemented during the course of the study.

2.3. Experimental Design

Seedlings were planted in two replicate plots approximately 150 m apart. Within each plot, seedlings were planted in an incomplete block design with four seedlings in each block: one small, one average, and one large size class from ten possible families grown following Kormanik et al.'s protocols [30] (therefore each block contained seedlings from multiple families), and one locally available seedling. There were between 10 and 73 seedlings within each size class (family) treatment, and 271 seedlings from the locally available treatment used in the study. The locally available seedlings were included to compare seedlings available at state tree nurseries in the northeast with seedlings grown under advanced nursery protocol designed specifically to yield high quality oak seedlings [30] (the half-sibling seedlings). A total of 1067 seedlings were planted in a 2.4 m × 2.4 m grid on each of the experimental sites between 12 and 13 April 2005. Three families with a total of 116 seedlings were excluded from the analysis due to low replication across seedling size classes, therefore 951 seedlings from seven families and one bulked seed lot (the locally available seedlings) were included in this study. Seedlings were hand planted using a Jim Gem KBC© bar, modified by adding 5 cm to each side of the 30 cm long blade, creating a blade 15 cm wide at the top, tapering to the tip.

2.4. Measurements

Height and root collar diameter of half-sibling seedlings were measured just after lifting. Height and ground level diameter of the seedlings from the locally available seedlings were measured directly following planting due to the timing of seedling availability. The term basal diameter (BD) is used to describe these baseline diameter measurements throughout the rest of the paper. Basal diameter measurements taken on unplanted bare-root seedlings may be larger than those taken on planted seedlings [37], which in this study may have slightly inflated BD at planting for half-sibling seedlings. Heights of all planted oaks were measured after bud set in 2005–2009, 2011, 2013, and 2015 and diameter at breast height (DBH) was measured following bud set in 2015. All height measurements were taken to the nearest centimeter and diameter measurements to the nearest tenth of a millimeter.

To characterize competing vegetation, a 2.6 m diameter competition plot was centered on each planted oak seedling and species, height and diameter of the tallest woody competitor within each plot was recorded in late 2015 [24,38]. This size plot was chosen because it approximates the space that a dominant or co-dominant tree occupies at crown closure [39]. Presence of stem forking at or below DBH was recorded in October, 2017.

2.5. Statistical Analysis

All analyses for this study were processed using SAS 9.3 software (SAS Institute 2011, Cary, NC, USA). Analysis of variance was used to detect differences among families and seedling size classes within family for initial height and basal diameter. A repeated-measures, linear mixed-model analysis of variance (LMM) with an autoregressive covariance structure was used to test the fixed effects of family (the bulked seed lot used for the locally available seedlings is considered one family for analysis), size class nested within family, year, and year interactions with family and size (family) on total tree heights for each year (2005–2009, 2011, 2013 and 2015). LMM was also used to evaluate the effects of family and size class within family on 2015 DBH. A DBH of '0' was scored for seedlings that were shorter than 1.37 m, the height at which DBH is taken. Stem forking at or below DBH (presence/absence) ($p < 0.0001$) was used as a covariate. Generalized LMM with binomial distribution was used to analyze 2015 survival (1 = alive, 0 = dead) and dominance probability of the seedlings. Surviving seedlings that attained at least 80 percent of the height of the tallest competitor within the competition plot were defined as dominant [21,23,24,40]. Data were checked for homogeneity of variance and normality. Least-significant-difference tests were performed to identify differences among means ($\alpha = 0.05$).

Rank correlations were used to evaluate the relationship between initial basal diameter and height each year it was measured, and correlation coefficients were squared to be expressed as R-square values. Logistic regression (Proc Logistic) was used to develop models, as previously described by Hosmer and Lemeshow [41], to study influences on eleventh-year (2015) survival and dominance probabilities. Initial height and basal diameter, family and size class were used as explanatory variables, as well as species of tallest competitor for dominance. Multiple regression was used to explain DBH in 2015 using the same variables, as well as height and DBH of the tallest competing woody stem. The log transformation of planted seedling height and DBH of the tallest competing woody stem were used to linearize relationships, however, untransformed values are reported in the tables. The most parsimonious model with the lowest corrected Akaike information criterion value was selected for each dependent variable. The Hosmer–Lemeshow goodness-of-fit statistic was used to test that the model adequately explained the data.

Height growth patterns over the length of the study were derived using quadratic regression for each seedling. Negative, positive, and zero slopes created nine different growth patterns, three linear patterns by three quadratic patterns. For example, a +Linear-Quadratic would indicate a seedling with positive growth but decreasing growth rate over time. Contingency tables were then used to compare the frequency of growth patterns across families and seedling size groups. To highlight differences, groups were pooled when no differences were found.

3. Results

3.1. Seedling Grading

Seedling height and basal diameter at the time of planting differed among families ($p < 0.0001$ for both, $F = 72.58$ and 29.51, respectively, Tables 1 and 2) and seedling size classes within family ($p < 0.0001$ for both, $F = 16.12$ and 25.39, respectively, Tables 1 and 2). Basal diameter at the time of planting differed among each size class for all families. Height was generally greater in the large size class than average or small for each family. The average size class was larger in height than the small size class in four of the seven half-sibling families. The locally available seedlings were similar in basal diameter to the small size class for all families and were shorter in height than small size class seedlings for all families except one (8).

3.2. Survival

After eleven years, sixty-eight percent of the planted oak trees were alive, with significant differences in survival among families ($p < 0.0001$, $F = 8.10$, Figure 1), but not among size classes within family (p = 0.1265, F = 1.71). Survival in 2015 ranged from forty-six percent for the locally available seedlings to ninety-three percent for family 8 (Figure 1). The logistic regression model that best explained 2015 survival included basal diameter at planting and family (Table 3, Figure 2). The max re-scaled R-square (adjusted to reach a maximum value of 1) value for this model was 0.20. Across families, the larger the seedling was at planting, the greater probability that it survived. A seedling that was 7.8 mm in BD at planting (the mean BD at planting) from family 8 had an 89 percent chance of survival in 2015, compared with 52 percent for a seedling of the same size from the locally available treatment (Figure 2). Two families (12 and 14) had survival rates similar to the locally available seedlings, lower than all other families, with the greatest mortality occurring in year 2 (Figure 1).

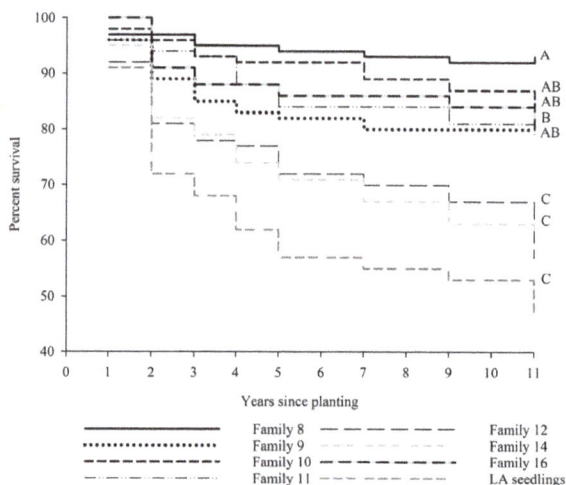

Figure 1. Mean survival among families over the 11-year study. Letters indicate differences in eleventh year survival (α = 0.05).

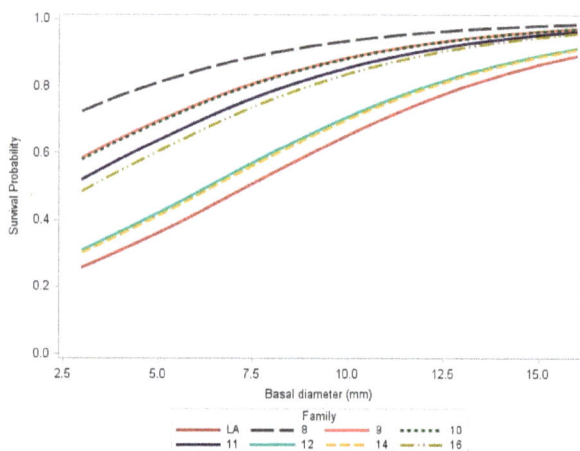

Figure 2. Regression model using family and basal diameter at planting to predict eleventh-year survival. LA indicates locally available seedling treatment.

Table 3. Coefficients and odds ratio values for regression models to predict survival (logistic regression), diameter at breast height (DBH, mm, multiple linear regression), and dominance (logistic regression) of the northern red oak trees eleven years after planting. The locally available seedling treatment (LA) is the reference for the family treatment and planted northern red oak is the reference for species of tallest competitor.

	Survival		DBH	Dominance	
	Coeff.	OR	Coeff.	Coeff.	OR
Intercept	−0.86			−2.02	
Independent variables					
Height at planting (cm)			0.32 ***		
Basal diameter at planting (mm) planting	0.24 ***	1.27		0.05	1.31
DBH of tallest competitor (mm)			0.04 ***		
Family					
8	1.10 **	7.59	12.16 **		
9	0.47 *	4.05	5.63 *		
10	0.45 *	3.98	6.43 **		
11	0.21	3.12	1.41		
12	−0.67 **	1.29	6.61 *		
14	−0.71 **	1.41	2.70		
16	0.07	2.71	6.36		
LA (reference)					
Species of tallest competitor					
Black cherry				−1.39 ***	0.16
Red maple				−0.51 **	0.38
Serviceberry				0.45	0.98
White oak				0.89 **	1.53
Wild northern red oak				0.18	0.75
Other				−0.09	0.58
Planted northern red oak (reference)					

$* p < 0.05, ** p < 0.01, *** p < 0.0001.$

3.3. Growth

The repeated measures height LMM found significant differences among family ($p < 0.0001$, $F = 14.54$, size class within family ($p = 0.0009$, $F = 2.67$), year ($p < 0.0001$, $F = 120.81$), family by year interaction ($p < 0.0001$, $F = 4.70$), and size class (family) by year interaction ($p < 0.0001$, $F = 2.08$). Families 8 and 16 had the greatest height over the course of the study (406 and 378 cm, respectively after 11 years) and trees from the locally available treatment remained the shortest (232 cm, Table 1, Figure 3). We present mean separation only for the height of families and size classes within family at year eleven (Table 1), while all means are displayed in Figure 3. Nearly all size classes were larger in height than seedlings in the locally available treatment after 11 years, except the small and average size classes for one family each (11 and 14, respectively). Large size classes were taller after eleven years than smaller and average size classes in four families, and two families had taller seedlings in the average size class than the small size class. One family (14), exhibited taller seedlings in the small size class than the average size classes.

Basal diameter at planting was positively related to height each year after planting ($p < 0.0001$ each year). Squaring the Pearson correlation coefficients produced R-square values of 0.35, 0.28, 0.24, 0.16, 0.14, 0.09, 0.10, and 0.07, for years 2005–2009, 2011, 2013, and 2015, respectively, exhibiting a relationship that diminished over time until, in year eleven, basal diameter at planting explained less than ten percent of the variation in height.

After eleven years, DBH differed among families and among seedling size classes within family ($p < 0.0001$, for each, $F = 6.52$ and $F = 3.27$, respectively, Table 2). There were no significant interactions between stem fork and eleventh year DBH. The fork was included as a covariate in the LMM model ($p < 0.0001$). Two families (8 and 16) were larger in DBH than the other families, while three families

(9, 11 and 14) were similar to the locally available seedling treatment. Trees in the large size class were larger than those in the average and small classes in three families (10, 11 and 14).

The final multiple linear regression model used to explain eleventh-year DBH included initial height, DBH of the tallest competing woody stem, and family (Table 3). Larger DBH of competing stems and larger initial height of the planted seedlings corresponded to greater DBH. On average, an increase in 10 mm in initial height of a seedlings led to an increase of 3.2 cm in eleventh-year DBH.

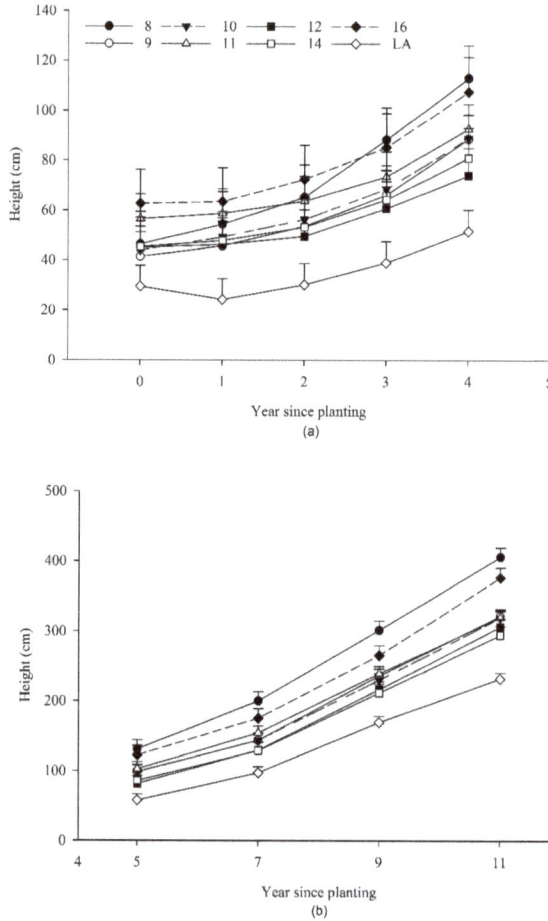

Figure 3. Height among families for years 0–4 (**a**) and 5–11 (**b**). Years are separated into two graphs to improve distinction among lines. Bars indicate standard error. LA indicates locally available seedling treatment.

3.4. Growth Pattern

Seedling height growth was described by six equations developed to model growth patterns over the eleven-year study from the nine possible combinations of positive, negative, or absent linear and quadratic terms. Each equation was developed using a minimum of thirty-one trees (Figure 4). PL-0Q (positive linear, no quadratic) and 0L-PQ (no linear, positive quadratic) growth patterns were pooled because there were no differences in frequency of seedlings found in each and both indicated a general positive growth trend over the length of the study. For the analysis evaluating distribution of families across the growth patterns, similar families were pooled, creating three family bins (Table 4). Pooling reduced the model chi-square from 116.75 with 35 degrees of freedom to 87.79 with 12 degrees of freedom ($p < 0.0001$ for each). Two families (12 and 14) were pooled with the locally available seedling treatment, four families were pooled into another bin (families 9, 10, 11 and 16), while family 8 remained in its own bin (Table 4). Fifty-five percent of seedlings in family 8 exhibited the PL-PQ growth pattern (steadily increasing growth rate); more than double the amount in the other two bins. Substantially more seedlings in the bin with the locally available seedlings exhibited the 0L-0Q (mortality or nominal growth) and NL-PQ growth patterns (top dieback, followed by re-sprouting), compared with seedlings in the other bins. For evaluating distribution of seedling size classes across growth patterns, small and average seedlings were similar and therefore pooled. Pooling these size classes and PL-0Q and 0L-PQ growth patterns reduced the model chi-square from 69.23 with 15 degrees of freedom to 56.95 with 8 degrees of freedom. There were significantly more seedlings from the locally available treatment in the 0L-0Q and fewer seedlings in the PL-PQ growth patterns than the other seedling size classes (Table 5).

Table 4. Frequency table showing growth patterns (depicted in Figure 4) displayed by families over the eleven-year study. LA indicates locally available seedling treatment. Families were pooled when their frequency across growth patterns did not differ, according to chi-square values. The percent of seedlings per group and chi-square value of each cell are listed. The overall chi-square value of the table is 87.79 ($p < 0.0001$) with twelve degrees of freedom.

Growth Pattern		Family			
		8	9, 10, 11, 16	12, 14 and LA	Total Seedlings (%)
0L-0Q	Percent	4	9	19	13
Trees died or added little growth	chi-square	3.50	4.19 *	9.25 *	
NL-PQ	Percent	8	28	34	29
Trees died back then grew	chi-square	8.50 *	0.16	2.59	
PL-NQ	Percent	4	5	9	7
Trees grew then died or died back	chi-square	0.70	1.11	2.30	
PL-PQ	Percent	55	23	10	19
Growth rate increased steadily over time	chi-square	34.32 *	2.47	16.73 *	
0L-PQ, PL-0Q (pooled)	Percent	30	34	29	32
No initial growth	chi-square	0.04	0.92	1.01	

Asterisks (*) indicates the observed number within a cell differs significantly from expected.

Table 5. Frequency table showing growth patterns (depicted in Figure 4) displayed by seedling size classes over the eleven-year study. LA indicates locally available seedling treatment. Small and average seedling size classes did not differ in frequency across the growth patterns and therefore were pooled. The percent of seedlings per group and chi-square value of each cell are listed. The overall chi-square value of the table is 56.95 ($p < 0.0001$) with eight degrees of freedom.

Growth Pattern		Seedling Size Class			
		Large	Small and Average	LA	Total Seedlings (%)
0L-0Q	Percent	6	11	24	13
Trees died or added little growth	Chi-square	6.11 *	1.6	17.44 *	
NL-PQ	Percent	26	29	33	29
Trees died back then grew	Chi-square	0.46	0.02	0.79	
PL-NQ	Percent	3	8	8	7
Trees grew and then died back	Chi-square	4.5 *	0.6	0.77	
PL-PQ	Percent	27	22	7	19
Growth increased over time	Chi-square	5.66 *	1.2	14.98 *	
0L-PQ, PL-0Q (pooled)	Percent	37	31	28	32
No initial growth	Chi-square	1.86	0.08	0.81	

Asterisks (*) indicates the observed number within a cell differs significantly from expected.

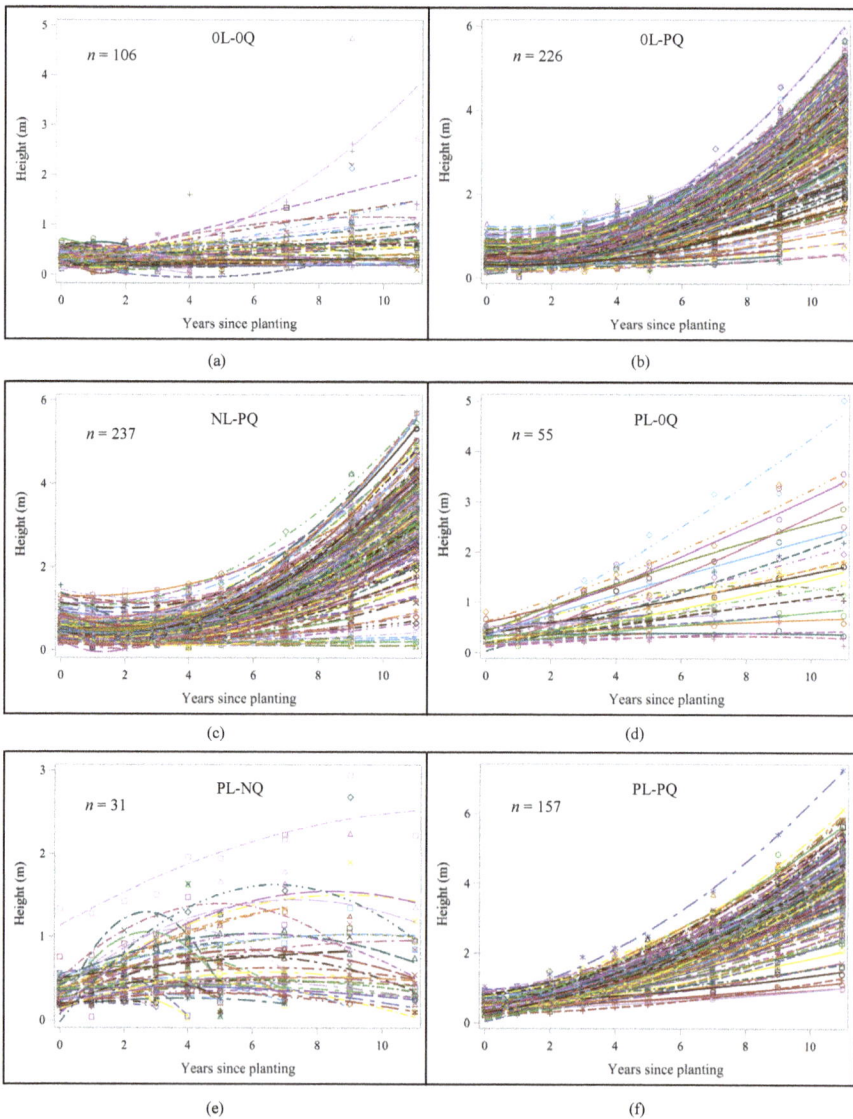

Figure 4. Linear (L) and quadratic (Q) terms were fit into regression models predicting growth patterns for each seedling. Each term was either not significant (0), or showed a positive (P) or negative (N) trend. Each seedling is represented by a single line within one of six growth patterns. (**a**) neither the linear nor quadratic terms were significant; seedlings grew nominally; (**b**) only the quadratic term was significant; seedlings grew slowly and then more rapidly over time; (**c**) negative linear and positive quadratic terms; seedlings experienced die-back before adding growth; (**d**) positive linear and no quadratic terms; seedlings grew linearly; (**e**) positive linear and negative quadratic terms; seedlings grew and then died-back; and (**f**) positive linear and quadratic terms; seedlings grew progressively more rapidly over time.

3.5. Dominance

Red maple was the tallest woody competitor in twenty-six percent of the competition plots. Serviceberry (*Amelanchier* spp. Medik.) was the next most abundant tallest competing species (19 percent of plots), followed by black cherry (*Prunus serotina*, Ehrh., 12 percent), planted northern red oak (11 percent; competition plot diameters were greater than the planting spacing, therefore adjacent planted seedlings were located in competition plots), white oak (9 percent), wild northern red oak (7 percent), and other species (7 percent). Black cherry was the tallest competing species, on average, with a mean height of 562 (±19) cm and 47.5 (±2.8) mm in DBH, compared to the average height of 321 (±6) cm and DBH of 26.5 (±0.5) mm for all planted northern red oak (Figure 5).

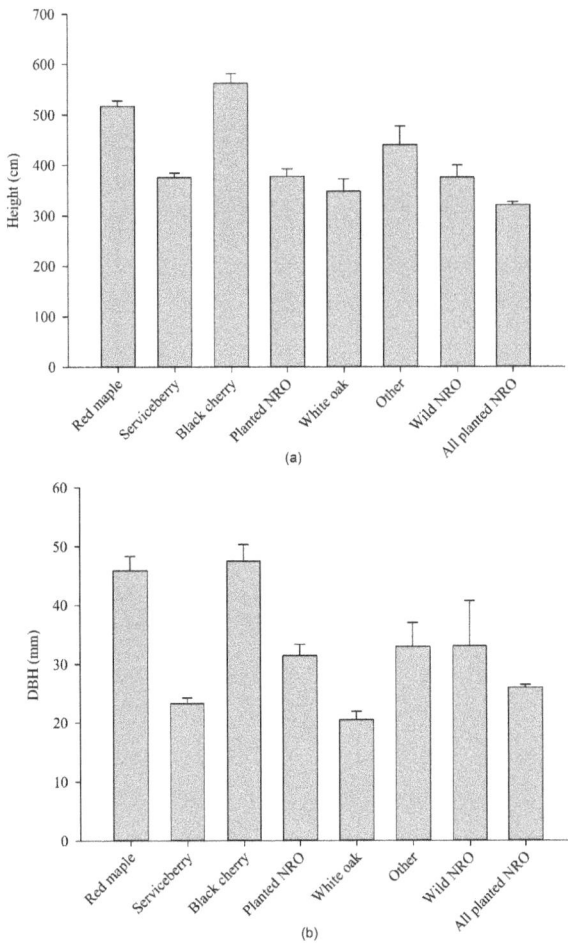

Figure 5. (**a**) Mean height of tallest woody stem found in competition plots, by species and mean height of the planted northern red oak trees. "Planted NRO" includes only planted northern red oak trees found in competition plots, while "All planted NRO" includes all planted northern red oak; (**b**) Mean DBH of tallest woody stem found in competition plots, by species and mean DBH of all planted northern red oak.

Dominance differed among families ($p = 0.0086$, $F = 2.83$, Figure 6), but not among seedling size classes ($p = 0.1152$, $F = 1.48$). Two half-sibling families, 8 and 16, had higher dominance than the locally available seedlings, while five were similar (Figure 6). The logistic regression model that best explained dominance included species of tallest competing woody stem and basal diameter at planting, with larger seedlings having a greater chance of being dominant (Table 3, Figure 7). According to the model, which had a max re-scaled R-square value of 0.18, an oak with a basal diameter of 7.8 mm at planting (the mean basal diameter at planting) had a 22 percent dominance probability when black cherry was the tallest competing species, compared with a 58 percent chance when a wild northern red oak, 74 percent when a white oak, 41 percent when a red maple, 64 percent when a serviceberry, and 52 percent when another species was the tallest competing stem in the competition plot (Figure 6).

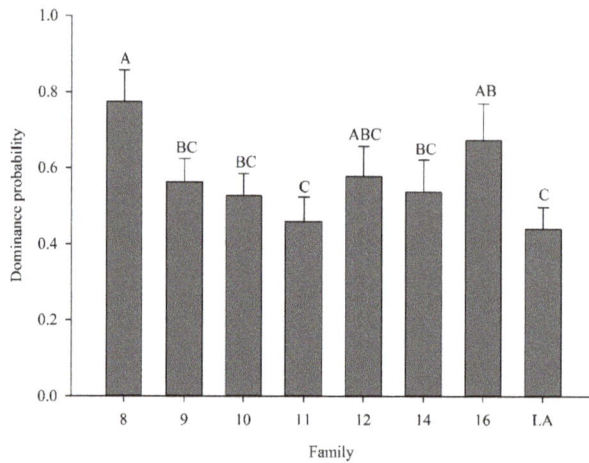

Figure 6. Mean eleventh-year dominance probability ($\pm$ standard error) among families. Letters indicate significant differences ($\alpha = 0.05$). LA indicates locally available seedling treatment.

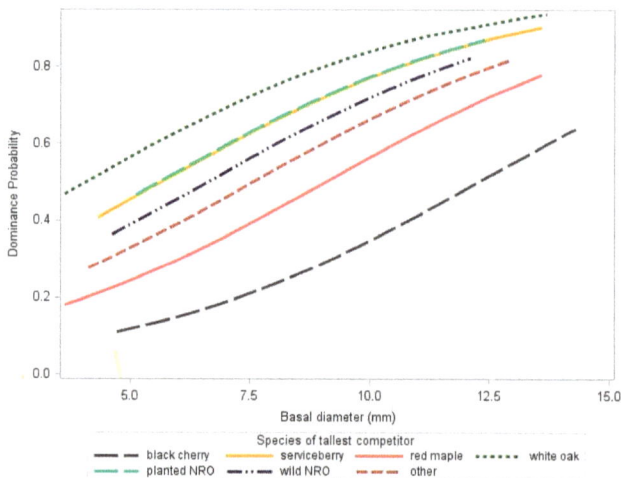

Figure 7. Regression model using basal diameter at planting and species of tallest competing stem to predict eleventh-year dominance probability. NRO stands for northern red oak.

4. Discussion

4.1. Family

The results from this eleven year study demonstrate the potential variability in survival, growth, and competitive ability among different northern red oak seed sources. One family (8) consistently showed improved performance across eleventh-year survival, height, DBH, dominance, and growth pattern. By 2015, family 8 was twenty-eight to forty-one percent larger in diameter and ten to thirty-four percent taller on average than the other families (except family 16). Two families (12 and 14) were similar in size to other families at the beginning of the study, but by year eleven, demonstrated inferior growth, survival, and competitive ability. These results show that superior phenotypes in the nursery may not be reflective of subsequent seedling performance, as Kriebel and others have also found [42]. Variation in acorn size among families as well as microsite variation in the nursery—either due to differences in soil or amounts of fertilization and irrigation—may have contributed to initial size differences among families, as families were not replicated in the nursery. Our results demonstrate the substantial variation in seedling performance among families and suggest genetic gains may be achieved through selection of mother trees with desirable phenotypic traits, such as faster juvenile growth. Genetic variation is essential for populations, in terms of adaptation to new stresses such as disease and climate change [43,44], therefore we do not recommend planting seedlings from only several parents. Rather, we do recommend identification of families that demonstrate repeated superior field growth, survival, and competitive ability through the development of tree improvement programs.

4.2. Seedling Quality

Size class within family differed in height and DBH for some families but did not affect dominance or survival, according to logistic regression models. Size classes did show differences in growth pattern over the eleven year study, with more locally available seedlings dying or adding little growth than the high-quality seedlings. Most other studies that have evaluated the effect of seedling size class on field performance have followed the seedlings for fewer than ten years [24,27,31,45], with the exception of [28], which evaluated performance 17 years after planting. Clark et al. [24] visually graded northern red oak seedlings into two categories and found, after seven years, that mean height and diameter of seedlings in the larger size class were 40 cm and 3.8 mm greater than the small size class, however there were no differences in survival. Ward et al. [45] graded bare-root northern red oak seedlings into four size classes based on the number of first order lateral roots (FOLR), and reported that after seven years, when protected from herbivory using tree shelters, seedlings in the largest class (>8 FOLR) were 82 cm taller than seedlings in smaller size classes. Zaczek et al. [28] evaluated 10 and 17 year survival and height among twenty stock types, including 1–0 bareroot seedlings, which were comparable in height and basal diameter to the locally available seedlings at planting in this study, and 2–0 bareroot seedlings, comparable to the large size class of the high-quality seedlings used in this study. Survival after 10 years was 44% for the 1–0 seedlings, similar to survival of the locally available seedlings in our study, and 77% for the 2–0 seedlings. Height after 10 years did not differ between the two stock types. None of these studies included a family treatment in their study; Clark et al. [24] used northern red oak seedlings derived from a single mother tree, Ward et al. [45] did not describe the origin of the seedlings used, and Zaczek [27,28] bulked seeds from four mother trees. While importance of seedling size class to outplanting success in our study was variable, the use of high-quality seedlings produced using improved nursery protocol as compared to seedlings derived from the locally available treatment conferred substantial increases in survival and growth. This suggests that improvements to seedling quality will improve success, but that planting only the very largest seedlings may not impart the same benefits for all families.

While the differences among size classes was variable for the measured traits, seedling size was a significant predictor of eleventh-year survival, DBH growth, and dominance probability. Initial basal diameter has consistently been found to be the most important predictor of future growth [25,46] and dominance [15,23]. Most studies followed growth for only one to seven years in the field ([23,28] are exceptions). Our study shows that over time, the relationship between initial diameter and height gradually weakened, similar to what [28] found, presumably due to the effects of other factors. The positive relationship between the DBH of the tallest competing stem and DBH of the planted northern red oak, for example, suggests that variation in microsite was an important driver of variation in growth. The importance of initial diameter to early height growth, however, should not be overlooked, as early height growth is vital for seedlings to become established and grow above the deer-browse line quickly.

At planting, seedlings from the locally available treatment were between thirty and fifty percent shorter and fifteen and twenty-five percent smaller in diameter than each of the half-sibling families grown under advanced nursery methods, likely due to differences in nursery protocols. After eleven years, locally available seedlings were still shorter—between twenty-two and thirty-four percent—versus the half-sibling families. They were similar in DBH to four of the seven families, and inferior in dominance probability only to two families (8 and 16). Far more of these seedlings died or grew slowly over the course of the study (24 percent), in comparison to seedlings in most of the half-sibling families (<9 percent, Table 4). Few locally available seedlings were large enough to measure DBH after eleven years. The number of families bulked into this seed lot is unknown, therefore, we cannot speculate on the contribution of genetics to field performance. Inferior size at planting undoubtedly played a major role in the poor survival and growth demonstrated by these seedlings. This inferior size can be explained by differences in nursery protocol and conditions; higher seed densities, shorter growing season, and different fertilization and irrigation regimes for the locally available seedling treatment.

4.3. Competitive Ability

By the end of the study, 58 percent of living seedlings were dominant. This number is substantially higher than what Morrissey and others [15] found for northern red oak seedlings planted in group selection harvests in Indiana, where only 20 percent of the seedlings were dominant after five years. Fast growing yellow poplar (*Liriodendron tulipifera* L.), a shade-intolerant species that commonly suppresses oak regeneration [47] was the predominant competitor in that study. Clark and others [24] found that 70 percent of planted northern red oak was dominant in shelterwood harvests after seven years, however competition control in year five likely contributed to the high dominance probability, as only 38 percent of the seedlings were dominant in year three. The xeric nature of our study site likely contributed to the relatively high number of trees that were competitive after eleven years. In our study, dominance was positively related to initial stem diameter, which is similar to what others have found [15,23,24]. The species of the tallest competing stem was also a strong predictor of eleven-year dominance, with black cherry posing the greatest challenge to planted oaks. For example, a tree 12 mm in basal diameter at planting, among the larger used in this study, would only have a thirty three percent chance of being dominant after eleven years when a black cherry was the tallest competitor (Figure 7).

4.4. Growth Patterns

We are not aware of any other study that has similarly modeled the individual growth of planted oak growth seedlings over time. A quadratic regression analysis provided the best comprehensive assessment of the eleven-year performance of the planted seedlings. While the LMM height analyses showed how each treatment level, family, or seedling size class, grew over time on average, the growth pattern analysis revealed a more nuanced assessment of growth, by showing the percentage of seedlings within each size class or family that followed certain growth trajectories. For example, far fewer seedlings from families 12 and 14 and the locally available seedling treatment showed a sustained positive growth rate in height over time. Rather, many died, grew slowly, or grew and then died back. However, the LMM analysis found no differences between these and three other families (9, 10 and 11). Likewise, fewer seedlings in the large size class experienced die-back after establishment than those in the smaller size classes and from the locally available seedling treatment, a trend the height LMM analysis was not able to detect. Particularly for multi-year datasets with multiple seedling treatments, this technique provides a comprehensive, more interesting analysis than simply evaluating mean size and survival.

5. Conclusions

This study demonstrates the importance of nursery practices and seedling quality to the survival, growth, and competitive ability of northern red oak seedlings planted on a xeric site. After eleven years, the high quality seedlings were taller than the lower quality locally available seedlings. Some families had larger DBH than seedlings from the locally available seedlings, which represented nursery seedlings used for outplantings in the Northeastern U.S. Seedlings in the large size class maintained their size advantage in some, but not all families, demonstrating the variation in growth among families. Overall success by using high-quality seedlings is substantial. While importance of initial size to height diminished over time, it remained an important predictor of eleven-year survival, DBH growth, and dominance. The differences among the half-sibling families show that pedigree can also influence success. The use of quadratic equations to model individual tree growth identified patterns of growth that may be characteristic of families or seedling size, enabling better choice of planting material.

Author Contributions: S.E.S., T.J.H., A.M.S. and J.K.B. conceived and designed the experiments; S.E.S., T.J.H. and C.C.P. performed the experiments, A.M.S. and C.C.P. analyzed the data, and C.C.P. wrote the paper.

Acknowledgments: We thank Christopher Prey, U.S. Military Academy Reservation for assistance with collection of acorns. We acknowledge Tim Dugan, District Forester of Delaware State Forest, PA and Tim Lander, District Forester of Weisler State Forest, PA, for their assistance with the establishment and monitoring of this study. We thank Fengyou Jia, Entomologist, Division of Forest Health, PA Bureau of Forestry, for technical assistance and data collection. Stacy Clark and Mark Coggeshall each provided useful edits to earlier versions of this manuscript. Resources for this study was provided by the USDA Forest Service, Northern Research Station, The Tennessee Tree Improvement Program, and The Pennsylvania Bureau of Forestry.

Conflicts of Interest: The authors declare no conflict of interest.

References

1. Aszalós, R.; Horváth, F.; Mázsa, K.; Ódor, P.; Lengyel, A.; Kovács, G.; Bölöni, J. First signs of old-growth structure and composition of an oak forest after four decades of abandonment. *Biologia (Bratisl)* **2017**, *72*, 1264–1274. [CrossRef]
2. Fei, S.; Kong, N.; Steiner, K.C.; Moser, W.K.; Steiner, E.B. Change in oak abundance in the eastern United States from 1980 to 2008. *For. Ecol. Manag.* **2011**, *262*, 1370–1377. [CrossRef]
3. Abrams, M.D. Where has all the White Oak gone? *Bioscience* **2003**, *53*, 927–939. [CrossRef]
4. Crow, T.R. Reproductive mode and mechanisms for of northern red oak (*Quercus rubra*)—A review. *For. Sci.* **1988**, *34*, 19–40.
5. Rooney, T.P.; Waller, D.M. Direct and indirect effects of white-tailed deer in forest ecosystems. *For. Ecol. Manag.* **2003**, *181*, 165–176. [CrossRef]
6. Abrams, M.D. Fire and the Development of Oak Forests. *Bioscience* **1992**, *42*, 346–353. [CrossRef]
7. Nowacki, G.J.; Abrams, M.D. The Demise of Fire and "Mesophication" of Forests in the Eastern United States. *Bioscience* **2008**, *58*, 123–138. [CrossRef]
8. Campbell, A.J.; Steiner, K.C.; Finley, J.J.; Leites, L. Limitations on regeneration potential after even-aged harvests in mixed-oak stands. *For. Sci.* **2015**, *61*, 874–881. [CrossRef]
9. Steiner, K.C.; Finley, J.C.; Gould, P.; Fei, S.; McDill, M. Oak Regeneration Guidelines for the Central Appalachians. *North. J. Appl. For.* **2008**, *25*, 5–16.
10. Dey, D.C.; Jacobs, D.; McNabb, K.; Miller, G.; Baldwin, V.; Foster, G. Artificial Regeneration of Major Oak (*Quercus*) Species in the Eastern United States—A Review of the Literature. *For. Sci.* **2008**, *54*, 77–106.

11. Brose, P.; Van Lear, D.; Cooper, R. Using shelterwood harvests and prescribed fire to regenerate oak stands on productive upland sites. *For. Ecol. Manag.* **1999**, *113*, 125–141. [CrossRef]

12. Miller, G.W.; Brose, P.H.; Gottschalk, K.W. Advanced Oak Seedling Development as Influenced by Shelterwood Treatments, Competition Control, Deer Fencing, and Prescribed Fire. *J. For.* **2017**, *115*, 179–189. [CrossRef]

13. Clark, F.B.; Watt, R.F. Silvicultural Methods for Regenerating Oaks. In Proceedings of the Oak Symposium, Upper Darby, PA, USA, 16–20 August 1971; pp. 37–42.

14. Loftis, D.L. Predicting post-harvest performance of advance red oak reproduction in the Southern Appalachians. *For. Sci.* **1990**, *36*, 908–916.

15. Morrissey, R.C.; Jacobs, D.F.; Davis, A.S.; Rathfon, R.A. Survival and competitiveness of *Quercus rubra* regeneration associated with planting stocktype and harvest opening intensity. *New For.* **2010**, *40*, 273–287. [CrossRef]

16. Schuler, J.L.; Robison, D.J. Performance of northern red oak enrichment plantings in naturally regenerating Southern Appalachian hardwood stands. *New For.* **2010**, *40*, 119–130. [CrossRef]

17. Struve, D.K.; Joly, R.J. Transplanted red oak seedlings mediate transplant shock by reducing leaf surface area and altering carbon allocation. *Can. J. For. Res.* **1992**, *22*, 1441–1448. [CrossRef]

18. Buckley, D.S.; Sharik, T.L.; Isebrands, J.G. Regeneration of northern red oak: Positive and negative effects of competitor removal. *Ecology* **1998**, *79*, 65–78. [CrossRef]

19. Ponder, F., Jr. Shoot and root growth of northern red oak planted in forest openings and protected by treeshelters. *North. J. Appl. For.* **1995**, *12*, 36–42.

20. Schlarbaum, S.E.; Kormanik, P.; Tibbs, T. Oak Seedlings: Quality Improved Available Now- Genetically Improved Available Soon. In Proceedings of the 25th Annual Hardwood Symposium, Cashiers, NC, USA, 7–10 May 1997; pp. 123–130.

21. Johnson, P.S.; Rogers, R. A method for estimating the contribution of planted hardwoods to future stocking. *For. Sci.* **1985**, *31*, 883–891.

22. Dey, D.C.; Parker, W.C. Morphological indicators of stock quality and field performance of red oak (*Quercus rubra* L.) seedlings underplanted in a central Ontario shelterwood. *New For.* **1997**, *14*, 145–156. [CrossRef]

23. Spetich, M.A.; Dey, D.C.; Johnson, P.S.; Graney, D.L. Competitive capacity of *Quercus rubra* L. planted in Arkansas' Boston Mountains. *For. Sci.* **2002**, *48*, 504–517.

24. Clark, S.L.; Schlarbaum, S.E.; Schweitzer, C.J. Effects of visual grading on northern red oak (*Quercus rubra* L.) seedlings planted in two shelterwood stands on the Cumberland Plateau of Tennessee, USA. *Forests* **2015**, *6*, 3779–3798. [CrossRef]

25. Davis, A.S.; Jacobs, D.F. Quantifying root system quality of nursery seedlings and relationship to outplanting performance. *New For.* **2005**, *30*, 295–311. [CrossRef]

26. Kormanik, P.P.; Sung, S.S.; Kormanik, T.L.; Zarnock, S.J.; Kormanik, P.P.; Oak, S.J.; Why, R. Oak Regeneration—Why Big Is Better. 1995, pp. 1–9. Available online: https://www.researchgate.net/profile/Shi_Jean_Sung/publication/237609808_Oak_Regeneration_Why_Big_Is_Better/links/542b29970cf29bbc126a800b/Oak-Regeneration-Why-Big-Is-Better.pdf (accessed on 1 September 2017).

27. Zaczek, J.J.; Steiner, K.C.; Bowersox, T.W. Northern red oak planting stock: 6-year results. *New For.* **1997**, *13*, 177–191. [CrossRef]

28. Zaczek, J.J.; Steiner, K.C. The influence of cultural treatments on the long-term survival and growth of planted *Quercus rubra*. In Proceedings of the 17th Central Hardwood Forest Conference, Lexington, KY, USA, 5–7 April 2010; pp. 294–305.

29. Sander, I.L. Height growth of new oak sprouts depends on size of advance reproduction. *J. For.* **1971**, *69*, 809–811.

30. Kormanik, P.P.; Sung, S.-J.S.; Kormanik, T.L. Toward a single nursery protocol for oak seedlings. In Proceedings of the 22nd Southern Forest Tree Improvement Conference, Atlanta, GA, USA, 14–17 June 1993; pp. 89–98.

31. Oswalt, C.M.; Clatterbuck, W.K.; Houston, A.E. Impacts of deer herbivory and visual grading on the early performance of high-quality oak planting stock in Tennessee, USA. *For. Ecol. Manag.* **2006**, *229*, 128–135. [CrossRef]

32. Kormanik, P.P.; Sung, S.-J.S.; Kass, D.J.; Schlarbaum, S. Effect of Seedling Size and First-Order-Lateral Roots on Early Development of Northern Red Oak on Mesic Sites. In Proceedings of the Ninth Biennial Southern Silvicultural Research Conference, Clemson, SC, USA, 25–27 February 1997; pp. 247–252.

33. Gregory, J.; Nowacki, M.D.A. Community, edaphic, and historical analysis of mixed oak forests of the Ridge and Valley Province in central Pennsylvania. *Can. J. For. Res.* **1991**, *22*, 790–800.

34. Clark, S.L.; Schlarbaum, S.E.; Keyser, T.L.; Schweitzer, C.J.; Spetich, M.A.; Simon, D.; Warburton, G.S. Response of planted northern red oak seedlings to regeneration harvesting, midstory removal, and prescribed burning. In *Proceedings of the 18th Biennial Southern Silvicultural Research Conference*; e–Gen. Tech. Rep. SRS–212; U.S. Department of Agriculture, Forest Service, Southern Research Station: Asheville, NC, USA, 2016; p. 8.

35. Clark, S.L.; Schlarbaum, S.E.; Kormanik, P.P. Visual grading and quality of 1-0 northern red oak seedlings. *South. J. Appl. For.* **2000**, *24*, 93–97.

36. Johnson, P.S. Responses of planted northern red oak to three overstory treatments. *Can. J. For. Res.* **1984**, *14*, 536–542. [CrossRef]

37. Clark, S.; Schweitzer, C.; Schlarbaum, S. Nursery quality and first-year response of American chestnut (*Castanea dentata*) seedlings planted in the southeastern United States. *Tree Plant. Notes* **2010**, *53*, 13–21.

38. Pinchot, C.C.; Schlarbaum, S.E.; Clark, S.L.; Saxton, A.M.; Sharp, A.M.; Schweitzer, C.J.; Hebard, F.V. Growth, survival, and competitive ability of chestnut (*Castanea* Mill.) seedlings planted across a gradient of light levels. *New For.* **2017**, *48*, 491–512. [CrossRef]

39. Sander, I.L.; Johnson, P.S.; Rogers, R. *Oak Advance Reproduction in the Missouri Ozarks*; North Central Forest Experiment Station, Forest Service, US Department of Agriculture: St. Paul, MN, USA, 1984.

40. Weigel, D.R. Oak planting success varies among ecoregions in the Central Hardwood Region. In Proceedings of the 12th Central Hardwood Forest Conference, Lexington, KY, USA, 28 February–2 March 1999; pp. 9–16.

41. Hosmer, D.W., Jr.; Lemeshow, S.; Sturdivant, R.X. *Applied Logistic Regression*, 3rd ed.; John Wiley & Sons: Honoken, NJ, USA, 2013.

42. Kriebel, H.B.; Merritt, C.; Stadt, T. Genetics of growth rate in *Quercus rubra*: Provenance and family effects by the early third decade in the North Central U.S.A. *Silvae Genet.* **1988**, *37*, 193–198.

43. Kitzmiller, J.H. Managing genetic diversity in a tree improvement program. *For. Ecol. Manag.* **1990**, *35*, 131–149. [CrossRef]

44. Johnson, R.; Lipow, S. Compatibility of Breeding for Increased Wood Production and Long-Term Sustainability: The Genetic Variation of Seed Orchard Seed and Associated Risks. United States Department of Agriculture Forest Service General Technical Report PNW. 2002, Volume 563, pp. 169–179. Available online: https://www.researchgate.net/profile/Randy_Johnson6/publication/237262080_Compatibility_of_breeding_for_increased_wood_production_and_long-term_sustainability_The_genetic_variation_of_seed_orchard_seed_and_associated_risks/links/53e012e50cf2aede4b4cbb24.pdf (accessed on 1 September 2017).

45. Ward, J.S.; Gent, M.P.N.; Stephens, G.R. Effects of planting stock quality and browse protection-type on height growth of northern red oak and eastern white pine. *For. Ecol. Manag.* **2000**, *127*, 205–216. [CrossRef]

46. Parker, W.C.; Dey, D.C. Influence of overstory density on ecophysiology of red oak (*Quercus rubra*) and sugar maple (*Acer saccharum*) seedlings in central Ontario shelterwoods. *Tree Physiol.* **2008**, *28*, 797–804. [CrossRef] [PubMed]

47. Loftis, D.L. Regenerating southern Appalachian mixed hardwood stands with the shelterwood method. *South. J. Appl. For.* **1983**, *7*, 212–217.

![forests logo] *forests*

MDPI

Article

Mechanized Tree Planting in Sweden and Finland: Current State and Key Factors for Future Growth

Back Tomas Ersson [1], Tiina Laine [2] and Timo Saksa [2,*]

[1]   School of Forest Management, SLU/Swedish University of Agricultural Sciences,
     73921 Skinnskatteberg, Sweden; back.tomas.ersson@slu.se
[2]   Luke/Natural Resources Institute Finland, Latokartanonkaari 9, 00790 Helsinki, Finland;
     tiinalaine86@gmail.com
*   Correspondence: timo.saksa@luke.fi; Tel.: +358-29-532-4834

Received: 29 March 2018; Accepted: 26 May 2018; Published: 21 June 2018

Abstract: In Fennoscandia, mechanized tree planting is time-efficient and produces high-quality regeneration. However, because of low cost-efficiency, the mechanization of Fennoscandian tree planting has been struggling. To determine key factors for its future growth, we compared the operational, planning, logistical, and organizational characteristics of mechanized planting in Sweden and Finland. Through interviews with planting machine contractors and client company foresters, we establish that mechanized tree planting in Sweden and Finland presently shares more similarities than differences. Some notable differences include typically longer planting seasons in Sweden, and a tendency towards two-shift operation and less frequent worksite pre-inspection by contractors in Finland. Because of similar challenges, mechanized planting in both countries can improve cost-efficiency through education of involved foresters, flexible information systems, efficient seedling logistics, and continued technical development of planting machines. By striving to have multiple client companies, contractors can reduce their operating radii and increase their machine utilization rates. Above all, our results provide international readers with unprecedented detailed and comprehensive figures and characteristics of Swedish and Finnish mechanized tree-planting activities. We conclude that cooperation between Sweden's and Finland's forest industries and research institutes could enhance the mechanization level of Fennoscandian tree planting.

Keywords: tree planting machine; contractor; mechanization; reforestation; silviculture; forestry; Fennoscandia

1. Introduction

Despite much effort over the last 50 years, tree planting in Fennoscandia has not successfully been mechanized on a large scale [1–3]. Mechanized tree planting has been shown to be time efficient [4], so the machines can potentially alleviate future labor shortages. In addition, because planting trees mechanically with today's machines produces high-quality regeneration (often with better quality than when trees are planted manually [2,5,6]), foresters in both Sweden and Finland are keen on this option, rather than the standard option of manual planting [7,8].

The prevalence of mechanized tree planting is quite low in both countries. In Sweden, the proportion of mechanically planted seedlings historically peaked at circa 12% during the late 1990s [9] with the highly productive, continuously advancing Silva Nova [10], but fell to <1% during the mid-2000s and has stayed there since [2]. In Finland, this figure is considered to be <5% [8], despite plenty of research and development work to promote mechanized planting. The proportion of mechanized planting in Finland has, in fact, stagnated or even slightly decreased over the last few years [11].

Previous research has highlighted some general reasons why the growth of mechanized tree planting in Fennoscandia is struggling. The main reason for this struggle has primarily been poor contractor profitability leading to reduced interest among contractors [2,3]. Poor profitability is a consequence of mechanized tree planting's relatively low cost-competitiveness, compared to manual tree planting. Cost-competitiveness is low, despite there being a relatively strong demand for mechanically planted trees in both countries [2,4]. Even though the time consumption of mechanized planting is lower compared to separate (spot) mounding followed by manual planting, the cost-efficiency has still been poor compared to manual methods [4]. Cost-efficiency is hampered by low productivity, which originates from operator inexperience, among other reasons [12]. The choice of worksites also strongly influences productivity, and it has been reported that the presence of stones, slash, and stumps decreases the planting machines' productivity [13–17]. The size and spatial distribution of worksites also affects cost-competitiveness [18]. The proportion of time spent relocating the machine decreases as worksite size increases, and the cost and time consumption of relocating increases as the distance between worksites increases [4,10].

Machine Utilization (MU) has a strong impact on contractor profitability, and MU is often comparatively low for planting machines [12]. In turn, factors like workplace organization and seedlings logistics have a strong impact on the planting machines' MU [10,19,20]. The theoretical MU of Finnish planting machines has been estimated to be potentially as high as 90% [21]. However, high MU rates can also potentially cause disadvantages, such as poorer worksite conditions and greater relocation distances.

The client for mechanized planting in Finland is usually a forest company, a local forest owners association (FOA), or a non-industrial private forest owner [8]. A client can be regarded as a service provider, and foresters working for a forest company/FOA are responsible for organizing mechanized planting activities (for example selecting the worksite), while the planting machine contractor is responsible for performing the actual planting work. Non-industrial private forest owners usually buy a planting service from a forest company/FOA rather than directly contract the planting machine contractor themselves. In Sweden, the clients are almost always forest companies; although sometimes, especially in connection to shutdown periods, such as July or winter, planting machines contractors can work directly for non-industrial private forest owners without using a client company as a middleman.

It is the nurseries' responsibility to grow seedlings for mechanized planting. The seedling types used during mechanized planting are generally the same as those used during manual planting [6,22]. In Finland, spring-planted seedlings are usually freezer-stored and packed in cardboard boxes. The summer-planted seedlings (since they are growing and need tending such as watering) are usually packed in open plastic trays, while the autumn-planted seedling are either packed in cardboard boxes or open plastic trays. In Sweden, from spring to autumn, each nursery sticks to only one seedling packaging system [23], and nurseries in southern Sweden stop delivering seedlings in July until the seedlings have hardened enough to withstand packing.

In Finland, mechanized tree planting is more common [3] and thus (despite using the same type of equipment) could be suspected to be relatively more cost-efficient than in Sweden. Accordingly, there might be factors like workplace organization, choice of worksites, and/or seedlings logistics that are coordinated differently within the two countries. Thus, there is a need to analyze how mechanized planting is organized in Sweden and Finland, and to compare and contrast their operational, planning, logistical, and organizational characteristics. This analysis just might identify solutions in one country that can help to increase the cost-competitiveness of mechanized tree planting in another.

The **objectives** of this study were to: (1) compare mechanized tree planting in Sweden and Finland; (2) identify factors that can increase the cost-competitiveness of Swedish and Finnish mechanized planting; (3) and determine key factors for the future growth of this reforestation tool in these two countries. Because of the business arrangements used in Swedish and Finnish mechanized tree planting, the data collection approach we used was to interview both contractors and client company foresters.

2. Materials and Methods

The development of mechanized planting has been a hot topic both in Finland and in Sweden over the last ten years [24]. Several studies have tried to establish the technical feasibility and effectivity of present planting devices and a couple of dissertations has also been published lately [2,3]. As a continuation of these earlier research activities, we conceived the idea to compare the current operational models for mechanized planting in Sweden and Finland.

A semi-structured thematic interview form based on previous studies was prepared for a field visit to southern Sweden conducted in August 2017. During this study trip, the authors had discussions/interviews with a planting machine contractor, with the forest company Södra's foresters at both the district, regional and company level, and with Södra's nursery professionals. These discussions provided the basic information of the current status of (as well as foresters' attitudes towards) mechanized planting in Sweden.

After the field study, four additional Swedish contractors were identified and interviewed (three more in southern Sweden and one in northern Sweden) in autumn 2017. The contractors were identified through contacts with Sweden's major forest companies. In conjunction with these semi-structured contractor interviews, additional semi-structured interviews were also held with those foresters (who were responsible for each planting machine contract) at the district level.

In Finland, some of the corresponding information could be found from an interview study conducted in 2014 [8]. This information was updated via a Metsäteho-maintained online catalogue of Finnish mechanized tree-planting contractors [25], and two typical contractors were interviewed using a semi-structured method during the winter of 2017–2018.

The interview questions were grouped into four categories. The first category concentrated on the operational characteristics and covered questions on equipment, operator, and production. The second category dealt with worksite characteristics like average size, selection criteria, selector, pre-planting inspection, and relocation. The third category comprised characteristics of seedling logistics, including number of planted seedlings, seedling packaging, seedling delivery method, equipment, seedling storage, and tending. The fourth category concentrated on organizational characteristics such as information systems and business arrangements.

The results are presented by showing the typical attributes (i.e., mode/mean for quantitative data, or type for qualitative data) and their range (i.e., variation among the attributes) in both Sweden and Finland. The range of the qualitative data provides an idea of the variation that exists among the contractors. This range was necessary to identify differences in work methods and novel or innovative solutions that may increase planting machine contractor profitability.

3. Results

3.1. Country Comparison

In general, mechanized tree planting was operationally quite similar in Sweden and Finland (Table 1), although there were some particular exceptions. Most notably, only one brand of planting device was used in Sweden (Bracke Planter), while three brands were used in Finland (Bracke Planter, M-Planter and Risutec); the typical operator experience level was twice as high in Finland as in Sweden (6 seasons vs. 3 seasons); the typical planting season was longer in Sweden than in Finland (7 months vs. 5 months); and two shifts were the norm in Finland, whereas one shift was typical in Sweden. The use of the M-Planter device in Finland led to a higher maximum productivity over a shift (360 pl/PMh; pl = seedlings; PMh = Productive Machine hour) compared to the maximum reported in Sweden with the Bracke Planter (240 pl/PMh). Typically, operators had no forestry education, although some Swedish operators had secondary forestry education and some Finnish operators had basic forestry education.

Table 1. Operational characteristics of mechanized tree planting in 2017 in Sweden and Finland (pl = seedlings; PMh = Productive Machine hour).

Category	Characteristic	Attribute	Sweden		Finland	
			Typical	Range	Typical	Range
Equipment	Base machine	Type	Tracked excavator	Tracked excavator	Tracked excavator	Tracked excavator, Wheeled harvester
		Mass (t)	20	16–22	16	14–22 *
	Planting device	Type	Bracke Planter	Bracke Planter	Bracke Planter	Bracke Planter, M-Planter, Risutec
		Seedling carousel capacity (pl)	71	70–196	72	60–242
		Number per contractor	1	1	1	1–2
Operator	Experience level (years/seasons)	Mechanized planting	3	1–23	6	1–24
		Other forestry experience	20	4–40	32	3–52
		Other excavator work	2	0–20	3	0–20
		Forestry education	None	None, Forestry secondary school	None	None, Basic forestry education
Production	Planting season	Length (months)	7	2–8	5	3–6
	Productivity (pl/PMh)	Average over a year	150	120–220	165	70–265
		Maximum over a shift	220	180–240	260	200–360
	Target production	Yearly (pl/year)	150,000	30,000–200,000	180,000	70,000–320,000
		Shift-wise (pl/shift)	1100	1000–1900	1200	1000–1800
	Average shift	Number per day	1	1–2	2	1–2
		Number per year	150	30–200	160	55–248
		Length (PMh)	6.5	6–8	7	6–8
	Non-planting work	Proportion of base machine's yearly PMh	20%	15–70%	38%	0–67%

* Data from 2015–2017.

Work sites, planning, and seedling logistics did not vary substantially between the countries either (Tables 2 and 3). However, typical work sites were larger in Finland than Sweden (4.7 ha vs. 3.5 ha); slash-harvested sites were typical in Finland while not in Sweden; contractors in Sweden always pre-inspected the work sites while this habit was rare in Finland; hired truck and trailer was typical when relocating in Sweden while contractor-owned truck and trailer was typical in Finland; relocations were typically faster in Finland than Sweden (1.5 h vs. 2.5 h); seedling packaging in Finland also sometimes entailed open plastic trays (during late-summer/early autumn planting); Swedish seedlings were typically stored on planting machines in enclosed metal boxes while Finnish seedlings were typically stored on covered racks.

Table 2. Work site characteristics of mechanized tree planting in 2017 in Sweden and Finland (pl = seedlings). Words enclosed by (brackets) in the 'Range' columns indicate that the circumstance sometimes occurs.

Characteristic	Attribute	Sweden		Finland	
		Typical	Range	Typical	Range
Average size	Area (ha)	3.5	2–6	4.7	2–10
	Seedling prescription (pl)	6500	5000–9000	8500	3600–18,000
Selection criteria	Requirement	Mesic to moist sites; not too stony/B.Q. ** <50%	Mesic to moist sites; not too stony; slash harvested; road closer than 300 m to edge of site	Mesic to moist sites, low stoniness/B.Q.** <60%	Low to medium stoniness
	Preference	Slash harvested; site > 3 ha	Site > 1–3 ha	Slash harvested; site >4.5 ha	Slash and stumps harvested; site >1–10 ha
Selector	of Sites	Forester	Contractor and/or forester	Forester	Contractor and/or forester
	of Route plan	Contractor	Forester and/or contractor	Contractor	Contractor, or in conjunction with forester, or forester only
Pre-planting inspection	Frequency; Assessor	Always; by contractor	Always—when in new area or involving new foresters; by contractor or operator	Rarely; by contractor	Always when new client company; by forester or contractor
Relocation	Method	Hired truck & trailer	Hired (own) truck or tractor & trailer	Own truck & trailer	Own(hired) truck or tractor & trailer
	Average distance (km)	30	20–40	22	5–60
	Maximum distance from contractor depot (km)	50	20–100	62	5–125
	Average time consumption (h)	2.5	2–3.5	1.5	1–3

** B.Q. = Boulder quota; see [26] for definition.

Table 3. Characteristics of seedling logistics for mechanized tree planting in 2017 in Sweden and Finland (pl = seedlings). Words enclosed by (brackets) in the 'Range' columns indicate that the circumstance sometimes occurs.

Characteristic	Attribute	Sweden		Finland	
		Typical	Range	Typical	Range
Planted seedling	Type (always container-grown)	*Picea abies*	*Picea abies* seedlings (and cuttings), *Pinus sylvestris, Larix* spp.	*Picea abies, Pinus sylvestris*	*Picea abies, Pinus sylvestris, Betula pendula*
	Average size (stem length/root plug volume)	30 cm/93 cm^3	20–35 cm/50–120 cm^3	25 cm/85 cm^3	15–30 cm/50–115 cm^3
Seedling packaging	Type	Cardboard box	Cardboard box	Cardboard box and/or Open plastic tray	Cardboard box and/or Open plastic tray
Seedling delivery	Capacity (pl/unit)	165	80–500	130	80–250
	Frequency	Weekly	From weekly to twice per season	Twice a month	From every few days to once per month

Table 3. *Cont.*

Characteristic	Attribute	Sweden		Finland	
		Typical	Range	Typical	Range
	Waypoints	Nursery—contractor's depot—roadside depot	Nursery—(contractor's depot)—roadside depot	Nursery—contractor's depot—roadside depot	Nursery—client's depot or contractor's depot—roadside depot
	Deliverer	Nursery-contracted courier to contractor's depot; contractor to roadside depot	Contractor or Nursery-contracted courier attends to the whole delivery	Nursery-contracted courier to contractor's depot	Nursery-contracted courier to client's depot or roadside depot
Contractor-owned equipment	Seedling storage at contractor's depot	Cooler storage	Uncooled storage hall or underground cellar or purchased (rented/shared) cooler storage	Semi-cooled storage	Uncooled storage hall or underground cellar or purchased (rented/shared) cooler storage
	Secondary seedling transport ***	Covered pickup truck	Covered pickup truck or covered(open) trailer	Covered pickup truck	Van or trailer or pickup truck
Storage on planting machine	Type	Enclosed metal box on the side of the crane pillar	(Ground-accessible) enclosed metal box on the side of the crane pillar	Covered or open rack on the back of the excavator	Covered or open rack on the base machine's back side, enclosed metal box on the side of the crane pillar
	Capacity (pl)	1800	1100–3000	1800	1000–4000
Seedling tending	Activities	Shading	Shading; watering; unstacking or moving (opening) boxes	Shading	Shading; watering; unstacking or moving (opening) boxes
	Average time consumption (min/shift)	30	0–30	30	0–60

*** Transporting seedlings from contractor's depot to roadside depot; c.f. definition in [19].

The typical information systems and business arrangements for mechanized tree planting were also quite similarly organized between Sweden and Finland (Table 4). Indeed, in both countries, work orders were typically to be delivered by the client company via an internet application minimum 2–2.5 months before planting. This application was also typically used to report the quality control sampling (usually with a sampling frequency of circa two plots per ha). Piece-rate remuneration was typical in both countries, although there was sometimes area-based payment in Finland.

Table 4. Organizational characteristics of mechanized tree planting in 2017 in Sweden and Finland (pl = seedlings).

Category	Characteristic	Attribute	Sweden		Finland	
			Typical	Range	Typical	Range
Information Systems	Work order (instructions and map)	Delivery method	Internet application	Visit to landowner or paper or email or Internet application	Internet application	Paper or email or Internet application
		Delivery deadline	2 months before arrival to site	0.5–4 months before arrival to site	2.5 months before arrival to site	0.5–5.5 months before arrival to site

Table 4. *Cont.*

Category	Characteristic	Attribute	Sweden		Finland	
			Typical	Range	Typical	Range
	Seedling ordering	Order method	Manually via forester	Manually via forester, Pre-season clump order	Manually via forester	Straight from nursery via contractor or forester, Pre-season clump order
		Minimum timespan from order to roadside depot	8 work days	0–10 work days	4 work days	0–7 work days
	Quality control	Reporting method	Internet application	None or Paper forms or Email or Internet application	Internet application	None or Paper forms or Email or Internet application
		Onsite sampling frequency	Two 25 m^2 sample plots per ha	One—three 25–50 m^2 sample plots per ha	One 50 m^2 sample plot per 500/1000 pl planted	One—four 50 m^2 sample plots per ha, Once per shift
	Productivity follow-up	Recipient	Contractor	None, Contractor	Contractor	None, Contractor or Forester
Business Arrangement	Between forest company and contractor	Remuneration	Piece-rate	Hourly compensation; Piece-rate	Piece-rate	Area-based or hourly compensation; Piece-rate
		Number of client forest companies	1	1–4	1	1–3
	Marketing towards landowners by forest companies	Method	Field demos	Field demos; Information dissemination to landowners	Field demos	Field demos; Information dissemination to landowners
		Frequency	Annually	Seasonally to None	Annually	Seasonally to Annually

3.2. Factors Leading to Cost-Competitive Mechanized Tree Planting

Based on the data in Tables 1–4, we identified 22 factors that lead to cost-competitive mechanized planting (Figure 1). Cost-competitiveness offers room for contractors to become profitable, and profitable contractors are necessary for the growth of mechanized tree planting in both countries. The 22 factors can be broadly defined as factors that: concern seedlings and the planting result; lower fixed costs; increase Machine Utilization (MU); decrease the negative consequences of relocating; and increase productivity and/or work quality (via, e.g., better site selection).

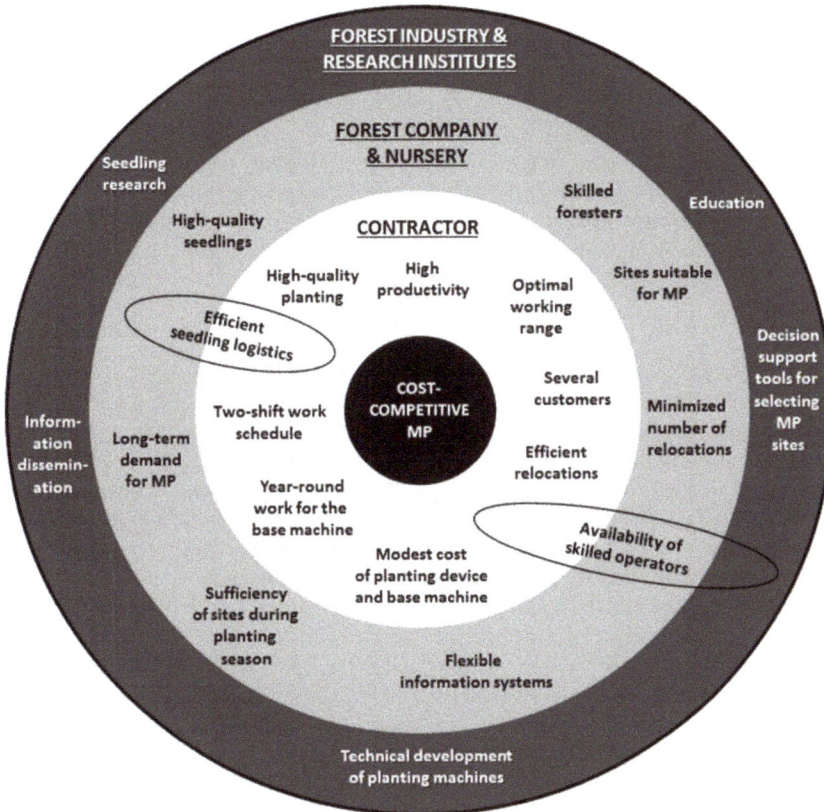

Figure 1. A conceptual framework showing the factors leading to cost-competitive mechanized tree planting (MP) in both Sweden and Finland. The factors in the outer darker ring are those chiefly shaped by the forest industry as a whole and by research institutes. The factors in the middle grey ring are those influenced by the forest company and the nursery supplying the seedlings. The factors in the inner white ring are those mainly affected by the planting machine contractor. The two encircled factors (efficient seedling logistics and availability of skilled operators) are influenced by two or three actors (contractor plus nursery, and contractor, forest company plus whole forest industry, respectively). Modified from reference [27].

3.3. Key Factors for Future Growth of Mechanized Tree Planting

Factors that were judged to be key (i.e., the lowest-hanging fruit) for the future growth of mechanized tree planting in both countries were as follows:

1. Education of foresters (to generate competent selectors of work sites, and acceptance of planting machines as a reliable reforestation tool), information to landowners (to create a higher demand for mechanized tree planting leading to a greater selection of suitable sites while reducing the working radius and increasing autumn planting opportunities), and education of operators (to ensure they know the best work methods leading to maximum productivity).

2. Flexible information systems that can help identify suitable planting sites, increase Machine Utilization (MU) through easier administration and seedling ordering, and increase the accuracy of follow-ups (e.g., using Risutec's ASTA system [28] or something similar).

3. Efficient logistics of suitable seedlings (which increases MU, and ensures high seedling vitality and post-planting performance).

4. High-quality planting work (which ensures continued demand for mechanized tree planting so that landowners receive added value as compensation for the machines' present-day higher planting costs vs. manual planting).

5. Contractors having several client forest companies as this arrangement supports efficient route planning, spreads risk and helps contractors leverage themselves against the (often so) larger forest companies, and reduces the working radius (which leads to, e.g., shorter commutes for the operator).

6. Continued technical development of planting machines (so as to ensure higher quality plantings, higher machine productivity, and/or lower costs in the future).

4. Discussion

According to our findings, mechanized tree planting in Sweden and Finland presently share more similarities than differences. Nonetheless, some Finnish contractors had two planting machines, while all Swedish contractors had only one. Also, Finnish contractors more often had two client companies and operated in two shifts per day, while Swedish contractors typically only had one client company and operated in a single shift per day. Still, annual use [29,30] (or more specifically Capacity Utilization [31] or Total Utilization [12]) of the Swedish planting machines was typically almost the same as the Finnish machines (in terms of average number of shifts per year: 150 vs. 160; Table 1) because of southern Sweden's longer planting season (typically 7 months vs. 5 months; Table 1).

In comparison to those of Fennoscandian harvesting contractors, our identified key factors share some similarities, like the need for route efficiency [32] and for contractors to find ways to leverage themselves against the larger client forest companies [33]. Our key factors "High quality planting work" and "Education of operators" have previously been identified by Laine et al. [8] as critical success factors for Finnish mechanized tree-planting contractors. Meanwhile, Mäkinen [34] concluded that for Finnish harvesting contractors, having only one client was a success factor. This is contrary to our key factor "Contractors having several client forest companies", but the harvesting contractors in Mäkinen's study worked year around and were reported to have Capacity Utilization rates up to 99%. Planting machines, in Finland especially, cannot work year around, and tend to have one-third as high Capacity Utilization rates [12]. Thus, during the short Fennoscandian planting seasons, efficient route planning and small operating radii (the latter which multiple clients can give rise to) become comparatively more important for planting machines if they are to secure enough productive hours to pay for their capital costs.

Swedish contractors always pre-inspected work sites, while Finnish contractors rarely did (Table 2). If the contractor is also the sole operator (as was the case for 3 of 5 Swedish contractors), contractor pre-inspection reduces the planting machines' number of productive machine hours (or at least their Machine Utilization rates), which is detrimental to their cost-efficiency. The lesser need for pre-inspection by Finnish contractors is probably a result of the Finnish foresters' comparatively greater experience in selecting sites suitable for mechanized planting (mechanized planting being at least five times more common in Finland than in Sweden).

In Finland, land-owning forest companies like UPM have embraced mechanized tree planting [3]. In Sweden, mechanized planting is demanded by non-industrial private forest owners but not by land-owning forest companies. This difference in acceptance might explain why Swedish contractors typically only had one client company, while Finnish contractors more often had two.

Efficient seedling logistics increases machine uptime, which is in turn a prerequisite for profitable mechanized tree planting [35]. Efficient seedling logistics requires cooperation between the forest company and contractor (Figure 1), but the terms of cooperation might be wholly dictated by the (generally) larger forest company (cf. Reference [36]), leading to contractor frustration and poor motivation. Nevertheless, simply understanding that efficient seedling logistics is also a responsibility of the forest company might lead to greater efforts by the forest company [18].

Similar to all mechanized forestry work [37,38], the operator's skill level has a profound effect on planting machine productivity [12,14]. Thus, profitable mechanized tree planting requires access to skilled operators. However, the onus of training operators to become skilled cannot be strictly delegated to the contractors (Figure 1). Instead, this responsibility must also be shared with the whole forest industry, as is done with the training of Fennoscandian harvester and forwarder operators [39,40]. Individual client forest companies can also help expand the pool of available, skilled operators by budgeting for competitive wages during pricing negotiations with the contractors.

Being a study based on interviews, some of the time-consumption figures are anecdotal and thus potentially erroneous. However, the figures were provided by the contractors themselves, so any errors are probably small. Moreover, figures for key characteristics like worksites and seedling logistics were double-checked with client company foresters.

We identified key factors for intermittently advancing planting machines, but they are probably relevant as well for continuously advancing planting machines. Because of their higher productivities, continuously advancing planting machines will most likely become dominant in the future in Fennoscandia (despite the prevalence of moraine/glacial till soils in the region) [2,3,21,41].

The cost-competitiveness of mechanized tree planting is certainly reliant on the cost-efficiency of the alternatives, specifically manual tree planting following mechanical site preparation [18]. However, further technical development of Fennoscandian manual tree planting has been judged to be poor [10], and the future supply of labor for manual planting is predicted to shrink [21], so significant cost decrease of the alternatives is unlikely. Additionally, any technical development regarding mechanical site preparation can potentially be transferred to planting machines as well [42].

Our identified key factors pave the way for future studies and development. For example, there is the need in both countries for decision support tools to select sites (based on rockiness/stoniness, etc.) and for tools to simplify quality management (e.g., follow-up planting using systems similar to ASTA). Such tools deserve to be developed and further studied. Considering the amount of time spent by contractors on seedling handling/tending (Table 3) and reloading [35], present-day seedling production and delivery methods are not good enough for today's intermittently advancing machines, and definitely not for future continuously advancing machines [43]. Thus, there is continued need for development of machine-specific seedlings and seedling packaging systems. Likewise is the need for continued technical development to increase the machines' productivity.

5. Conclusions

According to our observations, there are several well-known factors like the low Capacity Utilization and productivity of today's planting devices, which prevent the cost-competitiveness and growth of the Fennoscandian mechanized tree planting business. But there are also some factors that cannot be addressed by one counterpart alone. The availability of skilled operators and efficient seedling logistics are two such bottlenecks, which must be solved cooperatively.

In both countries, a large-scale breakthrough of mechanized planting is still waiting to happen. Neither in Finland nor in Sweden was mechanized planting on a routine-level similar to other forestry contracting. Indeed, there are many operational processes, which can be considered to be in their developmental infancy. This includes the process of choosing the work sites for mechanized planting, as well as the process of delivering seedlings throughout the whole planting season. Because mechanized planting in Sweden and Finland share many similar challenges, cooperation between the countries' forest industries and research institutes is both desirable and needed, if tree planting is to become mechanized in Fennoscandia.

Author Contributions: B.T.E. conceived and designed the experiments; B.T.E. and T.S. performed the experiments; B.T.E. and T.S. analyzed the data; B.T.E., T.L., and T.S. wrote the paper.

Acknowledgments: SLU and Luke provided the funding for this study. The travel costs for data gathering was funded by NordGen Forest-SNS scholarships 2017.

Forests **2018**, *9*, 370

Conflicts of Interest: The authors declare no conflict of interest. The founding sponsors had no role in the design of the study; in the collection, analyses, or interpretation of data; in the writing of the manuscript, and in the decision to publish the results.

References

1. Bäckström, P.-O.; Jansson, E.; Jeansson, E.; Sirén, G. *Comparative Study of Four Tree-Planting Machines*; Rapporter och Uppsatser Nr 21; Institutionen för Skogsföryngring; Skogshögskolan: Stockholm, Sweden, 1970.
2. Ersson, B.T. Concepts for Mechanized Tree Planting in Southern Sweden. Ph.D. Thesis, SLU, Umeå, Sweden, 2014.
3. Laine, T. Mechanized Tree Planting in Finland and Improving its Productivity. Ph.D. Thesis, University of Helsinki, Helsinki, Finland, 2017.
4. Hallongren, H.; Laine, T.; Rantala, J.; Saarinen, V.-M.; Strandström, M.; Hämäläinen, J.; Poikela, A. Competitiveness of mechanized tree planting in Finland. *Scand. J. For. Res.* **2014**, *29*, 144–151. [CrossRef]
5. Ersson, B.T.; Petersson, M. Återinventering av 2010 års Maskinplanteringar—3-års Uppföljning. [*Three-Year Follow-up of 2010's Mechanically Planted Seedlings*]; Rapport S048; Skoglig Service; Södra Skog: Växjö, Sweden, 2013.
6. Luoranen, J.; Viiri, H. Deep planting decreases risk of drought damage and increases growth of Norway spruce container seedlings. *New For.* **2016**, *47*, 701–714. [CrossRef]
7. Ersson, B.T.; Jundén, L.; Lindh, E.M.; Bergsten, U. Simulated productivity of conceptual, multi-headed tree planting devices. *Int. J. For. Eng.* **2014**, *25*, 201–213. [CrossRef]
8. Laine, T.; Kärhä, K.; Hynönen, A. A survey of the Finnish mechanized tree-planting industry in 2013 and its success factors. *Silva Fenn.* **2016**, *50*. [CrossRef]
9. Lindholm, E.-L.; Berg, S. Energy Use in Swedish Forestry in 1972 and 1997. *Int. J. For. Eng.* **2005**, *16*, 27–37.
10. Hallonborg, U.; von Hofsten, H.; Mattson, S.; Hagberg, J.; Thorsén, Å.; Nyström, C.; Arvidsson, H. Maskinell mlantering med Silva Nova -nuvarande status samt utvecklingsmöjligheter i jämförelse med manuell plantering [*Mechanized Planting with the Silva Nova Tree Planter—Recent State and Feasibility Compared with Manual Planting*]; Redogörelse Nr 6; Skogforsk: Uppsala, Sweden, 1995.
11. LUKE. *Statistics Database: Forest Statistics: Structure and Production: Silvicultural and Forest Improvement Work*; LUKE: Helsinki, Finland, 2018.
12. Rantala, J.; Laine, T. Productivity of the M-Planter Tree-Planting Device in Practice. *Silva Fenn.* **2010**, *44*, 859–869. [CrossRef]
13. Ersson, B.T.; Junden, L.; Bergsten, U.; Servin, M. Simulated productivity of one- and two-armed tree planting machines. *Silva Fenn.* **2013**, *47*. [CrossRef]
14. Laine, T.; Rantala, J. Mechanized tree planting with an excavator-mounted M-Planter planting device. *Int. J. For. Eng.* **2013**, *24*, 183–193. [CrossRef]
15. Rantala, J.; Harstela, P.; Saarinen, V.-M.; Tervo, L. A techno-economic evaluation of Bracke and M-Planter tree planting devices. *Silva Fenn.* **2009**, *43*, 659–667. [CrossRef]
16. St-Amour, M. *Reforestation Trials with the Bräcke P11.a Planter*; Advantage Report Volume 11, No. 19; FPInnovations-FERIC: Pointe-Claire, QC, Canada, 2009.
17. Von Hofsten, H. Hög kvalitet även på högkvaliteten med Öje-Planter [*The Öje Planter Machine—Good Performance at a Competitive Cost*]; Resultat NR 3; Skogforsk: Uppsala, Sweden, 1993.
18. Ersson, B.T. *Possible Concepts for Mechanized Tree Planting in Southern Sweden—An Introductory Essay on Forest Technology*; Arbetsrapport 269; Department of Forest Resource Management, SLU: Umeå, Sweden, 2010.
19. Ersson, B.T.; Bergsten, U.; Lindroos, O. The cost-efficiency of seedling packaging specifically designed for tree planting machines. *Silva Fenn.* **2011**, *45*, 379–394. [CrossRef]
20. Sønsteby, F.; Kohmann, K. Forsøk med maskinell planting på Østlandet [*Mechanized Planting Trials in Østlandet*]; Oppdragsrapport 3/03; Norsk Institutt for Skogforskning: Ås, Norway, 2003.
21. Strandström, M.; Hämäläinen, J.; Pajuoja, H. Metsänhoidon koneellistaminen—Visio ja T&K-ohjelma [*The Mechanization of Forestry—Vision and R&D Program*]; Metsätehon Raportti 206; Metsäteho: Helsinki, Finland, 2009.
22. Luoranen, J.; Rikala, R.; Smolander, H. Machine planting of Norway spruce by Bracke and Ecoplanter: An evaluation of soil preparation, planting method and seedling performance. *Silva Fenn.* **2011**, *45*, 341–357. [CrossRef]

23. Ersson, T. *Review of Transplanting and Seedling Packaging Systems in Swedish Tree Nurseries*; Technical Report; FPInnovations: Pointe Claire, QC, Canada, 2015; p. 22.

24. Nilsson, U.; Luoranen, J.; Kolström, T.; Örlander, G.; Puttonen, P. Reforestation with planting in northern Europe. *Scand. J. For. Res.* **2010**, *25*, 283–294. [CrossRef]

25. Metsäteho. Koneistuttaja palvelu [*Machine Service*]; Metsäteho: Helsinki, Finland, 2017.

26. Berg, S. Terrängtypschema [*Terrain Classification System for Forestry Work*]; Forskningsstiftelsen Skogsarbeten: Stockholm, Sweden, 1982.

27. Kärhä, K.; Hynönen, A.; Laine, T.; Strandström, M.; Sipilä, K.; Palander, T.; Rajala, P.T. Koneellinen metsänistutus ja sen tehostaminen Suomessa [*Mechanized Planting in Finland and Its Enhancement*]; Report NR 233; Metsäteho: Vantaa, Finland, 2014; p. 42.

28. Risutec. ASTA Documentation System. Available online: http://www.risutec.fi/en/products/softwood/asta (accessed on 12 March 2018).

29. Malinen, J.; Laitila, J.; Väätäinen, K.; Viitamäki, K. Variation in age, annual usage and resale price of cut-to-length machinery in different regions of Europe. *Int. J. For. Eng.* **2016**, *27*, 95–102. [CrossRef]

30. Spinelli, R.; Magagnotti, N.; Picchi, G. Annual use, economic life and residual value of cut-to-length harvesting machines. *J. For. Econ.* **2011**, *17*, 378–387. [CrossRef]

31. Ackerman, P.; Gleasure, E.; Ackerman, S.; Shuttleworth, B. *Standards for Time Studies for the South African Forest Industry*; ICFR/FESA: Scottsville, South Africa, 2014; p. 49.

32. Erlandsson, E.; Fjeld, D. Impacts of service buyer management on contractor profitability and satisfaction— A Swedish case study. *Int. J. For. Eng.* **2017**, *28*, 148–156. [CrossRef]

33. Mäkinen, P. Metsä-ja Puualan pk-Yritysten Menestystekijät [*Success Factors for SMEs in Forestry and Woodworking*]; Finnish Forest Research Institute Research Papers 869; Metla: Vantaa, Finland, 2002; pp. 1–52.

34. Mäkinen, P. Success Factors for Forest Machine Entrepreneurs. *J. For. Eng.* **1997**, *8*, 27–35.

35. Ersson, B.T.; Bergsten, U.; Lindroos, O. Reloading mechanized tree planting devices faster using a seedling tray carousel. *Silva Fenn.* **2014**, *48*. [CrossRef]

36. Eriksson, M. Developing Client-Supplier Alignment in Swedish Wood Supply. Ph.D. Thesis, SLU, Umeå, Sweden, 2016.

37. Lindroos, O. Scrutinizing the Theory of Comparative Time Studies with Operator as a Block Effect. *Int. J. For. Eng.* **2010**, *21*, 20–30.

38. Purfürst, F.T.; Erler, J. The Human Influence on Productivity in Harvester Operations. *Int. J. For. Eng.* **2011**, *22*, 15–22.

39. Ovaskainen, H. Timber harvester operators' working technique in first thinning and the importance of cognitive abilities on work productivity. *Diss. Fore.* **2009**. [CrossRef]

40. Valinger, K. Skogsbrukets Framtida arbetskraftsförsörjning—Skogsmaskinföraryrkets attraktionskraft [*The Future Labour in Swedish Forestry—The Attraction of Forest Machine Operation as a Profession*]; Arbetsrapport 244; Institutionen för Skoglig Resurshushållning, SLU: Umeå, Sweden, 2009.

41. Malmberg, C.-E. Mekanisering av skogsodling [*The Mechanization of Forest Cultivation*]; STU-info 783-1990; Styrelsen för Teknisk Utveckling: Stockholm, Sweden, 1990; p. 196.

42. Berg, S. Studier av mekaniserade system för markberedning och plantering. [*Studies of Mechanized Systems for Scarification and Planting*]; Meddelande Nr 19; Forskningsstiftelsen Skogsarbeten: Kista, Sweden, 1991.

43. Saarinen, V.-M.; Hyyti, H.; Laine, T.; Strandström, M. Kohti jatkuvatoimista koneistutusta. [*Towards Continuously Operating Planting Machines*]; Metsätehon Raportti 227; Metsäteho: Vantaa, Finland, 2013.

MDPI

St. Alban-Anlage 66

4052 Basel

Switzerland

Tel. +41 61 683 77 34

Fax +41 61 302 89 18

www.mdpi.com

Forests Editorial Office

E-mail: forests@mdpi.com

www.mdpi.com/journal/forests